Air Pollution Control and Sustainable Development: Innovative Methods and Policy Implications

Air Pollution Control and Sustainable Development: Innovative Methods and Policy Implications

Editors

Weixin Yang
Guanghui Yuan
Yunpeng Yang

Basel • Beijing • Wuhan • Barcelona • Belgrade • Novi Sad • Cluj • Manchester

Editors

Weixin Yang
Business School
University of Shanghai for
Science and Technology
Shanghai
China

Guanghui Yuan
School of Economics and
Management
Shanghai University of
Political Science and Law
Shanghai
China

Yunpeng Yang
Antai College of Economics
and Management
Shanghai Jiao Tong
University
Shanghai
China

Editorial Office
MDPI AG
Grosspeteranlage 5
4052 Basel, Switzerland

This is a reprint of articles from the Topical Collection published online in the open access journal *Sustainability* (ISSN 2071-1050) (available at: https://www.mdpi.com/journal/sustainability/topical_collections/APCSD).

For citation purposes, cite each article independently as indicated on the article page online and as indicated below:

Lastname, A.A.; Lastname, B.B. Article Title. *Journal Name* **Year**, *Volume Number*, Page Range.

ISBN 978-3-7258-2043-6 (Hbk)
ISBN 978-3-7258-2044-3 (PDF)
doi.org/10.3390/books978-3-7258-2044-3

© 2024 by the authors. Articles in this book are Open Access and distributed under the Creative Commons Attribution (CC BY) license. The book as a whole is distributed by MDPI under the terms and conditions of the Creative Commons Attribution-NonCommercial-NoDerivs (CC BY-NC-ND) license.

Contents

About the Editors vii

Preface ix

Weixin Yang, Hao Gao and Yunpeng Yang
Analysis of Influencing Factors of Embodied Carbon in China's Export Trade in the Background of "Carbon Peak" and "Carbon Neutrality"
Reprinted from: *Sustainability* **2022**, *14*, 3308, doi:10.3390/su14063308 1

Piotr Sekula, Zbigniew Ustrnul, Anita Bokwa, Bogdan Bochenek and Miroslaw Zimnoch
Random Forests Assessment of the Role of Atmospheric Circulation in PM_{10} in an Urban Area with Complex Topography
Reprinted from: *Sustainability* **2022**, , 3388, doi:10.3390/su14063388 21

Lu Wang, Xue Chen, Yan Xia, Linhui Jiang, Jianjie Ye, Tangyan Hou, et al.
Operational Data-Driven Intelligent Modelling and Visualization System for Real-World, On-Road Vehicle Emissions—A Case Study in Hangzhou City, China
Reprinted from: *Sustainability* **2022**, *14*, 5434, doi:10.3390/su14095434 64

Chuang Sun, Xuegang Chen, Siyu Zhang and Tianhao Li
Can Changes in Urban Form Affect $PM_{2.5}$ Concentration? A Comparative Analysis from 286 Prefecture-Level Cities in China
Reprinted from: *Sustainability* **2022**, *14*, 2187, doi:10.3390/su14042187 86

Pedro Jiménez-Guerrero
What Are the Sectors Contributing to the Exceedance of European Air Quality Standards over the Iberian Peninsula? A Source Contribution Analysis
Reprinted from: *Sustainability* **2022**, *14*, 2759, doi:10.3390/su14052759 106

Qin Liu, Ying Zhu, Weixin Yang and Xueyu Wang
Research on the Impact of Environmental Regulation on Green Technology Innovation from the Perspective of Regional Differences: A Quasi-Natural Experiment Based on China's New Environmental Protection Law
Reprinted from: *Sustainability* **2022**, *14*, 1714, doi:10.3390/su14031714 130

Jing Wang, Hui-Zhen Fu, Jiaqi Xu, Danqi Wu, Yue Yang, Xiaoyu Zhu and Jing Wu
Trends of Studies on Controlled Halogenated Gases under International Conventions during 1999–2018 Using Bibliometric Analysis: A Global Perspective
Reprinted from: *Sustainability* **2022**, *14*, 806, doi:10.3390/su14020806 153

Alok Tiwari and Mohammed Aljoufie
Modeling Spatial Distribution and Determinant of $PM_{2.5}$ at Micro-Level Using Geographically Weighted Regression (GWR) to Inform Sustainable Mobility Policies in Campus Based on Evidence from King Abdulaziz University, Jeddah, Saudi Arabia
Reprinted from: *Sustainability* **2021**, *13*, 12043, doi:10.3390/su132112043 170

Ju Wang, Yue Zhong, Zhuoqiong Li and Chunsheng Fang
Temporal and Spatial Analysis of $PM_{2.5}$ and O_3 Pollution Characteristics and Transmission in Central Liaoning Urban Agglomeration from 2015 to 2020
Reprinted from: *Sustainability* **2022**, *14*, 511, doi:10.3390/su14010511 184

Jie Zhou, Hanlin Lan, Cheng Zhao and Jianping Zhou
Haze Pollution Levels, Spatial Spillover Influence, and Impacts of the Digital Economy:
Empirical Evidence from China
Reprinted from: *Sustainability* **2021**, *13*, 9076, doi:10.3390/su13169076 . . . . **202**

About the Editors

Weixin Yang

Weixin Yang received his Ph.D. in Economics from Shanghai Academy of Social Sciences. He is currently teaching at the University of Shanghai for Science and Technology. He has published over 100 papers, book chapters, and conference proceedings. He is also an invited reviewer and academic editor of well-known journals. His main research interests are the digital economy and sustainable development.

Guanghui Yuan

Guanghui Yuan received his Ph.D. from Shanghai University of Finance and Economics. He is currently teaching at Shanghai University of Political Science and Law. He has published more than 50 research papers. His main research interests include system modeling, information economy, data mining, and algorithm design.

Yunpeng Yang

Yunpeng Yang received his Ph.D. from the University of Shanghai for Science and Technology. He is currently teaching at Antai College of Economics and Management, Shanghai Jiao Tong University. He has published more than 30 research papers. His main research interests include resource and environmental management, economic policy analysis, and game theory.

Preface

We are pleased to present the topical collection "Air Pollution Control and Sustainable Development: Innovative Methods and Policy Implications." The topical collection brings together a collection of research that addresses one of the most pressing challenges of our time—the control of air pollution in the context of sustainable development. The included contributions span diverse regions and approaches, reflecting the global nature of air pollution and its far-reaching impacts. Our aim is to explore and promote innovative methods that not only improve air quality but also contribute to broader goals of sustainability.

The scope of this topical collection includes studies that focus on advanced modeling techniques, such as data-driven intelligent systems and geographically weighted regression, as well as analyses of the effects of urban planning, environmental regulations, and technological innovations on air quality control. Through these multidisciplinary perspectives, we seek to provide valuable insights for researchers, policymakers, and practitioners who are designing effective and sustainable solutions to air pollution.

This topical collection is the outcome of the collaborative efforts of the contributing authors. We would like to express our deepest thanks to all of the researchers who submitted their work, as well as the reviewers who provided invaluable comments. Special thanks are also due to the editors of *Sustainability*, whose support and guidance have been instrumental in bringing this topical collection to print.

We hope that this topical collection will serve as a useful resource for anyone interested in the intersection of air pollution control and sustainable development and that it inspires further research and policy development in this critical area.

Weixin Yang, Guanghui Yuan, and Yunpeng Yang
Editors

Article

Analysis of Influencing Factors of Embodied Carbon in China's Export Trade in the Background of "Carbon Peak" and "Carbon Neutrality"

Weixin Yang [1,†], Hao Gao [1,†] and Yunpeng Yang [2,*,†]

[1] Business School, University of Shanghai for Science and Technology, Shanghai 200093, China; iamywx@outlook.com (W.Y.); gaohao0302@outlook.com (H.G.)
[2] Antai College of Economics and Management, Shanghai Jiao Tong University, Shanghai 200030, China
* Correspondence: yang_yunpeng@outlook.com; Tel.: +86-21-5596-0082
† All the authors contributed equally to this work.

Abstract: Since China's reform and opening up, especially after its accession to the World Trade Organization, its foreign trade has achieved fruitful results. However, at the same time, the extensive foreign trade growth model with high energy consumption and high pollution has also caused a rapid increase in carbon emissions. There is a large amount of embodied carbon emissions in the export trade. In order to achieve the strategic goals of "Carbon Peak" and "Carbon Neutrality", and at the same time build a green trading system to achieve coordinated development of trade and the environment, it is of great significance to study embodied carbon emissions and how to decouple them with China's foreign trade. This paper uses the Logarithmic Mean Divisia Index method to decompose the influencing factors of the embodied carbon in China's export trade in order to study the impact of three factors: export scale, export structure, and carbon emission intensity. The results show that the change in export scale is the most important factor affecting the embodied carbon of China's export trade, and the expansion of export scale has caused the growth of trade embodied carbon. Carbon emission intensity is the second influential factor, and the decline in carbon intensity would slow down the growth of trade embodied carbon, while changes in the export structure have the smallest impact on trade embodied carbon. The high carbonization of the overall export structure will cause growth of trade embodied carbon, but the tertiary industry has seen some improvement in the export structure, which could facilitate the decline of trade embodied carbon.

Keywords: carbon peak; carbon neutrality; export trade; embodied carbon; Logarithmic Mean Divisia Index

Citation: Yang, W.; Gao, H.; Yang, Y. Analysis of Influencing Factors of Embodied Carbon in China's Export Trade in the Background of "Carbon Peak" and "Carbon Neutrality". *Sustainability* **2022**, *14*, 3308. https://doi.org/10.3390/su14063308

Academic Editor: Silvia Fiore

Received: 25 January 2022
Accepted: 10 March 2022
Published: 11 March 2022

Publisher's Note: MDPI stays neutral with regard to jurisdictional claims in published maps and institutional affiliations.

Copyright: © 2022 by the authors. Licensee MDPI, Basel, Switzerland. This article is an open access article distributed under the terms and conditions of the Creative Commons Attribution (CC BY) license (https://creativecommons.org/licenses/by/4.0/).

1. Introduction

Since China's reform and opening up, its foreign trade has achieved tremendous growth. According to statistical yearbook released by the National Bureau of Statistics, in 2001, China's total foreign trade of goods imports and exports was only $509.65 billion U.S. dollars, while by 2020, this number has increased to $4655.91 billion U.S. dollars, with an increase of 813.55% and a compound annual growth rate of over 11% [1,2]. On the other hand, China's total foreign trade of service imports and exports has increased from 78.45 billion U.S. dollars in 2001 to $661.72 billion U.S. dollars, with an increase of 743.49% and a compound annual growth rate of 11.26% [1,2]. By 2020, China had become the world's largest trader, as well as a major trading partner of more than 100 countries [3]. Its total foreign trade has accounted for more than 13% of total global trade, and its growth rate is much higher than that of total global trade [4–6]. The growth in foreign trade has played a huge role in stimulating China's economic development, but the extensive growth model of foreign trade has also caused a huge negative impact on the environment [7,8]. China's over-reliance on factors such as labor and resources in global trade has resulted in China staying at the low end of the global trade value chain for a long time [9–11]. On the one hand, in China's export structure, resource-intensive and labor-intensive products account for more than 50% of total exports, and this percentage has shown an increasing trend [12,13]. On the other hand, processing trade still occupies a large proportion in China's export trade, especially in the first few years after joining the WTO, during which the proportion of processing trade was once over 50% [14,15]. The foreign trade growth model discussed above has resulted in China's large export scale with a low added value of export. In addition, China's export is heavily dependent on consumption of resources, causing environmental pollution as well as continuous growth of carbon emissions [16].

According to statistics from the Global Carbon Budget Database, China's total domestic carbon emissions were 3.51 billion tons in 2001, which has increased to 10.67 billion tons by 2020 with an increase of 203.56%. The proportion of China's carbon emissions in total global carbon emissions has also increased from 13.62% in 2001 to 30.64% in 2020, making China the world's largest carbon emitter [17]. Excessive carbon emissions will not only have a negative impact on China's economic development, but will also threaten people's health and even survival [18]. As carbon is being the main greenhouse gas, the increase of its concentration in the atmosphere has led to global warming, resulting in temperature rise, sea level rise, and various extreme climates, which could cause immeasurable damage to food production, the ecological environment, infrastructure construction, and the safety of people's lives and property [19–21]. According to the fifth assessment report of the Intergovernmental Panel on Climate Change (IPCC), the current impact of human activities on climate change is negative and large, and we cannot let it continue to develop [22]. In response to this severe situation, the United Nations adopted the "United Nations Framework Convention on Climate Change", with the goal of controlling global temperature changes within a safe range [23]. As a supplement to this framework, the "Kyoto Protocol" adopted in 1997 put forward emission reduction requirements for some countries [24]. Since China's total carbon emissions were limited at the time, it was not bound by mandatory emission reduction requirements. However, with the continuous growth of carbon emissions, China's carbon emissions have attracted more and more attention from the international community. In the "Paris Agreement" signed in 2015, China has been designated as one of the main countries for carbon emission reduction, and all countries are required to set emission reduction targets by 2030 by means of "independent contributions" [25]. In this regard, China's leader Xi Jinping solemnly pledged at the 75th UN General Assembly in 2020 to achieve the peak of carbon emissions by 2030 and achieve carbon neutrality by 2060 [26].

As China being a major trading country and a major carbon emitter, its trade growth has not only driven economic growth, but also continuously increased carbon emissions, resulting in a large amount of embodied carbon in export trade [27]. In order to achieve the strategic goals of "Carbon Peak" and "Carbon Neutrality", it is of great theoretical and practical importance to study embodied carbon emissions and how to decouple them with China's foreign trade.

2. Literature Review

When studying the influencing factors of carbon emissions, the academic community often use the structural decomposition analysis method based on the input-output analysis model [28–30]. This method is based on the input-output table and fully considers the relationship between sectors. This method decomposes the changes in carbon emissions into the sum of changes caused by different independent variables, and analyzes the contribution of changes in each independent variable to changes in carbon emissions. For example, Ali et al. (2020) designed an emission multiplier product matrix to estimate the carbon emissions generated by British industrial activities and decomposed the factors affecting carbon emissions. They found that technological progress had played a key role in reducing carbon emissions in the UK. The final demand structure achieved through technological progress could help reduce carbon emissions [31]. Araujo et al. (2020) conducted a quantitative study on the influencing factors of carbon emissions of countries that newly joined the EU. The results of structural decomposition showed that their total amount of carbon emissions had increased due to the expansion of the trade scale and changes in their industrial structure [32]. Engo et al. (2021) used the structural decomposition method and decoupling model to analyze the carbon emissions of North African countries. They found that the effects of scale, energy intensity and economic structure are different among those countries [33]. Kim and Tromp (2021) used the multi-region input-output method and the structural decomposition model to calculate the embodied carbon in trade between Brazil and China. The study found that the changes in China's final demand and export structure are the main factors accounting for the increase of embodied carbon emissions [34].

However, the structural decomposition analysis method has the issue of incomplete decomposition when decomposing variables, that is, there could be some "decomposition residual" [35,36]. In recent researches, the academic circle often handles the decomposition residual by taking the average of the positive and negative extreme values [37–39]. However, when researchers adopt different decomposition orders, the results obtained are not always consistent [40–42].

In view of this, many scholars have adopted the Logarithmic Mean Divisia Index (LMDI) method of the Divis Decomposition method. For examples, Raza and Lin (2020) applied the LMDI method to study carbon emissions in Pakistan's transport sector. Their findings suggest that economic growth was the main factor responsible for the increase in carbon emissions from Pakistan's transport sector during the 1984–2018 period [43]. Pita et al. (2020) used the LMDI-I index method to study the influencing factors of carbon emissions from road transport in Thailand. The results show that the type of fuel and energy efficiency are the main factors affecting carbon emissions in this sector in Thailand. By increasing the proportion of biofuels used and further improving energy efficiency, the carbon emission level in this sector will be significantly reduced [44]. Chontanawat et al. (2020) used the LMDI method to study the carbon emission levels and influencing factors of various industries in Thailand from 2005 to 2017. The results show that the upgrading of industrial structure has reduced carbon emissions, while the increase in energy intensity of some industries has led to their carbon emissions rising [45]. Hasan and Wu (2020) investigated carbon emissions from the power sector in Bangladesh from 1979 to 2018 and used the LMDI method to analyze the industry's future emissions levels. The research results show that CO_2 intensity and power intensity are the main factors leading to the increase of carbon emissions, and the widespread application of renewable energy technologies in the future will be an important policy tool to reduce carbon emissions [46].

This method is very robust. It can deal with zero and negative values very well, and can achieve complete decomposition without residuals so that the decomposition results are more reliable. The following studies demonstrate this advantage: Taka et al. (2020) used Kaya identity and LMDI method to study carbon emissions in Ethiopia's energy sector. The results they obtained show that economics, population, and fossil fuel were the main contributors to the increase in carbon emissions, while the increase in energy intensity would significantly reduce the increase in carbon emissions [47]. Yasmeen et al. (2020) used the LMDI method to assess Pakistan's carbon emissions during 1972–2016. Their results also show that economic development is the main factor for the increase in per capita carbon emissions in Pakistan, while improving the energy structure and improving energy efficiency can help reduce per capita carbon emissions [48]. Ozturk et al. (2021) used the Tapio decoupling index and LMDI method to study carbon emissions of three typical representatives of emerging economies, Pakistan, India, and China. The results show that although the energy intensity of the above three countries has reduced carbon emissions, their economic development, population, energy structure, and other factors have increased carbon emissions [49]. Using the LMDI-I model and the Innovative Accounting Approach, Cansino et al. (2021) studied Ecuador's carbon emissions from 2000–2014. They found that the most important factors affecting Ecuador's carbon emissions are carbon intensity, population growth and economic development, with the country's energy and transportation sectors being the most sensitive to increases in carbon emissions [50].

According to this latest trend in the academic circle, we utilized the LMDI method to decompose the changes in the embodied carbon of China's export trade into the export scale effect, export structure effect, and the carbon intensity effect, in order to analyze the impact of changes in the export scale, export structure, and carbon intensity on the embodied carbon of China's export trade in depth.

In the following sections of this paper, Section 3 introduces the calculation method of real trade volume and the decomposition model of influencing factors of export trade embodied carbon. Section 4 calculates and discusses the influencing factors of embodied carbon in China's export trade by three major industries by using the World Input-Output Database (WIOD) (2016 edition). Section 5 provides the conclusions of this paper.

3. Materials and Methods

3.1. Real Trade Volume Calculation

In view of the possibility of double counting in traditional trade statistics, this paper has conducted research based on trade value-added, and converted the trade value-added of each year to comparable prices of 2010. In the accompanying economic and social accounts of the 2016 version of the WIOD database, price indices of value added by sector for each country are available each year. Since the value-added price of each sector in 2010 has been used as the reference price in the account, and the value-added price index of each sector in 2010 is set to be 100, we convert the value-added of each sector in other years into 2010 comparable prices. Similarly, indicators such as trade scale, trade structure and carbon emission intensity are also calculated based on the above converted trade added value. The main steps of real trade volume calculation of this paper are as follows.

3.1.1. Trade Value Added

The trade value-added is divided into export trade value-added and import trade value-added. The former refers to the trade value-added created by domestic production while the latter refers to the trade value-added created by foreign production in imports.

Due to economic globalization, the production of a final product is often not completed within one country, but has undergone production and processing in multiple countries. For example, the primary products are manufactured in one country, and then exported to another country as intermediate products for further processing, and eventually exported to a third country as final products for consumption. The international division of labor is conducive to each country's comparative advantages and factor endowment advantages to participate in the production of final products, which is of great significance to promoting the economic development of all countries [51]. Developing countries in particular can participate in international trade by virtue of their comparative advantages in resources and labor and benefit from the international trade. Even trade in intermediate products could help developing countries mitigate the distortion of factor markets and optimize the allocation of resources.

However, the rapid growth of intermediate products trade has caused statistical problems. Since traditional trade statistics normally focus on the total value of commodities, and do not consider the trade of intermediate products. Therefore, the multiple flows of intermediate products between countries will cause statistical duplication, thereby "inflating" the trade volume [52–54]. Taking mobile phone production as an example, assume that the design and production of core components such as mobile phone chips and integrated circuits are completed in the United States, and then the components are exported to China as intermediate products. Chinese companies would process and assemble the mobile phones, and then the finished products are exported to Japan. In the above-mentioned trade process, the actual trade volume of China is only the added value of mobile phones during processing and assembly, and should not be calculated based on the export value of finished mobile phones according to customs statistics. Therefore, in the context of rapid growth of international trade in intermediate products, the concept of trade value added should be adopted to accurately calculate the real trade volume, trade gains and value flow in order to identify and measure a country's real trade competitive advantage [55–57].

3.1.2. Calculation of Trade Value Added

This paper has adopted the input-output method to measure the trade added value of China's export. Please refer to Table A1 in Appendix A to find the meaning of variables used in calculation. First, this paper calculates the direct value-added coefficient v_i^r, which represents the value-added contained in the total output of sector i in country r, that is, the total input of sector i after removing the intermediate input (that is, the initial input) [58]. The calculation method is shown in the following equation:

$$v_i^r = \frac{I_i^r}{Y_i^r} \quad (r = 1, 2, \ldots, M, i = 1, 2, \ldots, N) \tag{1}$$

Express the above direct value-added coefficient in a matrix form, as shown in the following equation:

$$v = \begin{bmatrix} v^1 \\ \ldots \\ v^M \end{bmatrix} \tag{2}$$

Equation (3) can be obtained by establishing a connection with the Leontief model.

$$v \odot X = v \odot (I - A)^{-1} F \tag{3}$$

where $v \odot X$ represents the vector obtained by multiplying the corresponding position of vector v and vector X. A is the direct consumption coefficient matrix, and its element A_{ij}^{rs} is the direct consumption coefficient, representing consumption per unit of output in sector j in country s to sector i in country r. In the input-output table, from the perspective of the relationship between the rows, the intermediate output Z plus the final output F equals the total output X, that is, $Z + F = X$. The intermediate output can be expressed by the total output and the direct consumption coefficient, that is, $Z = AX$. So we can get $AX + F = X$. After matrix inversion, we can further get $X = (I - A)^{-1} F$, which is the Leontief model [59,60].

Moreover, $v \odot (I - A)^{-1}$ is the complete value-added coefficient, representing the total value added for each unit of final product produced by different sectors in different countries. Here, the final demand matrix F is expressed as a block matrix for illustration, as shown in equation:

$$F = \begin{bmatrix} F^{11} & \ldots & F^{1M} \\ \ldots & \ldots & \ldots \\ F^{M1} & \ldots & F^{MM} \end{bmatrix} = \begin{bmatrix} F^1 & \ldots & F^M \end{bmatrix} \tag{4}$$

The final demand matrix F is divided by columns into the demand vector of each economy for the final product produced by different sectors in different countries, where F^s is an $(M \times N) \times 1$ dimensional vector, representing the demand of country s for the final product of different sectors in different countries. By multiplying with the inverse Leontief matrix [61,62], that is, $(I - A)^{-1} F^s$, the demand of country s for all products of various sectors in each country can be obtained. By multiplying with the corresponding position of the direct value-added coefficient, that is, $v \odot (I - A)^{-1} F^s$, the calculation method of trade value added can be obtained, as shown in this equation:

$$V = v \odot (I - A)^{-1} F$$

$$= \begin{bmatrix} v \odot (I-A)^{-1} F^1 & \cdots & v \odot (I-A)^{-1} F^s & \cdots & v \odot (I-A)^{-1} F^M \end{bmatrix} \quad (5)$$

where $v \odot (I-A)^{-1} F^s$ represents the total added value of different sectors in different countries in order to meet the final demand of country s. Based on that, the matrix is further divided by country, and thus Equation (6) can be obtained:

$$V = \begin{bmatrix} V^{11} & \cdots & V^{1M} \\ \cdots & V^{rs} & \cdots \\ V^{M1} & \cdots & V^{MM} \end{bmatrix} \quad (6)$$

in which V^{rs} is the vector of added value of various sectors in country r in order to meet the final demand of country s. Thus, the export value-added vector of country r can be obtained, as shown in this equation:

$$EV^r = \sum_{s, s \neq r} V^{rs} \quad (7)$$

in which EV^r is an $N \times 1$ dimensional vector, which represents the added value of export trade of various sectors in country r. Similarly, the import value-added vector of country r can be calculated, as shown in this equation:

$$IV^r = \sum_{s, s \neq r} V^{sr} \quad (8)$$

where IV^r is an $N \times 1$ dimensional vector, which represents the import value-added of country r from various foreign sectors.

3.1.3. Data Source and Calculation Process

We used MATLAB (the software of MathWorks, Inc. Natick, MA, USA. Version: r2018b) to perform calculations. The multi-region input-output table needs to be established in order to calculate the trade value added. This paper has established a multi-region input-output table based on the WIOD database. The value-added price indices of various sectors in China are from the supporting economic and social accounts of the 2016 version of the WIOD [63]. The economic and social accounts include data such as the number of employees, compensation of labor and capital, and price indices of 56 sectors in 44 economies during the 2000–2014 period, providing a sound data support for the multi-region input-output table. In the process of trade value added calculation, inflation must be dealt with. Due to the existence of inflation, the same goods or services have different prices in different years, which makes it impossible to directly compare the trade value added of different years. When dealing with inflation, this paper first uses the international input-output tables of current prices to calculate trade value-added, and then uses the value-added price indices of various sectors in the economic and social accounts to obtain the trade value added at comparable prices. In actual calculation, this paper converts the trade value-added obtained in the first step with the 2010 comparable prices. Therefore, unless otherwise specified, the trade value-added data in this paper are the trade value-added calculated based on 2010 comparable prices rather than current prices.

3.2. Decomposition Model of Influencing Factors of Embodied Carbon in Export Trade

This paper has used the factor decomposition method to decompose the embodied carbon in China's export trade. This method decomposes the change of the target variable into changes of specific influencing factors to study the role of each factor. As mentioned in Sections 1 and 2 above, this paper has adopted the LMDI method according to the latest trends in the academic circle in order to analyze the influencing factors of embodied carbon in China's export trade. This method has strong robustness. It can deal with zero and negative values very well, and can achieve complete decomposition without residuals [64,65].

The calculation method of the LMDI model is shown in the following equation:

$$EC = \sum_i EC_i = \sum_i EV \frac{EV_i}{EV} \frac{EC_i}{EV_i} = \sum_i QS_i I_i \qquad (9)$$

The above equation decomposes the total embodied carbon in export trade into the sum of embodied carbon of various sectors, which is further expressed as the sum of the product of export scale, export structure and carbon intensity. In Equation (9), $EC\left(=\sum_i EC_i\right)$ is the total embodied carbon in export trade; EC_i represents the embodied carbon of sector i; $EV\left(=\sum_i EV_i\right)$ is the total export value-added; EV_i represents the export value-added of sector i. $Q(=EV)$ represents the total export scale. $S_i(=EV_i/EV)$ is the ratio of the export value added of sector i to the total export value added, representing the structural composition of the total export value added. $I_i(=EC_i/EV_i)$ is the ratio of the export trade embodied carbon of sector i to the export value added of sector i, representing the embodied carbon intensity of sector i. This indicator reflects the production technology and energy technology levels of sector i.

Further, this paper has decomposed the changes in the total embodied carbon in export trade from period 0 to t according to the decomposition method shown in this equation:

$$\Delta EC = EC^t - EC^0 = \Delta EC_Q + \Delta EC_S + \Delta EC_I \qquad (10)$$

In the above equation, ΔEC_Q is the export scale effect, representing the changes in embodied carbon in export trade caused by changes in export value added. ΔEC_S is the export structure effect, representing the changes in embodied carbon in export trade caused by changes in the export structure. ΔEC_I is the carbon intensity effect, implying the changes in embodied carbon in export trade caused by technological changes of various sectors. According to the LMDI model, the equations of the three variables above are as follows:

$$\Delta EC_Q = \sum_i \frac{EC_i^t - EC_i^0}{\ln EC_i^t - \ln EC_i^0} \ln\left(\frac{Q^t}{Q^0}\right) \qquad (11)$$

$$\Delta EC_S = \sum_i \frac{EC_i^t - EC_i^0}{\ln EC_i^t - \ln EC_i^0} \ln\left(\frac{S_i^t}{S_i^0}\right) \qquad (12)$$

$$\Delta EC_I = \sum_i \frac{EC_i^t - EC_i^0}{\ln EC_i^t - \ln EC_i^0} \ln\left(\frac{I_i^t}{I_i^0}\right) \qquad (13)$$

4. Results and Discussion

4.1. Analysis of Overall Influencing Factors of Embodied Carbon in China's Export Trade

Based on the LMDI model and the WIOD data, this paper has decomposed the influencing factors of embodied carbon in China's export trade. The results obtained are shown in Figure 1:

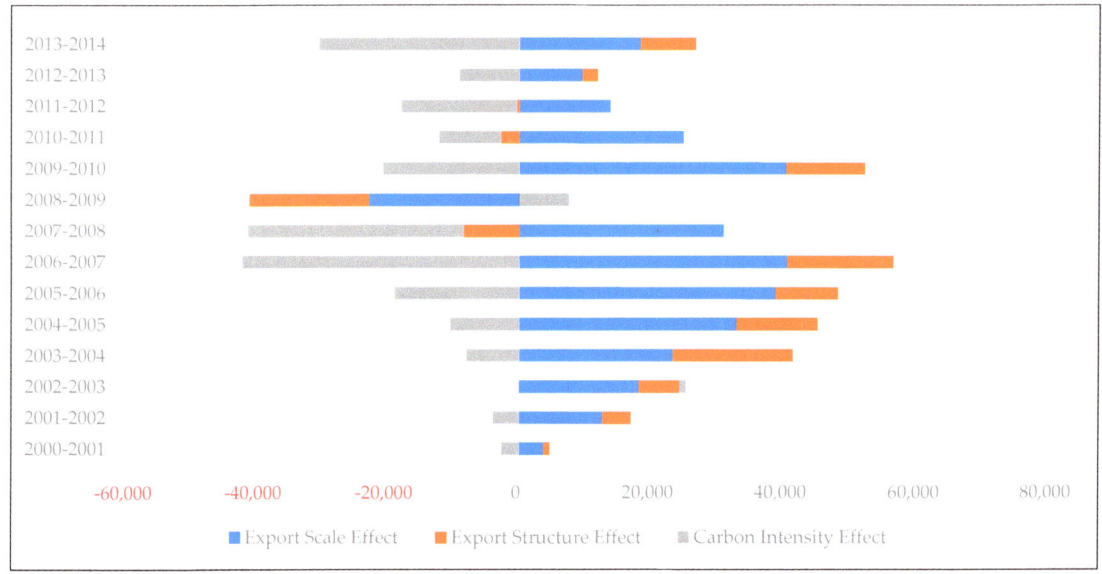

Figure 1. The specific effect value of influencing factors of embodied carbon in China's export trade (unit: 10,000 tons).

In the graph above, the bars represent the change in carbon embodied in export trade between two years. In terms of the export scale effect, the change in the embodied carbon in export trade caused by the export scale effect from 2000 to 2001 was 37.18 million tons, and it has been increasing since then. Between 2006 and 2007, the export scale effect reached a local maximum of 405 million tons. From 2008 to 2009, the export scale declined due to the financial crisis, and the export scale effect showed a negative value of −226 million tons during the research period. Between 2009 and 2010, the export scale effect recovered to 403 million tons and then declined, but remained at a level above 90 million tons. Therefore, the changes in the embodied carbon of export trade caused by the export scale effect are generally positive and the value of the export scale effect is relatively large, and only has a negative value in very few cases.

In terms of the export structure effect, the change in the embodied carbon in export trade caused by the export structure effect from 2000 to 2001 was 9.64 million tons, which has increased since then. During the period of 2003–2004, the export structure effect was 180 million tons. After a short period of decline, it rose to 159 million tons during the period of 2006–2007, and then dropped to a negative value. From 2008 to 2009, the export structure effect reached its lowest value of −183 million tons, and rebounded to 118 million tons during the period of 2009–2010. During 2010-2011 and 2011–2012, the export structure effect dropped to a negative value again, but rose to a positive value after that and remained below 100 million tons. Therefore, the export structure effect in the embodied carbon in export trade is generally positive, with negative values appear from time to time. The value of the export structure effect is relatively small compared to the export scale effect. However, from the perspective of reducing the embodied carbon in export trade as well as lowering the environmental cost of international trade, China's export structure has been deteriorating during most of the research period. Only for a few years has the embodied carbon declined due to the optimization of the export structure.

In terms of the carbon intensity effect, the change in the embodied carbon in export trade caused by the carbon intensity effect from 2000 to 2001 was −26.36 million tons, which showed a downward trend thereafter. From 2006 to 2007, the carbon intensity effect reached its lowest value of −420 million tons, and then rebounded. During the period of

2008–2009, the carbon intensity effect showed a positive value of 75.59 million tons, and then dropped to a negative value, and fluctuated at around −100 million tons. Therefore, the impact of the carbon intensity effect on the embodied carbon in China's export trade is generally negative, which indicates that during most of the research period, the use of clean energy and technological progress helped reduce the embodied carbon in China's export trade.

In summary, during the period of 2000–2014, the cumulative change in embodied carbon in export trade caused by the export scale effect was 2.85 billion tons; the cumulative change caused by the export structure effect was 595 million tons; the cumulative change caused by the carbon intensity effect was −1.96 billion tons. The total cumulative change caused by these three types of effects was 1.49 billion tons (as shown in Figure 2).

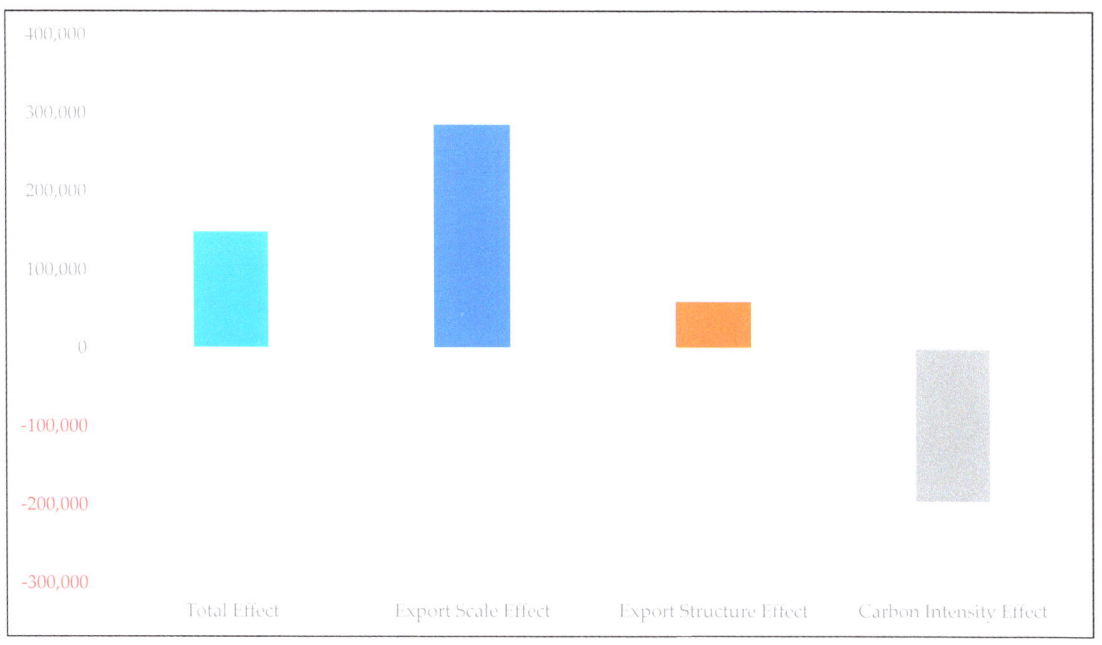

Figure 2. Cumulative change of embodied carbon in China's export trade caused by different influencing factors (unit: 10,000 tons).

In Figure 2, the proportions of the export scale effect, the export structure effect, and the carbon intensity effect were 191.56%, 40.03%, and −131.60%, respectively. This indicates that on the one hand, the expansion of the export scale and the deterioration of the export structure caused the embodied carbon in China's export trade to increase. The deterioration here is from the perspective of carbon emissions. It refers to the increase in the proportion of high energy consuming, high carbon emission, and low value-added sectors in the export structure, while the proportion of clean and high value-added sectors industries declines. The export scale effect was the most important driver of such growth. On the other hand, the use of clean energy and the decline of carbon intensity brought by technological progress helped reduce the embodied carbon in China's export trade [66].

4.2. Analysis of Influencing Factors of Embodied Carbon in the Export Trade of the Primary Industry

Based on the LMDI model and the WIOD data, this paper has decomposed the influencing factors of embodied carbon in the export trade of China's primary industry. The results obtained are shown in Figure 3:

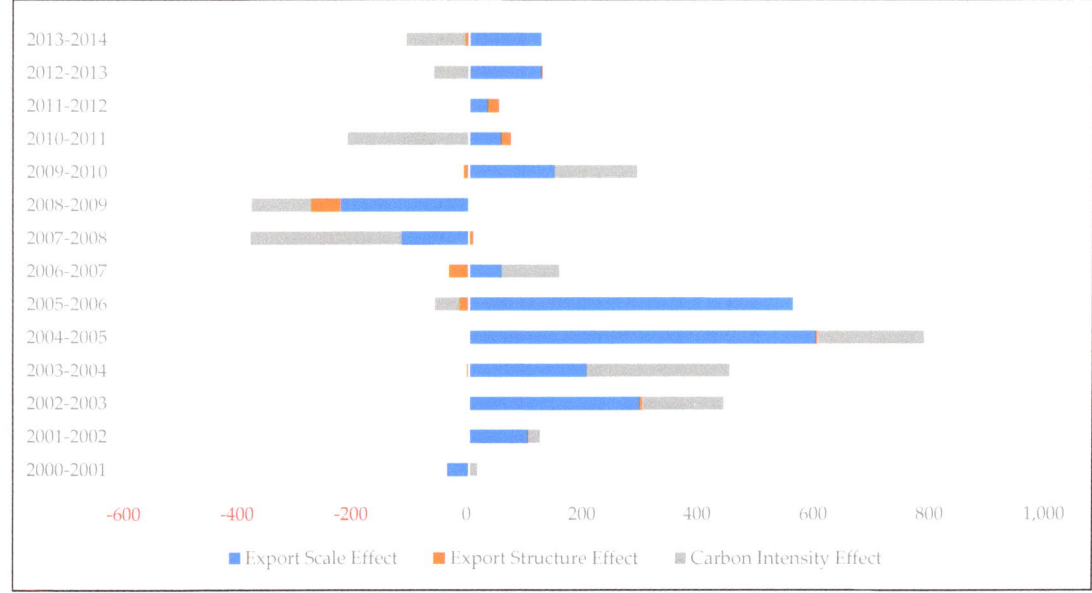

Figure 3. Influencing factors of embodied carbon in the export trade of China's primary industry (unit: 10,000 tons).

In terms of the export scale effect of the primary industry, during the period of 2000–2001, the change of the embodied carbon caused by the export scale effect was −375.2 thousand tons, which experienced a gradual increase thereafter. During the period of 2004–2005, the export scale effect increased to a maximum value of 6.00 million tons, but continued to decline thereafter and reached a minimum value of −2.24 million tons during the period of 2008–2009. The export scale effect of the primary industry rebounded to 1.47 million tons between 2009 and 2010, and then fluctuated around 1 million tons. Overall speaking, the changes of the embodied carbon in the export trade of the primary industry caused by the export scale effect are basically positive, with negative values appearing only in a few years. During the research period, the export scale effect has shown large positive values during the early stage, and its absolute value has decreased in the later stage.

In terms of the export structure effect of the primary industry, the changes of the embodied carbon caused by the export structure effect were relatively small, and mostly negative. During the research period, the export structure of China's primary industry has not experienced major changes, so the changes of embodied carbon caused by the export structure effect were relatively small.

In terms of the carbon intensity effect of the primary industry, during 2000–2001, the change of embodied carbon in the export trade of the primary industry caused by the carbon intensity effect was 142.4 thousand tons, which showed some increase thereafter. Between 2003 and 2004, the change of embodied carbon caused by the carbon intensity effect was 2.48 million tons, which has declined since then. The carbon intensity effect is mostly negative in terms of the embodied carbon in the export trade of the primary industry.

In summary, during the period of 2000–2014, the cumulative change of embodied carbon in the export trade of the primary industry caused by the export scale effect was 19.23 million tons; the cumulative change caused by the export structure effect was −722.3 thousand tons; the cumulative change caused by the carbon intensity effect was 604.7 thousand tons. The total cumulative change caused by these three types of effects was 19.11 million tons (as shown in Figure 4).

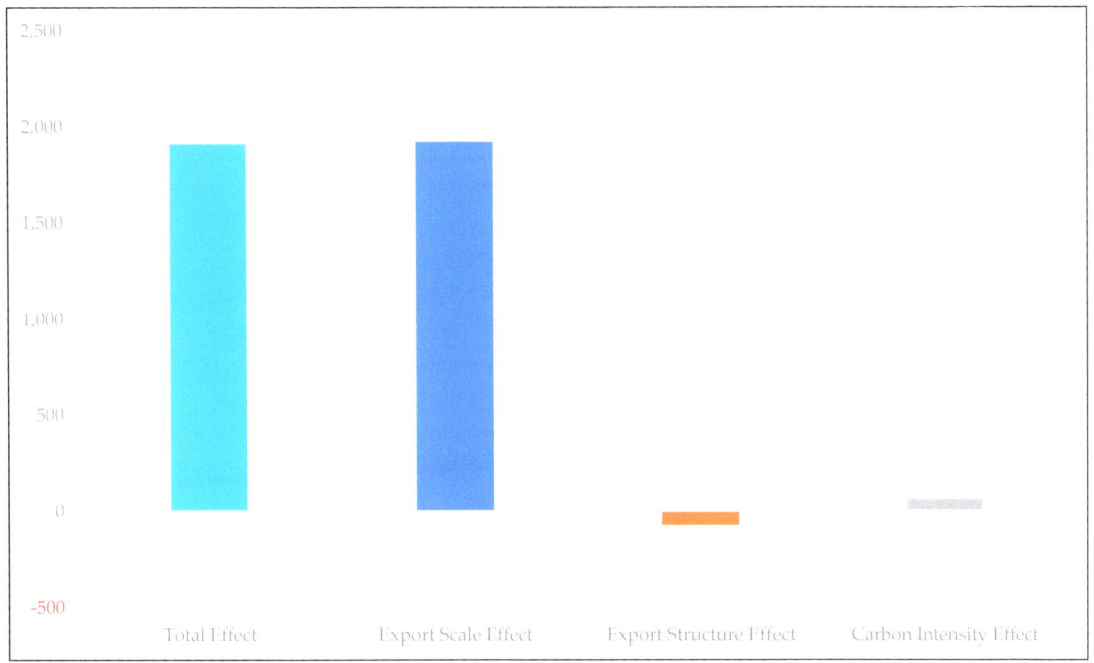

Figure 4. Cumulative change of embodied carbon in China's export trade of the primary industry caused by different Influencing Factors (unit: 10,000 tons).

In the above Figure 4, the proportions of the export scale effect, the export structure effect, and the carbon intensity effect were 100.62%, −3.78%, and 3.16%, respectively. This indicates that on the one hand, the expansion of the export scale is the most important driver of the embodied carbon in the export trade of the primary industry. The improvement of the export structure has helped reduce the embodied carbon to a certain extent. However, since the export structure of the primary industry has not changed much, the impact of the export structure effect is relatively small. On the other hand, the carbon intensity effect has led to an increase in the embodied carbon in the export trade of the primary industry. This is mainly due to the fact that the production technology of China's primary industry is not advanced and the emissions from the consumption of non-clean energy would also lead to the growth of embodied carbon in the export trade [67,68].

4.3. Analysis of Influencing Factors of Embodied Carbon in the Export Trade of the Secondary Industry

Based on the LMDI model and the WIOD data, this paper has decomposed the influencing factors of embodied carbon in the export trade of China's secondary industry. The results obtained are shown in Figure 5:

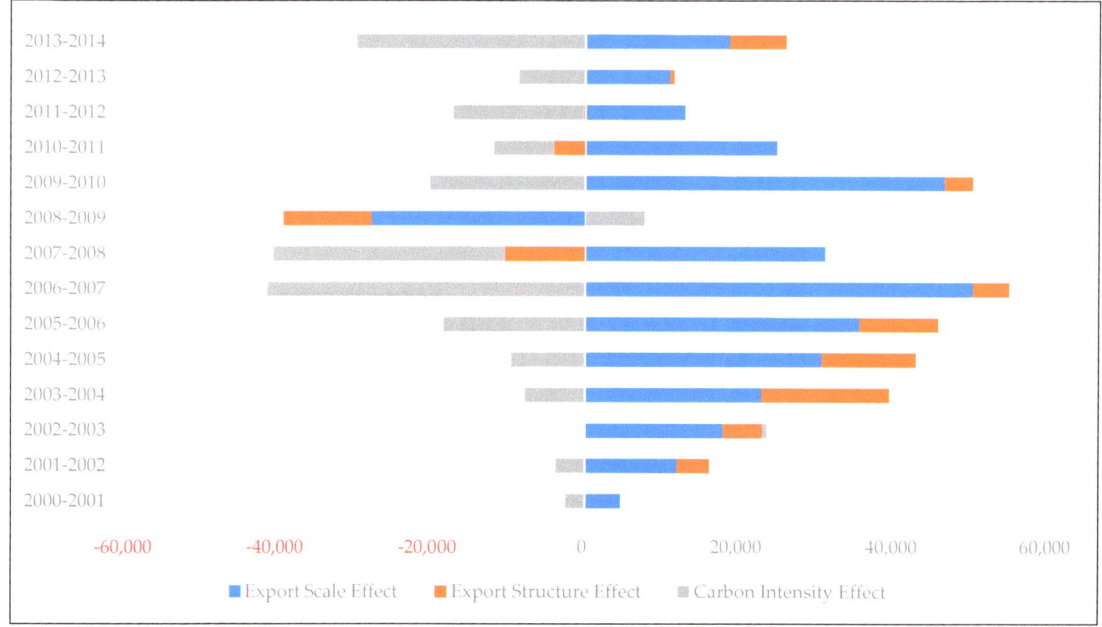

Figure 5. Influencing factors of embodied carbon in the export trade of China's secondary industry (unit: 10,000 tons).

In terms of the export scale effect of the secondary industry, during the period of 2000–2001, the change of the embodied carbon caused by the export scale effect was 45.47 million tons, which continued to increase thereafter. Between 2006 and 2007, the export scale effect reached 500 million tons. The export scale effect only turned negative (−277 million tons) once during the period of 2008–2009. After that, it rebounded to 463 million tons between 2009 and 2010 with a downward trend, and remained above 100 million tons. Overall, the changes of the embodied carbon in the export trade of the secondary industry caused by the export scale effect are basically positive, and the value of the export scale effect is relatively large.

In terms of the export structure effect of the secondary industry, the changes of the embodied carbon caused by the export structure effect first increased, and then decreased, and then rebounded. The export structure effect remained positive during most of the research period, which indicates that the export structure of the secondary industry is deteriorating, thus leading to the growth of the embodied carbon in the export trade of the secondary industry.

In terms of the carbon intensity effect of the secondary industry, the changes of the embodied carbon caused by this effect showed the trend of decline, a short rise, and then decline again. This effect remained negative for most of the research period, which indicates that the use of clean energy and technological progress have reduced the embodied carbon in the export trade of the secondary industry.

In summary, during the period of 2000–2014, the cumulative change of embodied carbon in the export trade of the secondary industry caused by the export scale effect was 2.9 billion tons; the cumulative change caused by the export structure effect was 376 million tons; the cumulative change caused by the carbon intensity effect was −1.87 billion tons. The total cumulative change caused by these three types of effects was 1.41 billion tons (as shown in Figure 6).

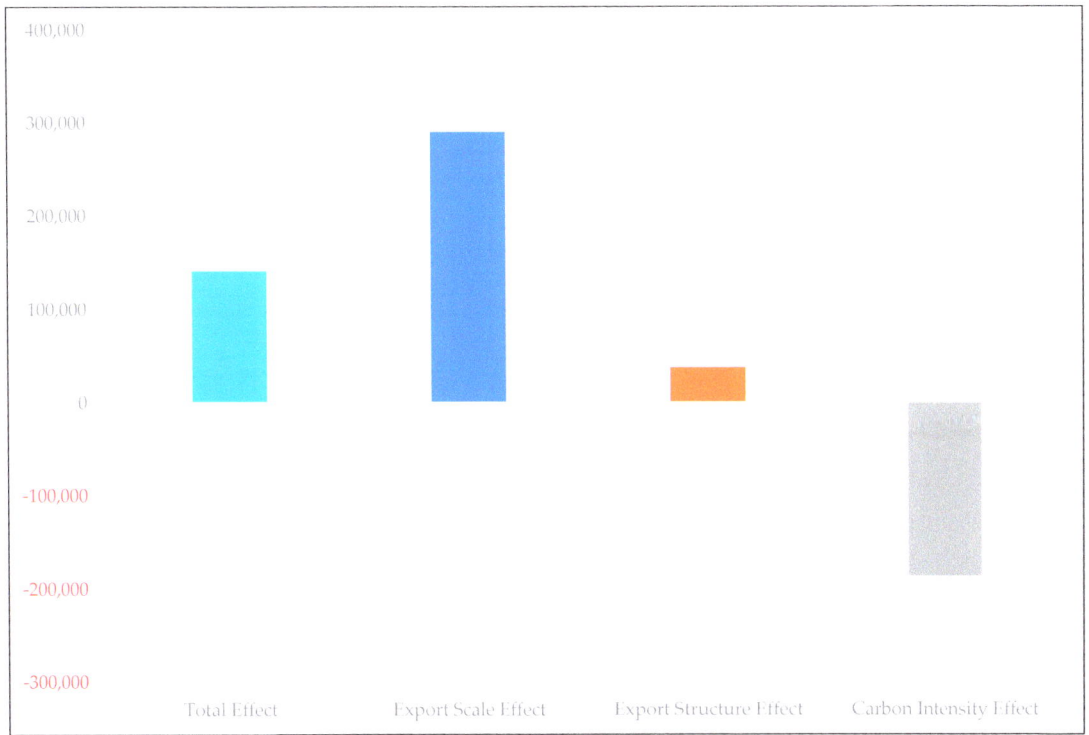

Figure 6. Cumulative change of embodied carbon in China's export trade of the secondary industry caused by different influencing factors (unit: 10,000 tons).

In the above Figure, the proportions of the export scale effect, the export structure effect, and the carbon intensity effect were 205.67%, 26.67%, and −132.62%, respectively. This indicates that the export scale effect is the most important driver of embodied carbon increase in the secondary industry. Secondly, the export structure effect has also caused growth of the embodied carbon in the export trade, which indicates that from the perspective of environmental costs, the export structure of China's secondary industry is deteriorating. Finally, the decrease of the carbon intensity effect has caused a decline of the embodied carbon in the export trade of the secondary industry, with a relatively significant impact.

4.4. Analysis of Influencing Factors of Embodied Carbon in the Export Trade of the Tertiary Industry

Based on the LMDI model and the WIOD data, this paper has decomposed the influencing factors of embodied carbon in the export trade of China's tertiary industry. The results obtained are shown in Figure 7:

In terms of the export scale effect of the tertiary industry, the change of the embodied carbon in the export trade of the tertiary industry caused by this effect repeated the patterns of increase first and then decrease during the research period. During most of the research period, the export scale effect remained positive, which indicates that the expansion of the export scale has caused the growth of the embodied carbon in the export trade of the tertiary industry. One exception is that during the period of 2008–2009, due to the impact of the financial crisis, the decrease of the export scale led to a decline in the embodied carbon in the export trade of the tertiary industry.

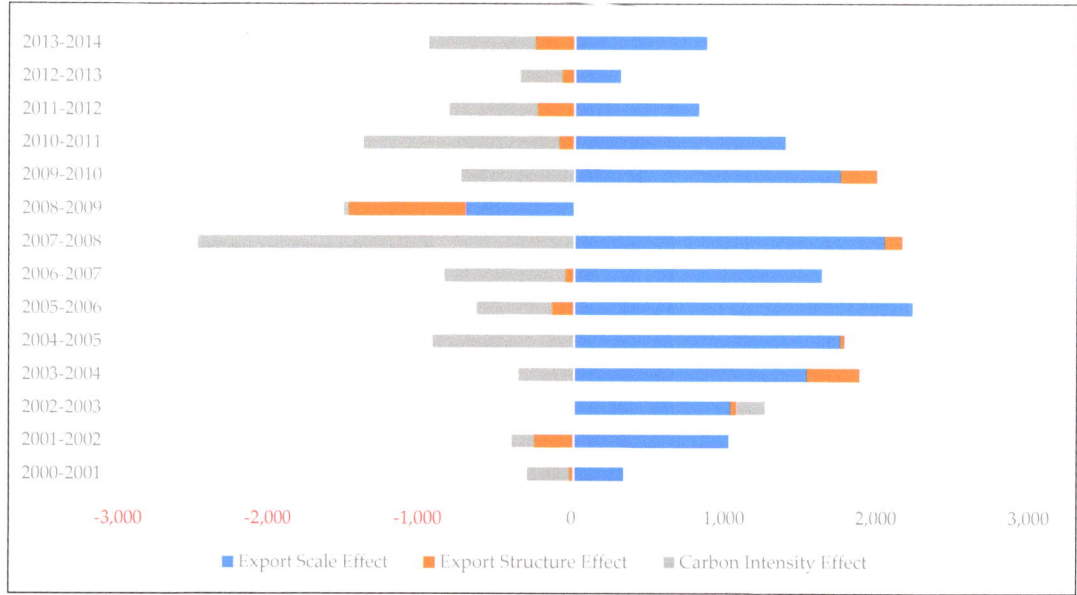

Figure 7. Influencing factors of embodied carbon in the export trade of China's tertiary industry (unit: 10,000 tons).

In terms of the export structure effect of the tertiary industry, the changes of the embodied carbon caused by this effect showed large fluctuations during the research period, with no obvious trend that can be identified. Overall speaking, during most of the research period, the export structure effect was negative, which indicates that the improvement of the export structure of the tertiary industry has caused a decline in the embodied carbon of export trade.

In terms of the carbon intensity effect of the tertiary industry, the changes of the embodied carbon in export trade caused by this effect remained negative for most of the research period and their values were relatively large, which indicates that the decrease of the carbon intensity effect has caused a large decline of the embodied carbon in the export trade of the tertiary industry.

In summary, during the period of 2000–2014, the cumulative change of embodied carbon in the export trade of the tertiary industry caused by the export scale effect was 1.59 million tons; the cumulative change caused by the export structure effect was −11.83 million tons; the cumulative change caused by the carbon intensity effect was −89 million tons. The total cumulative change caused by these three types of effects was 58.53 million tons (as shown in Figure 8).

In the Figure above, the proportions of the export scale effect, the export structure effect, and the carbon intensity effect were 272.28%, −20.21%, and −152.06%, respectively. This indicates that the export scale effect is the most important driver of embodied carbon increase in the tertiary industry. Secondly, the cumulative change caused by the export structure effect was negative, indicating that the tertiary industry has experienced export structure optimization during foreign trade, which has led to the decrease of the embodied carbon in export trade. Finally, technological progress and the use of clean energy have led to a decline in the carbon intensity effect, as well as a decline in the embodied carbon of the export trade of the tertiary industry [69,70].

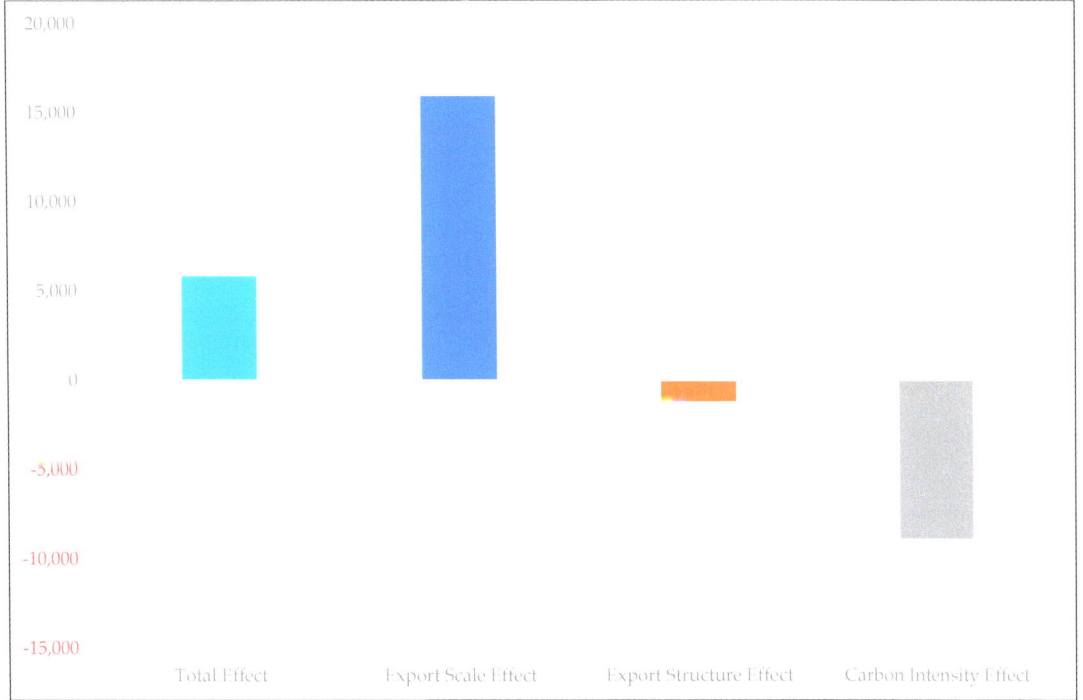

Figure 8. Cumulative change of embodied carbon in China's export trade of the tertiary industry caused by different influencing factors (unit: 10,000 tons).

4.5. Possible Strategies to Reduce the Impact

Based on the above analysis results of China's export trade embodied carbon impact factors, the following strategies may be effective means to reduce the impact:

(1) Calculate the carbon footprint of the relevant sectors. A carbon footprint is a collection of greenhouse gas emissions caused by an organization, business, product or individual through various production and consumption processes. It describes the carbon emissions impact of an individual's awareness and behavior on the natural world. In order to reduce the impact of carbon embodied in export trade, China needs to start calculating the carbon footprint of relevant sectors included in export trade [71].

(2) Promote the development of circular economy. China needs to improve resource conservation and recycling in export trade, and organize export trade into a circular process of "resources-products-renewable resources", so that all materials and energy can be rationally and lastingly utilized in this continuous cycle to reduce carbon emissions and the impact on the natural environment [72].

5. Conclusions

This paper has adopted the LMDI method to decompose the influencing factors of the embodied carbon in China's export trade, and studies the changes of the embodied carbon from the perspectives of export scale effect, export structure effect and carbon intensity effect in order to discuss the impact of the export scale, export structure and carbon emission intensity of each sector on the embodied carbon of export trade. The calculation results show that overall speaking, the expansion of the export scale is the most important driver of embodied carbon growth. The cumulative impact of the export scale effect was 2.85 billion tons. The deterioration of the export structure is a secondary factor causing the growth of the embodied carbon in China's export trade. The cumulative impact of the export structure effect was 595 million tons. Meanwhile, the decline of the carbon intensity was an important factor leading to the decrease of the embodied carbon in China's export trade. The cumulative impact of the carbon intensity effect was -1.96 billion tons.

In China's national economy, the primary industry refers to agriculture, forestry, animal husbandry and fishery. The secondary industry refers to mining, manufacturing, electricity, heat, gas and water production and supply, and construction. The tertiary industry is the service industry [73].

In terms of the three industries of the national economy, the continuous expansion of the export scale of the primary industry was the most important driver of its embodied carbon growth. The cumulative impact of the export scale effect in the primary industry was 19.23 million tons. During the research period, the export structure of the primary industry has been improved, resulting in a small decline in the embodied carbon of the export trade of the primary industry. The cumulative impact of the export structure effect in the primary industry (such as agriculture, forestry, etc.) was 722.3 thousand tons. Meanwhile, the increase of the carbon intensity has caused increase of the embodied carbon in the export trade of the primary industry. The cumulative impact of the carbon intensity effect in the primary industry was 604.7 thousand tons.

As for the secondary industry, the continuous expansion of its export scale was also the most important driver of embodied carbon increase. The cumulative impact of the export scale effect in the secondary industry was 2.9 billion tons. The deterioration of the export structure of the secondary industry was the secondary factor of embodied carbon increase in the export trade of the secondary industry. The cumulative impact of the export structure effect in the secondary industry was 376 million tons. However, the decrease of carbon intensity played an important role in the reduction of embodied carbon in the export trade of the secondary industry. The cumulative impact of the carbon intensity effect in the secondary industry was -1.87 billion tons.

As for the tertiary industry, the continuous expansion of its export scale was also the most important driver of embodied carbon increase in the export trade. The cumulative impact of the export scale effect in the tertiary industry was 159 million tons. The optimization of the export structure and the decline of carbon intensity have played an important role in the reduction of the embodied carbon in the export trade of the tertiary industry. The cumulative impacts of the export structure effect and the carbon intensity effect in the tertiary industry during the research period were -11.83 million tons and -89 million tons, respectively.

Since the statistical period of the WIOD database ends in 2014, the data analysis and calculation after 2014 need to be explored in future research to reflect the latest changes in the field of embodied carbon in export trade. This is one limitation of our study.

In addition, with the enrichment of research methods and the continuous updating of research tools, we will further tackle the technical problems of incomplete decomposition in the structural decomposition analysis method in the future research, in order to further improve the existing literature on the analysis of factors affecting the embodied carbon in export trade.

Author Contributions: Conceptualization, W.Y. and H.G.; methodology, H.G.; software, H.G.; validation, W.Y., H.G., and Y.Y.; formal analysis, H.G.; resources, W.Y.; data curation, W.Y.; writing—original draft preparation, H.G.; writing—review and editing, W.Y.; visualization, H.G.; supervision, W.Y.; project administration, Y.Y.; funding acquisition, W.Y. and Y.Y. All authors have read and agreed to the published version of the manuscript.

Funding: Weixin Yang was financially supported by the General Project of Shanghai Philosophy and Social Science Planning (2021BGL014). Yunpeng Yang was financially supported by the Youth Project of Shanghai Philosophy and Social Science Planning (2021EJB006).

Institutional Review Board Statement: Not applicable.

Informed Consent Statement: Not applicable.

Data Availability Statement: The calculation data used in this paper come from the WIOD database, which have been explained in the main text.

Conflicts of Interest: The authors declare no conflict of interest.

Appendix A. The Meaning of Variables Used in Calculation

Table A1. The meaning of variables used in Section 3.

Variable	Meaning
M	Number of countries
N	Number of sectors
I	The initial input vector of each sector in every country
I_i^r	Initial input in sector i in country r
v	The vector of direct value added coefficients of each sector in every country
v_i^r	The direct value − added coefficient in sector i in country r
A	Direct consumption coefficient matrix
A_{ij}^{rs}	Consumption per unit of output in sector j in country s to sector i in country r
F	Final demand matrix
F_i^{rs}	Demand of country s for final product of sector i in country r
X	The total output matrix obtained by multiplying the inverse Leontief matrix $(I-A)^{-1}$ by the final demand matrix F
X_i^{rs}	The demand of country s for the total output of sector i in country r
V	Value added trade flow matrix
V_i^{rs}	The export value − added transferred from sector i in country r to country s through trade
EV^r	Value − added export vector of sectors in country r
IV^r	Value − added import vector of sectors in country r
EC	The total embodied carbon in export trade
EC_i	The embodied carbon in export trade of sector i
EV	The total export value-added
EV_i	The export value − added of sector i
Q	The total export scale
S_i	The ratio of the export value added of sector i to the total export value added
I_i	The ratio of the export trade embodied carbon of sector i to the export value added of sector i
ΔEC	Total change of embodied carbon in export trade
ΔEC_Q	The export scale effect
ΔEC_S	The export structure effect
ΔEC_I	The carbon intensity effect

References

1. National Bureau of Statistics of China. *China Statistical Yearbook 2002*; China Statistic Press: Beijing, China, 2002; ISBN 9787503738265.
2. National Bureau of Statistics of China. *China Statistical Yearbook 2021*; China Statistic Press: Beijing, China, 2021; ISBN 9787503796258.
3. Gil, J. The Language Comprehensive Competitiveness of Chinese: The Objective Perspective. In *The Rise of Chinese as a Global Language: Prospects and Obstacles*; Springer International Publishing: Cham, Switzerland, 2021; pp. 51–71. ISBN 978-3-030-76171-4.
4. Yang, C.; Tsou, M. Exports and the demand for skilled labor in China: Do foreign ownership and trade type matter? *Econ. Model.* **2022**, *106*, 105692. [CrossRef]
5. Wang, Z.; Sun, Z. From Globalization to Regionalization: The United States, China, and the Post-COVID-19 World Economic Order. *J. Chin. Polit. Sci.* **2021**, *26*, 69–87. [CrossRef] [PubMed]
6. Kan, S.; Chen, B.; Han, M.; Hayat, T.; Alsulami, H.; Chen, G. China's forest land use change in the globalized world economy: Foreign trade and unequal household consumption. *Land Use Policy* **2021**, *103*, 105324. [CrossRef]
7. Yang, W.; Li, L. Energy Efficiency, Ownership Structure, and Sustainable Development: Evidence from China. *Sustainability* **2017**, *9*, 912. [CrossRef]
8. Li, Y.; Yang, W.; Shen, X.; Yuan, G.; Wang, J. Water Environment Management and Performance Evaluation in Central China: A Research Based on Comprehensive Evaluation System. *Water* **2019**, *11*, 2472. [CrossRef]
9. Shi, B.; Wang, X.; Gao, B. Transmission and Diffusion Effect of Sino-US Trade Friction along Global Value Chains. *Financ. Res. Lett.* **2021**, 102057. [CrossRef]

10. Bown, C.P.; Erbahar, A.; Zanardi, M. Global value chains and the removal of trade protection. *Eur. Econ. Rev.* **2021**, *140*, 103937. [CrossRef]
11. Cheng, D.; Wang, J.; Xiao, Z. Global value chain and growth convergence: Applied especially to China. *Pacific Econ. Rev.* **2021**, *26*, 161–182. [CrossRef]
12. Bhowmik, R.; Zhu, Y.; Gao, K. An analysis of trade cooperation: Central region in China and ASEAN. *PLoS ONE* **2021**, *16*, e0261270. [CrossRef]
13. Miao, M.; Liu, H.; Chen, J. Factors affecting fluctuations in China's aquatic product exports to Japan, the USA, South Korea, Southeast Asia, and the EU. *Aquac. Int.* **2021**, *29*, 2507–2533. [CrossRef]
14. Li, Y.; Yang, M.; Zhu, L. FDI, Export Sophistication, and Quality Upgrading: Evidence from China's WTO Accession. *Jpn. World Econ.* **2021**, *59*, 101086. [CrossRef]
15. Kim, M.; Xin, D. Export spillover from foreign direct investment in China during pre- and post-WTO accession. *J. Asian Econ.* **2021**, *75*, 101337. [CrossRef]
16. Jiang, B.; Li, Y.; Yang, W. Evaluation and Treatment Analysis of Air Quality Including Particulate Pollutants: A Case Study of Shandong Province, China. *Int. J. Environ. Res. Public Health* **2020**, *17*, 9476. [CrossRef] [PubMed]
17. Global Carbon Project. Global Carbon Budget. Available online: https://www.globalcarbonproject.org/carbonbudget/ (accessed on 18 January 2022).
18. Lu, S.; Zhao, Y.; Chen, Z.; Dou, M.; Zhang, Q.; Yang, W. Association between Atrial Fibrillation Incidence and Temperatures, Wind Scale and Air Quality: An Exploratory Study for Shanghai and Kunming. *Sustainability* **2021**, *13*, 5247. [CrossRef]
19. Shen, X.; Yang, W.; Sun, S. Analysis of the impact of China's hierarchical medical system and online appointment diagnosis system on the sustainable development of public health: A case study of Shanghai. *Sustainability* **2019**, *11*, 6564. [CrossRef]
20. Liu, H.; Liu, J.; Yang, W.; Chen, J.; Zhu, M. Analysis and Prediction of Land Use in Beijing-Tianjin-Hebei Region: A Study Based on the Improved Convolutional Neural Network Model. *Sustainability* **2020**, *12*, 3002. [CrossRef]
21. Yang, W.; Yang, Y. Research on Air Pollution Control in China: From the Perspective of Quadrilateral Evolutionary Games. *Sustainability* **2020**, *12*, 1756. [CrossRef]
22. Intergovernmental Panel on Climate Change. Fifth Assessment Report. Available online: https://www.ipcc.ch/assessment-report/ar5/ (accessed on 18 January 2022).
23. United Nations Climate Change. What is the United Nations Framework Convention on Climate Change? Available online: https://unfccc.int/process-and-meetings/the-convention/what-is-the-united-nations-framework-convention-on-climate-change (accessed on 18 January 2022).
24. United Nations Climate Change. Kyoto Protocol—Targets for the First Commitment Period. Available online: https://unfccc.int/process-and-meetings/the-kyoto-protocol/what-is-the-kyoto-protocol/kyoto-protocol-targets-for-the-first-commitment-period (accessed on 18 January 2022).
25. United Nations Climate Change. The Paris Agreement. Available online: https://unfccc.int/process-and-meetings/the-paris-agreement/the-paris-agreement (accessed on 18 January 2022).
26. The State Council of the People's Republic of China. Xi's Statements at UN Meetings Demonstrate China's Global Vision, Firm Commitment. Available online: http://english.www.gov.cn/statecouncil/wangyi/202010/02/content_WS5f771a17c6d0f7257693d023.html (accessed on 18 January 2022).
27. Akbar, U.; Li, Q.; Akmal, M.A.; Shakib, M.; Iqbal, W. Nexus between agro-ecological efficiency and carbon emission transfer: Evidence from China. *Environ. Sci. Pollut. Res.* **2021**, *28*, 18995–19007. [CrossRef]
28. Hossain, M.A.; Chen, S.; Khan, A.G. Decomposition study of energy-related CO_2 emissions from Bangladesh's transport sector development. *Environ. Sci. Pollut. Res.* **2021**, *28*, 4676–4690. [CrossRef]
29. Malik, A.; Egan, M.; du Plessis, M.; Lenzen, M. Managing sustainability using financial accounting data: The value of input-output analysis. *J. Clean. Prod.* **2021**, *293*, 126128. [CrossRef]
30. Hastuti, S.H.; Hartono, D.; Putranti, T.M.; Imansyah, M.H. The drivers of energy-related CO_2 emission changes in Indonesia: Structural decomposition analysis. *Environ. Sci. Pollut. Res.* **2021**, *28*, 9965–9978. [CrossRef] [PubMed]
31. Ali, Y.; Pretaroli, R.; Sabir, M.; Socci, C.; Severini, F. Structural changes in carbon dioxide (CO_2) emissions in the United Kingdom (UK): An emission multiplier product matrix (EMPM) approach. *Mitig. Adapt. Strateg. Glob. Chang.* **2020**, *25*, 1545–1564. [CrossRef]
32. Araújo, I.; Jackson, R.; Borges Ferreira Neto, A.; Perobelli, F. European union membership and CO_2 emissions: A structural decomposition analysis. *Struct. Chang. Econ. Dyn.* **2020**, *55*, 190–203. [CrossRef]
33. Engo, J. Driving forces and decoupling indicators for carbon emissions from the industrial sector in Egypt, Morocco, Algeria, and Tunisia. *Environ. Sci. Pollut. Res. Int.* **2021**, *28*, 14329–14342. [CrossRef]
34. Kim, T.; Tromp, N. Carbon emissions embodied in China-Brazil trade: Trends and driving factors. *J. Clean. Prod.* **2021**, *293*, 126206. [CrossRef]
35. Vervliet, N.; Debals, O.; Sorber, L.; De Lathauwer, L. Breaking the Curse of Dimensionality Using Decompositions of Incomplete Tensors: Tensor-based scientific computing in big data analysis. *IEEE Signal Process. Mag.* **2014**, *31*, 71–79. [CrossRef]
36. Jackson, J.W.; VanderWeele, T.J. Decomposition Analysis to Identify Intervention Targets for Reducing Disparities. *Epidemiology* **2018**, *29*, 825–835. [CrossRef]

37. Leal, P.A.; Marques, A.C.; Fuinhas, J.A. Decoupling economic growth from GHG emissions: Decomposition analysis by sectoral factors for Australia. *Econ. Anal. Policy* **2019**, *62*, 12–26. [CrossRef]
38. Wang, S.; Zhu, X.; Song, D.; Wen, Z.; Chen, B.; Feng, K. Drivers of CO_2 emissions from power generation in China based on modified structural decomposition analysis. *J. Clean. Prod.* **2019**, *220*, 1143–1155. [CrossRef]
39. Riener, M.; Kainulainen, J.; Henshaw, J.D.; Orkisz, J.H.; Murray, C.E.; Beuther, H. GAUSSPY+: A fully automated Gaussian decomposition package for emission line spectra. *Astron. Astrophys.* **2019**, *628*, A78. [CrossRef]
40. Sun, T.; Hobbie, S.E.; Berg, B.; Zhang, H.; Wang, Q.; Wang, Z.; Hättenschwiler, S. Contrasting dynamics and trait controls in first-order root compared with leaf litter decomposition. *Proc. Natl. Acad. Sci. USA* **2018**, *115*, 10392–10397. [CrossRef] [PubMed]
41. Towne, A.; Schmidt, O.T.; Colonius, T. Spectral proper orthogonal decomposition and its relationship to dynamic mode decomposition and resolvent analysis. *J. Fluid Mech.* **2018**, *847*, 821–867. [CrossRef]
42. Herviou, L.; Bardarson, J.H.; Regnault, N. Defining a bulk-edge correspondence for non-Hermitian Hamiltonians via singular-value decomposition. *Phys. Rev. A* **2019**, *99*, 52118. [CrossRef]
43. Raza, M.Y.; Lin, B. Decoupling and mitigation potential analysis of CO_2 emissions from Pakistan's transport sector. *Sci. Total Environ.* **2020**, *730*, 139000. [CrossRef] [PubMed]
44. Pita, P.; Winyuchakrit, P.; Limmeechokchai, B. Analysis of factors affecting energy consumption and CO_2 emissions in Thailand's road passenger transport. *Heliyon* **2020**, *6*, e05112. [CrossRef] [PubMed]
45. Chontanawat, J.; Wiboonchutikula, P.; Buddhivanich, A. An LMDI decomposition analysis of carbon emissions in the Thai manufacturing sector. *Energy Rep.* **2020**, *6*, 705–710. [CrossRef]
46. Hasan, M.M.; Wu, C. Estimating energy-related CO_2 emission growth in Bangladesh: The LMDI decomposition method approach. *Energy Strateg. Rev.* **2020**, *32*, 100565. [CrossRef]
47. Taka, G.N.; Huong, T.T.; Shah, I.H.; Park, H.S. Determinants of Energy-Based CO_2 Emissions in Ethiopia: A Decomposition Analysis from 1990 to 2017. *Sustainability* **2020**, *12*, 4175. [CrossRef]
48. Yasmeen, H.; Wang, Y.; Zameer, H.; Solangi, Y.A. Decomposing factors affecting CO_2 emissions in Pakistan: Insights from LMDI decomposition approach. *Environ. Sci. Pollut. Res.* **2020**, *27*, 3113–3123. [CrossRef]
49. Ozturk, I.; Majeed, M.T.; Khan, S. Decoupling and decomposition analysis of environmental impact from economic growth: A comparative analysis of Pakistan, India, and China. *Environ. Ecol. Stat.* **2021**, *28*, 793–820. [CrossRef]
50. Cansino, J.M.; Sánchez Braza, A.; Espinoza, N. Moving towards a green decoupling between economic development and environmental stress? A new comprehensive approach for Ecuador. *Clim. Dev.* **2021**, 1–19. [CrossRef]
51. Zhang, Y. Basic Theory and Discipline System of World Economy. *World Econ. Stud.* **2020**, *7*, 3–16+135.
52. Daudin, G.; Rifflart, C.; Schweisguth, D. Who produces for whom in the world economy? *Can. J. Econ. Can. D'économique* **2011**, *44*, 1403–1437. [CrossRef]
53. Balié, J.; Del Prete, D.; Magrini, E.; Montalbano, P.; Nenci, S. Does Trade Policy Impact Food and Agriculture Global Value Chain Participation of Sub-Saharan African Countries? *Am. J. Agric. Econ.* **2019**, *101*, 773–789. [CrossRef]
54. Liu, Q.; Zhu, Y.; Yang, W.; Wang, X. Research on the Impact of Environmental Regulation on Green Technology Innovation from the Perspective of Regional Dif-ferences: A Quasi-natural Experiment Based on China's New Environmental Protection Law. *Sustainability* **2022**, *14*, 1714. [CrossRef]
55. Johnson, R.C. Measuring Global Value Chains. *Annu. Rev. Econ.* **2018**, *10*, 207–236. [CrossRef]
56. Linsi, L.; Mügge, D.K. Globalization and the growing defects of international economic statistics. *Rev. Int. Polit. Econ.* **2019**, *26*, 361–383. [CrossRef]
57. Syverson, C. Macroeconomics and Market Power: Context, Implications, and Open Questions. *J. Econ. Perspect.* **2019**, *33*, 23–43. [CrossRef]
58. Koopman, R.; Wang, Z.; Wei, S. Tracing Value-Added and Double Counting in Gross Exports. *Am. Econ. Rev.* **2014**, *104*, 459–494. [CrossRef]
59. Leontief, W. *Input-Output Economics*; Oxford University Press: New York, NY, USA, 1966; ISBN 9780196315690.
60. Leontief, W. Environmental repercussions and the economic structure: An input-output approach. *Rev. Econ. Stat.* **1970**, *52*, 262–271. [CrossRef]
61. Montibeler, E.E.; de Oliveira, D.R.; Cordeiro, D.R. Fundamental economic variables: A study from the leontief methodology. *EconomiA* **2018**, *19*, 377–394. [CrossRef]
62. Mardones, C.; Silva, D. Evaluation of Non-survey Methods for the Construction of Regional Input–Output Matrices When There is Partial Historical Information. *Comput. Econ.* **2022**, 1–33. [CrossRef]
63. University of Groningen. WIOD 2016 Release. Available online: https://www.rug.nl/ggdc/valuechain/wiod/wiod-2016-release (accessed on 18 January 2022).
64. Goh, T.; Ang, B.W. Tracking economy-wide energy efficiency using LMDI: Approach and practices. *Energy Effic.* **2019**, *12*, 829–847. [CrossRef]
65. Alajmi, R.G. Factors that impact greenhouse gas emissions in Saudi Arabia: Decomposition analysis using LMDI. *Energy Policy* **2021**, *156*, 112454. [CrossRef]
66. Yang, Y.; Yang, W.; Chen, H.; Li, Y. China's energy whistleblowing and energy supervision policy: An evolutionary game perspective. *Energy* **2020**, *213*, 118774. [CrossRef]

67. Doytch, N.; Narayan, S. Does transitioning towards renewable energy accelerate economic growth? An analysis of sectoral growth for a dynamic panel of countries. *Energy* **2021**, *235*, 121290. [CrossRef]
68. Gao, H.; Yang, W.; Wang, J.; Zheng, X. Analysis of the Effectiveness of Air Pollution Control Policies based on Historical Evaluation and Deep Learning Forecast: A Case Study of Chengdu-Chongqing Region in China. *Sustainability* **2021**, *13*, 206. [CrossRef]
69. Wang, S.; Zeng, J.; Liu, X. Examining the multiple impacts of technological progress on CO_2 emissions in China: A panel quantile regression approach. *Renew. Sustain. Energy Rev.* **2019**, *103*, 140–150. [CrossRef]
70. Khan, A.N.; En, X.; Raza, M.Y.; Khan, N.A.; Ali, A. Sectorial study of technological progress and CO_2 emission: Insights from a developing economy. *Technol. Forecast. Soc. Change* **2020**, *151*, 119862. [CrossRef]
71. Marrucci, L.; Marchi, M.; Daddi, T. Improving the carbon footprint of food and packaging waste management in a supermarket of the Italian retail sector. *Waste Manag.* **2020**, *105*, 594–603. [CrossRef]
72. Marrucci, L.; Daddi, T.; Iraldo, F. The integration of circular economy with sustainable consumption and production tools: Systematic review and future research agenda. *J. Clean. Prod.* **2019**, *240*, 118268. [CrossRef]
73. National Bureau of Statistics of China. The Regulation of Three Industries Division. Available online: http://www.stats.gov.cn/tjsj/tjbz/201301/t20130114_8675.html (accessed on 28 February 2022).

Article

Random Forests Assessment of the Role of Atmospheric Circulation in PM$_{10}$ in an Urban Area with Complex Topography

Piotr Sekula [1,2,*], Zbigniew Ustrnul [2,3], Anita Bokwa [3], Bogdan Bochenek [2] and Miroslaw Zimnoch [1]

1. Faculty of Physics and Applied Computer Science, AGH University of Science and Technology, 30-059 Kraków, Poland; zimnoch@agh.edu.pl
2. Institute of Meteorology and Water Management, National Research Institute, IMGW-PIB, 01-673 Warszawa, Poland; zbigniew.ustrnul@imgw.pl (Z.U.); bogdan.bochenek@imgw.pl (B.B.)
3. Institute of Geography and Spatial Management, Jagiellonian University, 30-387 Kraków, Poland; anita.bokwa@uj.edu.pl
* Correspondence: piotr.sekula@fis.agh.edu.pl; Tel.: +48-516-467-918

Abstract: This study presents the assessment of the quantitative influence of atmospheric circulation on the pollutant concentration in the area of Kraków, Southern Poland, for the period 2000–2020. The research has been realized with the application of different statistical parameters, synoptic meteorology tools, the Random Forests machine learning method, and multilinear regression analyses. Another aim of the research was to evaluate the types of atmospheric circulation classification methods used in studies on air pollution dispersion and to assess the possibility of their application in air quality management, including short-term PM$_{10}$ daily forecasts. During the period analyzed, a significant decreasing trend of pollutants' concentrations and varying atmospheric circulation conditions was observed. To understand the relation between PM$_{10}$ concentration and meteorological conditions and their significance, the Random Forests algorithm was applied. Observations from meteorological stations, air quality measurements and ERA-5 reanalysis were used. The meteorological database was used as an input to models that were trained to predict daily PM$_{10}$ concentration and its day-to-day changes. This study made it possible to distinguish the dominant circulation types with the highest probability of occurrence of poor air quality or a significant improvement in air quality conditions. Apart from the parameters whose significant influence on air quality is well established (air temperature and wind speed at the ground and air temperature gradient), the key factor was also the gradient of relative air humidity and wind shear in the lowest troposphere. Partial dependence calculated with the use of the Random Forests model made it possible to better analyze the impact of individual meteorological parameters on the PM$_{10}$ daily concentration. The analysis has shown that, for areas with a diversified topography, it is crucial to use the variability of the atmospheric circulation during the day to better forecast air quality.

Keywords: random forests; atmospheric circulation; air quality; machine learning; complex topography

1. Introduction

The abundant air pollution with particulate matter (PM) is a serious environmental and social problem in many regions all over the world [1–4]. Exposure to ambient PM concentration with a diameter below 10 μm (PM$_{10}$) increases the possibility of preterm birth [5], deaths from respiratory disease [6] and also causes lung irritation, cellular damage, coughing asthma, and cardiovascular diseases [7]. High PM concentrations in urbanized areas are the consequence of the interaction of many factors, including anthropogenic and natural sources of air pollution, chemical and physical reactions between primary and secondary pollutants, and dispersion conditions determined by atmospheric circulation types, meteorological conditions, and meso- and microclimatic features of the analyzed

Citation: Sekula, P.; Ustrnul, Z.; Bokwa, A.; Bochenek, B.; Zimnoch, M. Random Forests Assessment of the Role of Atmospheric Circulation in PM$_{10}$ in an Urban Area with Complex Topography. *Sustainability* **2022**, *14*, 3388. https://doi.org/10.3390/su14063388

Academic Editors: Weixin Yang, Guanghui Yuan and Yunpeng Yang

Received: 20 January 2022
Accepted: 10 March 2022
Published: 14 March 2022

Publisher's Note: MDPI stays neutral with regard to jurisdictional claims in published maps and institutional affiliations.

Copyright: © 2022 by the authors. Licensee MDPI, Basel, Switzerland. This article is an open access article distributed under the terms and conditions of the Creative Commons Attribution (CC BY) license (https://creativecommons.org/licenses/by/4.0/).

area [1,8,9]. Numerous studies confirm that atmospheric circulation is an important factor determining the level of air pollution in the lower troposphere, especially in urbanized and industrial areas, which are characterized by elevated pollution emissions [3,10–13]. The atmospheric circulation processes contribute not only to the dispersion of pollution but also to its transport over great distances from emission sources [14]. Previous research indicated that the duration of air pollution episodes is mainly influenced by the processes of atmospheric blocking and atmospheric stagnation, which contribute to the accumulation of pollutants near the ground, especially during wintertime periods [3,15,16]. Analyses of future climate change indicate that the occurrence of an increase in air stagnation cases is expected [17,18]. Studies performed for Thessaloniki showed that smog episodes can also occur often under weak flow conditions, with warm air advection, as a consequence of stabilization of the lower troposphere and limited vertical mixing [19]. A similar effect of reducing the available mixing volume caused by warm air advection occurs often in mountain valleys and is linked to foehn occurrence [20,21]. It is worth mentioning that similar local weather conditions can occur under very different rearrangements of the large-scale flow; therefore, there is a need to study the relations between unfavorable local pollutant dispersion conditions and large-scale atmospheric circulation, with the impact of particular environmental features.

The problem of the increased PM_{10} concentration level in the European Union is common for all the nations which are members; in 2018, the daily PM_{10} concentration limit (50 µg·m^{-3}) was exceeded in numerous cities in Poland, Bulgaria, the Czech Republic, Croatia, Hungary, Italy and Slovenia [22]. In Poland, poor air quality is a problem, especially in southern regions, both in urban agglomerations and many small localities, where the highest number of days with exceedance of the daily limit of PM_{10} concentration on the national scale is noted [23]. The problem of air quality in Southern Poland also concerns the area of Kraków, the second largest city in Poland in terms of the number of inhabitants, where air pollution has been a serious and unsolved environmental and social problem for several decades [24–26]. In the Małopolska region, where Kraków is located, the main source of PM_{10} is the emission from the municipal and housing sector (78.9% of the annual emission), from transportation (5%), and from industry (7.8%). During recent decades, there have been many actions aimed to reduce local emissions of PM_{10} and SO_2 from different sectors. Those actions include liquidation of solid fuel boilers, thermal modernization of buildings, installation of renewable energy sources, modernization of public transport and heating networks or the expansion of bicycle routes. As a result, the air quality in the city has gradually improved, although the PM_{10} daily limits are still exceeded during the cold seasons [23]. In addition, on 1 September 2019, the prohibition of solid fuels usage in individual heating devices in Kraków was introduced, which could partially contribute to the reduction of PM concentration level during the cold season. The atmospheric circulation conditions play a crucial role in determining the air quality in the city, as it is located in the Wisła River valley, in an area of very diversified relief. The properties of planetary boundary layer (PBL) are strongly modified both by the relief and the synoptic situation, and so are the air pollution's dispersion conditions which in turn affect the concentration of pollutants. Studies of fog occurrence for Kraków city for the period from 1965 to 2015 indicated that fog occurred usually on days with non-advective anticyclonic types Ca and Ka or cyclonic and anticyclonic advection from sector S-SW (types SWa and Sc) according to Niedźwiedź classification [27]. This indicated that the majority of winter fogs at Kraków might be related to air pollution from heating during frosty anticyclonic winter weather. Research of long-term variability of the cloud for Kraków has shown that the greatest cloudiness and one of the smallest variabilities are associated with cyclonic situations involving northerly and northeasterly advection. The relationships between the cloudiness and atmospheric circulations were stronger during the cold half of the year than during the warm half, when the radiation factor plays a major role [28]. One of the situations when the influence of atmospheric circulation on air pollution dispersion is well visible is the occurrence of foehn winds, which can worsen or improve the air quality in the city [21].

High weather variability in Poland is associated with frequent movement of low and high-pressure systems [29]. Studies of circulation types for Kraków in the 20th century were summarized by Z. Ustrnul [30]. Significant variation in the annual incidence of individual circulation types according to the Niedźwiedź classification was found; the most frequent were anticyclonic non-directional types (high-pressure center and anticyclonic wedge or ridge); they constituted 15% of all cases during the year. The second most common types were those with the advection of air masses from the western sector (SW-W-NW) (both during anticyclonic and cyclonic situations), with the frequency reaching a total of 40% during the year. The least frequent were the cyclonic types, with the advection of air from the North and the eastern sector. In the individual seasons (from spring to winter), there was quite large variability in dominant circulation types. A slight positive trend was observed for circulation types with air advection from the West. In the 20th century, circulation types were characterized by high inter-annual frequency variability and the absence of distinct, characteristic periods, with the prevalence of certain types.

Assessment of the role of atmospheric circulation on PM_{10} concentration level during a particular period is highly challenging because many factors including emission level, changeable weather conditions, microclimatic features and chemical and physical processes affect the air quality levels.

Recently, there has been a growing interest in the application of machine learning techniques in statistical analysis [31,32] and forecasting air quality over the wide temporal and spatial scale [33,34]. The most commonly used machine learning tools include Artificial Neural Network [35], Deep Neural Network, Extreme-Gradient Boosting [36] and Random Forests [33]. Machine learning techniques have been successfully applied to assess population exposure to poor air quality in metropolitan areas [33], downscaling of air pollutants at a higher resolution [37], and also meteorological normalization used in air quality trend analysis [31]. Random Forests, which is a machine learning method based on constructing decision trees, is widely used for regression and classification. One of the main advantages of this method, besides its being accurate and straightforward in implementation, is the simple and intuitive way of accessing variables that are important in the process of training the model in complex and nonlinear problems. The research was undertaken in order to assess the quantitative influence of atmospheric circulation on the pollutant concentration in the area of Kraków, and to compare the results for the study period mentioned with the research from earlier decades. The research was also aimed to evaluate two main groups of the classification methods of atmospheric circulation types' used in studies on air pollution dispersion (described in Section 2.5), and to assess the possibility of their application in air quality management. In our study, we were focused more on the interpretation of the importance of variables by the Random Forests technique than predictions of the model itself. Kraków is an adequate study area for such considerations as, on one hand, it is located in diversified environmental conditions, and on the other hand, relatively long series of air quality measurements are available. Two different classifications of atmospheric circulation were used in the present study: a manual classification by Niedźwiedź [29] and an automatic classification by Lityński [38,39]; those two different classifications were used with the aim of minimizing the risk of misinterpretation of the results, and both classification methods were widely used by different groups of researchers in studies of atmospheric types over Central Europe [14,40,41]. Previous studies concerning the influence of atmospheric circulation on air quality in Krakow, with the application of Niedźwiedź classification, have been summarized in the monograph by J. Godłowska [42]. The research indicated that during air masses advection from S-SW sector and non-directional circulation types (high-pressure center and anticyclonic wedge or ridge), wind speed near the ground was reduced, which in consequence could lead to the occurrence of a high-level PM_{10} concentration during the cold season. Studies of atmospheric stability in Kraków using SODAR for the period 1994–1999 showed that, for types with air masses advection from a direction between 135° and 225°, the anticyclonic wedge and the high-pressure center, the duration of elevated thermal inversions in the cold season was the longest.

Understanding the relationship between the pollutants' concentration and the atmospheric circulation, at the synoptic and local scale, is crucial for forecasting air pollution episodes and minimizing the negative impact of air pollution on the health of city residents and on the condition of the natural environment.

2. Material and Methods

Data used in the present study come from different sources and cover the period October 2000–September 2020 (additionally, the period October 2021–December 2021 was selected for operational tests of the predictive model). Data analyses were realized for two sub-periods: cold half-year (October–March) and warm half-year (April–September), owing to significant differences in air pollution emissions and concentrations, which have been observed in Kraków in those sub-periods. First, data sets are described, and their basic statistical features are shown, then the methods combining data from different data sets are presented. Air quality in Kraków was characterized with data on PM_{10} as the allowed concentrations of that pollutant are exceeded much more frequently than the concentrations of other pollutants. The authors are aware that the division of the year proposed above is only one of the options available, as, in particular years, the frequency of circulation types and meteorological conditions may change significantly; however, such a division was also used in other climatological studies [43,44].

2.1. Study Area

Kraków is the second largest city in Poland, located in the Małopolska (Lesser Poland) region, with an area of 326.8 km^2 and the number of inhabitants reaching almost 800,000 [45]. The Kraków agglomeration consists of the city itself and the highly populated towns and villages which surround it, and the total number of inhabitants is estimated to exceed 1 million. The city's area belongs to three different geographical regions and geological structures presented at Figure 1, i.e., the Polish Uplands, the Western Carpathians, and the basins of the Carpathian Foredeep in between. The central part of the city is located in the Wisła River valley, at an altitude of about 200 m a.s.l. In the western part of Kraków, the valley is as narrow as 1 km; however, in the eastern part of the city, the valley widens to about 10 km and there is a system of river terraces (Figure 1b). The hilltops bordering the city to the north and the south reach about 100 m above the river valley floor, similar to the hilltops in the western part of the valley which means that the city is located in a semi-concave landform (open only to the east) and sheltered from the prevailing western winds (Figure 1b). The local scale processes linked to the impact of relief include, for example, katabatic flows, cold air pool formation, frequent air temperature inversions, much lower wind speed in the valley floor than at the hilltops [46]. All the factors mentioned contribute to the poor natural ventilation of the city, and one of its consequences is the occurrence of high PM_{10} concentration levels, especially during heating seasons.

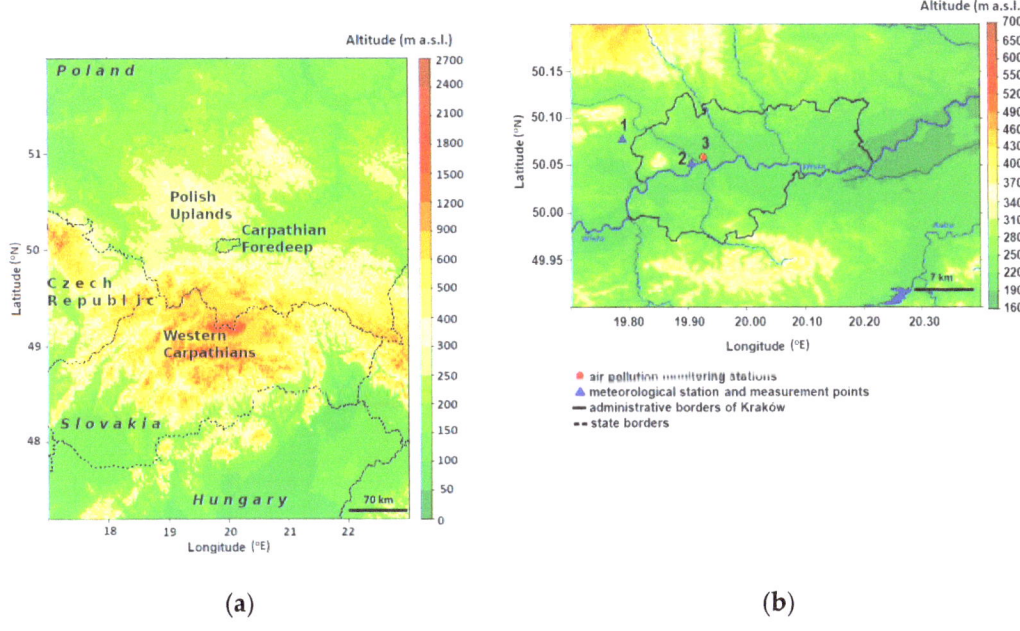

Figure 1. Location of the region studied: (**a**). in Central Europe, (**b**). at the junction of the Wisła River valley, Polish Uplands and the Western Carpathian Foothills. Numbers included in Figure 1b are described in Table 1.

Table 1. Location of meteorological and air quality stations in Kraków and its vicinities, and elements used in the study.

No.	Station	Lat N	Lon E	Altitude (m a.s.l.)	Manager of the Station	Landform	Parameters	Data Availability Period	Data Resolution
1	Balice	50.08	19.80	237	IMWM-NRI	Valley bottom	V, D, T, RH, C, PP	1960–currently	1 h, 3 h and 1 day
2	TV mast: 2 m a.g.l. 100 m a.g.l.	50.05	19.90	222 272 322	JU	Valley bottom	T	1.01.2010–currently	3 h
3	Krasińskiego St	50.06	19.93	207	NIEP	Valley bottom	PM_{10}	1.01.2000–currently	1 day

Explanations: V—wind speed (m·s^{-1}), D—wind direction, T—air temperature (°C), RH—relative humidity (%), C—cloudiness (oktas), PP—atmospheric precipitation (mm), PM_{10}—mean daily PM_{10} concentration (µg·m^{-3}).

2.2. Instrumental Meteorological Data

Weather data for Kraków were obtained from the meteorological station located in the Wisła River valley (Balice). The station is administered by the Institute of Meteorology and Water Management—National Research Institute (IMWM-NRI). Measurements of air temperature in the vertical profile of the valley were performed by the Jagiellonian University (JU) at the television mast of EMITEL company, located in the western part of the city (Bokwa, 2010). Measurements of meteorological parameters at the point administered by JU and IMWM-NRI were realized in accordance with WMO guidelines [47]. The location of measurement points and details on weather data used in the study are included in Figure 1 and Table 1. The measurements from the TV mast were crucial in the analysis of ground thermal stratification in the Wisła River valley at the local scale. These measurements were not available for the whole study period.

2.3. Atmospheric Reanalysis

In order to analyze the stratification of the lower troposphere for the period October 2000–September 2020, ERA5 reanalysis provided by European Center for Medium Range

Weather Forecasts [48] was used in this study. Air temperature, relative humidity and wind components data from pressure levels 975, 925 and 850 hPa were applied, at 00:00, 6:00, 12:00 and 18:00 UTC, respectively, for grid point representing Kraków (geographical coordinates 50° N and 20° E). The pressure level 1000 hPa was not used in the analysis owing to the fact that, in some cases, this level could be below the ground level.

2.4. Air Quality Measurements

Data on PM_{10} concentrations in Kraków come from the databases of the National Inspectorate of Environmental Protection (NIEP) [49]. The methodology for measuring PM_{10} concentration was realized in accordance with the guidelines of the European Parliament and of the Council included in Directive 2008/50/EC [50]. Daily data from the measurement point located in Krasińskiego St. for the period October 2000–September 2020 were used (Table 1). The measurement point is located in a street canyon, in the city center, at the bottom of the Wisła River valley, with a very busy municipal transportation route and intensive traffic. A comparison of mean daily PM_{10} concentration from Krasińskiego St. and two other air quality stations: Kurdwanów district and Bulwarowa St., located in the eastern and northern part of the city, for the common period 2010–2020 (3556 days), confirmed a high correlation between measurements from all those points. For the analysis of daily PM_{10} concentration the Pearson correlation coefficient was used, for three pairs of stations: Krasińskiego–Bulwarowa, Krasińskiego–Kurdwanów and Bulwarowa–Kurdwanów, where correlation coefficients were close to 0.93. The station in Krasińskiego St. is characterized by an increased level of daily PM_{10} concentration in comparison with other measurement points in Kraków during the year.

The period October 2000–September 2020 is suitable for showing the seasonal, long-term variability of the pollutants' concentration resulting from changes in the level of air pollutants emissions as well as fluctuations in circulation conditions.

Appendix A summarized air quality measurements used in this study, with special focus on the variability of PM_{10} daily concentration in cold and warm half-years, number of days with exceedance of selected concentration limits and deseasonalized trend observed in the multiyear period.

2.5. Atmospheric Circulation Classification

According to the suggestions of many authors [29,51–53], more than one classification of circulation types was used. Owing to the methodological approach, two classifications of circulation types have been chosen. Each of them represents a different group of atmospheric circulation classifications. Therefore, they differ essentially in many features and, above all, in the method of distinguishing individual types. The first classification included is the traditional, manual approach often used in Poland and developed by T. Niedźwiedź [29]. The second one is an objective classification according to Lityński's original concept [38]. Taking into account these 2 different classifications allowed for a more objective look at the impact of circulation and its changes in the analyzed 20-year period on the state of the atmosphere, including the concentration of pollution in the study area.

Classification by Lityński is based on an automatic approach, which may be considered, in simplified terms, as the objective one. In the literal sense, it is not like that, because it is based on arbitrarily imposed criteria; however, this approach is different from the manual and obviously subjective one proposed by Niedźwiedź. Both classifications are based on different input data sources. The division of Lityński uses numerical data (currently grid data), while the division of Niedźwiedź is based on the assessment of synoptic daily maps (charts). The spatial scale is also different in both classifications. The Niedźwiedź classification is a typical mesoscale one, while the Lityński classification characterizes the circulation on a larger scale. To sum up, both classifications have a different synoptic approach, and their application seems advisable.

Detailed information circulation classification characteristics for both approaches with analysis of multiyear trends for individual atmospheric patterns are summarized in Appendix B.

2.6. Atmospheric Stratification Determination

Data on air temperature and relative humidity provided by the European Center for Medium Range Weather Forecasts for atmospheric pressure levels 975, 925 and 850 hPa representing the point with geographical coordinates 50° N and 20° E were used to determine the presence of low (layer 975–925 hPa) or upper (925–850 hPa) inversion layers. The atmospheric stratification gradient was determined as the difference between the two nearest levels (layers 975–925 hPa and 925–850 hPa). Lower limits for the occurrence of air temperature inversion and air relative humidity inversion in the lower troposphere were set to 0 °C and 10%, respectively.

Additionally, with the aim of analyzing the near-ground thermal inversion layer, the measurements in the vertical profile (2 m to 100 m a.g.l.) obtained from the TV mast are used. The period of the day has been divided into two sub-periods of equal length:
- daytime period: from 6 to 17 UTC;
- nighttime period: from 18 to 5 UTC the next day.

The near-ground thermal gradient was calculated as the difference between lower and upper measurement points. The lower limit of the occurrence of thermal inversion between two levels of the TV mast was set equal to +1 °C. The condition was checked for each time period separately, and then summed up for day and night periods for individual days. Data for the TV mast station were available for the period from January 2010 to September 2020.

2.7. Data Analysis

The data set created to assess the influence of meteorological conditions on air quality includes:
- Meteorological observations from Balice synoptic station with 6-h resolution: air temperature, relative air humidity, wind speed and direction, cloudiness, the 6-h sum of atmospheric precipitation;
- air temperature, relative air humidity and wind speed and direction at three pressure levels obtained from ERA5 reanalysis (975, 925 and 850 hPa); differences between neighboring pressure levels of air temperature, relative air humidity, wind speed and wind direction (layers 975–925 hPa and 925–850 hPa) with 6-h resolution;
- mean daily PM_{10} concentration from previous day;
- difference of mean daily PM_{10} concentration between current day and previous day (used for determining PM_{10} decrease);
- day of week;
- atmospheric circulation types on a certain day according to Niedźwiedź and Lityński classification.

With the aim of investigating the relation between PM_{10} concentration and meteorological conditions, the Random Forests algorithm was used, which is an ensemble machine learning method based on constructing many decision trees. This method combines a large number of small decision trees into new predictors, and therefore is able to make a better prediction. By using this method, it is possible to assess which variables have the highest importance in machine learning. In our study, we compared results from multilinear regression with stepwise selection and the Random Forests method. Studies of variable selection for Random Forests models were conducted with use of the Boruta method available in package Pomona on GitHub repository [54,55]. In order to provide the best of hyperparametric values, repeated leave-group-out cross-validation (LGOCV) was used. The resampling method LGOCV was available in the function trainControl in the caret R package. For the multilinear regression model, the stepwise Akaike Information Criterion (AIC) algorithm was used [56], which is available in the function stepAIC in the MASS R package. The meteorological database from Balice station and ERA-5 reanalysis and PM_{10} daily concentration at the previous day were used as an input to models that were trained to predict daily PM_{10} concentration and its day-to-day changes on a randomly sampled 75% of data. The remaining 25% was used to validate models. Two sets of input data, which differ in the time resolution of meteorological parameters listed above (6-h resolution data and daily averages obtained from 6-h resolution data), were used in the studies. This analysis aimed to answer the question of whether the increase of the temporal resolution of parameters describing weather conditions during the day would improve model accuracy. The hyperparameters tuning and selection of crucial variables was done separately for each Random Forests and multilinear regression model. The plots with variable importance are presented, for clarity only, for the most important parameters. With the aim of determining the partial relationship between daily averages of individual meteorological parameters and the level of the daily PM_{10} concentration, the optimized Random Forests model was used. Partial dependencies were obtained with use of the function partial_dependence available in the open-source package edarf in the R environment.

In both half-years, days with the worst air quality and days with a significant improvement of air quality in relation to the previous day were selected. That choice of this selection of analyses is due to the fact that such situations are important in terms of the inhabitants' health protection, but also for various environmental effects. In both groups of the cases selected, very high concentrations of PM_{10} occur, so the analyses should support the assessment of the atmospheric circulation and weather conditions which contribute to such situations. In the second group of cases, the analyses should additionally support the assessment of the conditions favorable for a sudden decrease of the PM_{10} concentrations, due to the change of the dispersion conditions. Owing to the fact that the distribution of PM_{10} daily concentration differs significantly between both half-years (see Figure 2), some criteria of the cases delimitation in both sub-periods differ, too.

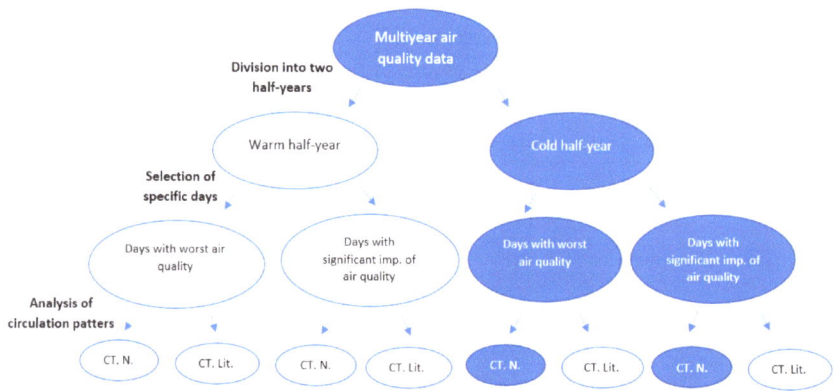

Figure 2. Procedure of data and analyses selection; elements with blue background represent research steps described in detail in the article. Explanation: CT. N.—atmospheric circulation types by Niedźwiedź; CT. Lit—atmospheric circulation types by Lityński.

The criteria used to distinguish the two groups of days are the following:

- Group 1: days with high PM_{10} concentration against the background of a particular half-year, which meet two conditions: daily PM_{10} concentration is greater than the upper quartile in the selected half-year (see Table A3) and greater than 50 or 40 $\mu g \cdot m^{-3}$ during cold or warm half-year, respectively. The number of days meeting the above conditions is 842 and 837 for cold and warm half-years, respectively.
- Group 2: days characterized by significant PM_{10} concentration decrease in relation to the previous day, which meet three conditions: the decrease is greater than 25% of the concentration on the previous day, the decrease of PM_{10} daily concentration is

equal at least to 20 or 10 µg·m^{-3}, in cold or warm half-years, respectively, and days assigned to Group 1 are omitted. The number of days meeting the above conditions is 634 and 461 for cold and warm half-years, respectively.

With the aim of better understanding the role of atmospheric circulation and local topography on the weather conditions over the analyzed region the distribution of the daily meteorological parameters from Balice station and vertical profiles from ERA5 reanalysis have been analyzed for individual circulation types for cold and half-years separately.

In the next step, the types of atmospheric circulation were assigned to the days from both groups and half-years), and then weather conditions for particular atmospheric circulation types were analyzed. The meteorological parameters used in this part of the study were selected based on the parameter importance obtained from the machine learning analyses. Data used in the study include wind speed and air temperature near the ground, atmospheric precipitation, vertical gradient of air temperature and relative air humidity, air temperature, relative humidity and wind speed at pressure level 925 hPa and wind speed difference in layer between 925 and 975 hPa. The aim of that step was to see whether there are significant differences in weather conditions during a certain atmospheric circulation type occurring in the two groups of days described above and the remaining days.

Selection of half-year for further analyses.

Comprehensive research for both half-years indicated that the problem of air quality in the warm half-year is insignificant compared to the cold half-year. In the analyzed multi-year period, an average share of days with exceedance of the PM_{10} daily limit in warm half-year was twice as low as in the cold half-year (34% and 67%, respectively). Furthermore, during the period 2016–2020, the average share of days with exceedance of the PM_{10} daily limit in the warm half-year was equal to 14%, with the lowest values in 2017 and 2020 (8% and 2%, respectively). Days with exceedance of daily PM_{10} limit occurred mostly in April and September (early spring and early autumn periods), while in June-July such cases almost did not occur. Therefore, only the cold half-year has been described in detail in this paper.

Selection of atmospheric circulation classification.

The analysis of the influence of atmospheric circulation on the dispersion conditions and the level of PM_{10} concentration was performed for both circulation type classifications (Lityński and Niedźwiedź classifications) for two half-years with particular attention to selected days with the worst air quality and a significant improvement in air quality.

The aim of the research was to determine which circulation type classification better separates the circulation patterns that negatively affect dispersion conditions from the patterns, from those which positively influence the air quality in an urbanized valley. The analysis with the use of both circulation classifications for both half-years and for both groups showed similar dependencies. In order to determine which type of circulation classification is more appropriate for the analysis of air quality in the cold half-year, the Gini coefficient [57] was determined for Niedźwiedź classification (11 and 21 types) and for Lityński classification (27 types). The Gini coefficient has been widely used to measure the inequality among values of a frequency distribution [58,59], the value ranges from 0 to 1. The zero value of the Gini coefficient indicates full uniformity of the distribution. The zero value of the Gini coefficient indicates the perfect equality of the distribution, while the greater the Gini coefficient refers to the greater spread of the distribution. The value of the Gini coefficient was similar for the two Niedźwiedź classifications (0.346 and 0.347), a slightly lower value was obtained for the Lityński classification, equal to 0.338. Owing to this fact and for the better clarity of the article, the paper presents only the analysis for 11 types of Niedźwiedź classification. Similar studies concerning the relation between circulation type classifications and smog days, using Gini coefficient, were conducted in COST Action 733 for air pollution in winter in Polish urban areas [12]. The analysis of the Gini coefficient for individual cold half-years also showed that the variability variation in air quality for individual types of circulation was greater for the types of Niedźwiedź classification (11 and 21 types) than for the Lityński classification (maximum difference was equal 0.037 and average difference equal to 0.010). An additional argument in favor of the selected circulation classification by Niedźwiedź is that it was designed to be most suitable for Southern Poland, while the Lityński circulation classification describes the atmospheric circulation in Central Europe [38].

Schematic representation of the scientific analysis has been presented in Figure 2.

3. Results

The analysis was focused on two groups of days determined for cold half-year:
- Group 1—days with the highest daily concentration; of PM_{10};
- Group 2—days with the greatest decrease day by day in the concentration of PM_{10}.

For both groups of days, the most frequent circulation types according to the Niedźwiedź classification were selected. Weather conditions during days from both groups were compared with remaining days for selected atmospheric circulation patterns.

With the aim of estimating the impact of individual meteorological parameters on air quality, the results of ensemble machine learning methods were used.

All the analyses of the influence of atmospheric circulation on air quality in the light of PM_{10} confirmed the significant role of circulation types. In the last 20 years, despite a significant reduction in emissions, which is the result of administrative pro-ecological activity, there are still serious smog episodes, when the average daily concentration of PM_{10} in the cold half-year period may exceed 100 $\mu g \cdot m^{-3}$. This, of course, applies to non-advective circulation types, although surprisingly high dust concentrations may also occur during the types from southerly advection. The results of the performed analyses, i.e., circulation type vs. PM_{10} concentration is included in the Appendix D.

Random Forests Analyses

At the first step the Random Forests and multilinear regression models were built to predict daily PM_{10} concentration with the use of two different meteorological data resolution sets (6-h resolution and daily averages from 6-h resolution data). The Boruta variable selection method was applied for Random Forests models, and it showed that for the model which uses daily averages, the following parameters: daily cloudiness, wind direction changes in the layer between 925 hPa and 850 hPa and wind direction at 850 hPa were unnecessary. For the Random Forests model which uses meteorological parameters with 6-h resolution 28 of 111 selected variables were rejected by the Boruta method, including both circulation types, wind direction at 850 hPa, wind direction and wind speed change in layer between 925 and 850 hPa, relative air humidity at 2 m a.g.l. at 0, 12 and 18 UTC, wind direction at 925 hPa at 0, 6 and 18 UTC and day of week. The results of both Random Forests models were similar, the average value of mean absolute error (MAE) and root-mean square error (RMSE) for both models were equal 19.6 $\mu g \cdot m^{-3}$ and 26.9 $\mu g \cdot m^{-3}$, respectively. The Random Forests models analysis for specific measured PM_{10} concentration ranges 0–50 (25% testing data), 50–100 (40% testing data) and 100–200 (27% testing data) indicated that the RMSE error was equal respectively to 18, 20, and 32 $\mu g \cdot m^{-3}$. The group of observations with PM_{10} concentration exceeding 200 $\mu g \cdot m^{-3}$ was relatively small (5% testing data), and the RMSE for this group was the largest equal to 42 $\mu g \cdot m^{-3}$. An example plot presenting comparison observations with the Random Forests model forecast is included in Appendix C, Figure A6a). In Figure 3 the most important parameters affecting daily PM_{10} concentration are presented. Air quality on the previous day (PM_{10} daily concentration) was the most important parameter for both models. For the sake of clarity of Figure 3, this parameter was not presented in the chart owing to the large differences between the GINI Index for this parameter and the next one. The analysis of variable importance for both Random Forests models confirmed the similarity of the results. Apart from the parameters whose significant influence on air quality is well established (air temperature and wind speed at the ground and air temperature gradient), the key factor was also the gradient of relative air humidity and wind shear in the lowest troposphere (layer between 975 and 925 hPa; Figure 3a).

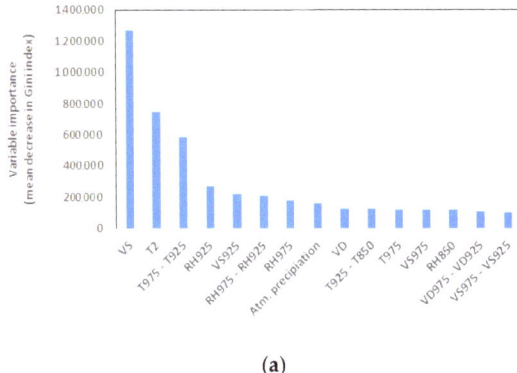 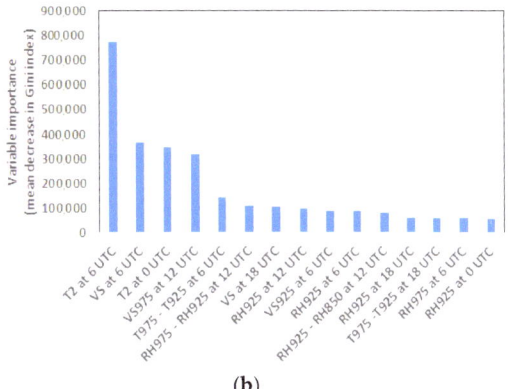

(a) (b)

Figure 3. Variable importance plots for Random Forests models trained to predict PM$_{10}$ daily levels in Krasińskiego air quality station with use (**a**) daily averages (**b**) 6-h resolution of meteorological parameters in cold half-years. Explanation: T2—temperature at 2 m a.g.l.; VS—wind speed at 10 m a.g.l.; VD—wind direction at 10 m a.g.l.; RH975, RH925 and RH850—relative air humidity at 975, 925 and 850 hPa, respectively; T975, T925 and T850—air temperature at 975, 925 and 850 hPa; VS975 and VS925—wind speed at 975 and 925 hPa; VD975 and VD925—wind direction at 975 and 925 hPa.

With the aim of analyzing the significance of individual parameters concerning the model accuracy, tests by removing a single variable from the model were performed (Figure 4). These studies also confirmed that the most important parameter was the PM$_{10}$ concentration level on the previous day. The lack of this variable in the model affected on MAE increased by almost 25% compared to the forecast results where all parameters were included.

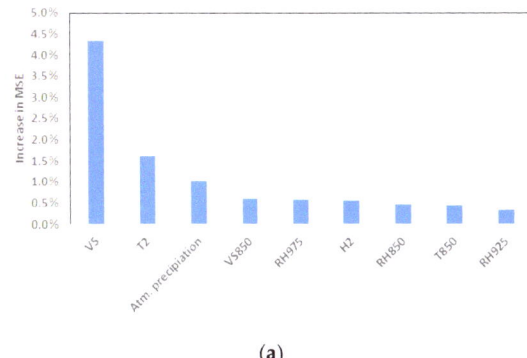 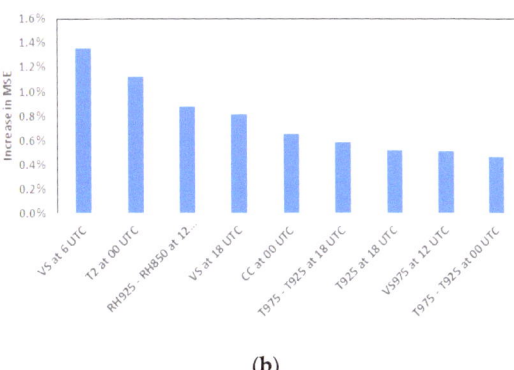

(a) (b)

Figure 4. Increase in mean square error (MSE) of predicted PM$_{10}$ daily levels by Random Forests models with use (**a**) daily averages (**b**) 6-h resolution of meteorological parameters in cold half-years. Explanation: T2—temperature at 2 m a.g.l.; VS—wind speed at 10 m a.g.l.; CC—cloudiness; RH975, RH925 and RH850—relative air humidity at 975, 925 and 850 hPa, respectively; T975, T925 and T850—air temperature at 975, 925 and 850 hPa; VS975—wind speed at 975 hPa.

Studies of multilinear regression for both data groups were done for the same teaching and testing sets. Variable criterion with use of Akaike algorithm showed that for data with daily averages of meteorological parameters, six variables were excluded in the analysis: relative air humidity at 2 m a.g.l., wind speed at three pressure levels (975 hPa, 925 hPa and 850 hPa), relative humidity gradient in layer between 925 and 850 hPa and relative humidity at 850 hPa. The results obtained for multilinear models were slightly worse for the Random Forests model which used the same data set, RMSE and MAE for multilinear models were equal to 29 µg·m^{-3} and 21.4 µg·m^{-3}, respectively. A comparison of results obtained from four Random Forests and multilinear regression models is presented in the Taylor diagram in Figure 5. The analysis of results presented at Figure 5 indicates that there are slight differences between the two model groups in Pearson correlation coefficient and standard deviation of predicted values. In the case of multilinear regression and meteorological parameters with 6-h resolution, the results of the multilinear model were close to the previous multilinear model. The application of the Akaike method to select the best parameters for the multilinear model caused a reduction of the parameters from 111 to 44 variables. The selection of crucial variables did not improve model performance, RMSE and MAE were equal to 29 µg·m^{-3} and 21.4 µg·m^{-3}, respectively. In this case the application of the Random Forests model for numerous variables showed better results than multilinear regression. The comparison of observations with the multilinear model forecast with the use of daily averages of meteorological parameters presented at Figure A6b) in Appendix C indicates that the model underestimates PM$_{10}$ daily concentration for values below 25 µg·m^{-3}. In contrast, Random Forests models more often overestimate PM$_{10}$ concentration than multilinear regression in the range between 0 and 50 µg·m^{-3} (Figure A7 in Appendix C).

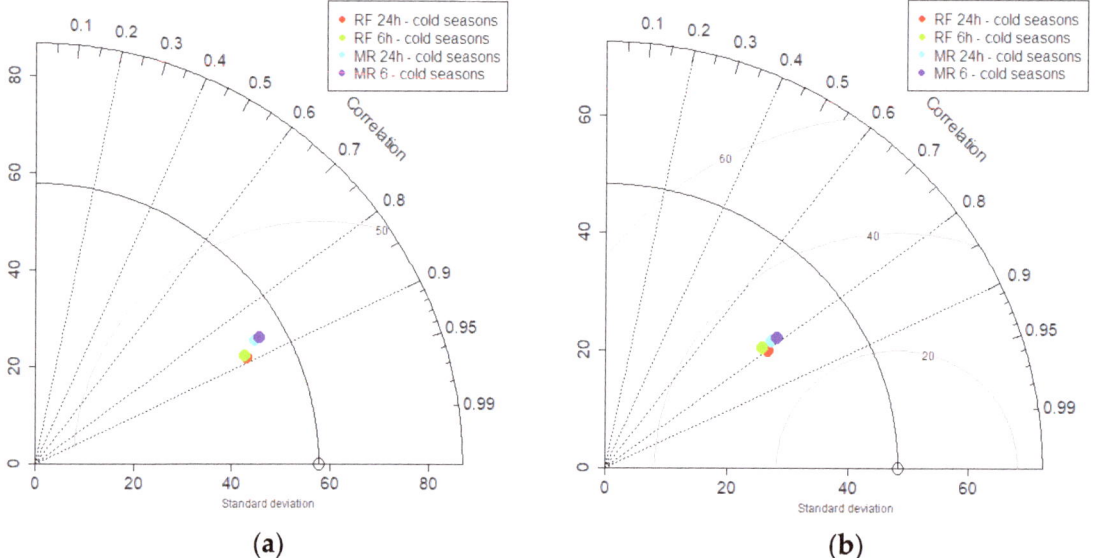

Figure 5. Taylor diagram for (**a**) predicted PM10 daily concentration and (**b**) day-to-day PM$_{10}$ daily concentration changes for Random Forests (RF) and multilinear regression models (MR) in cold half-years.

In the second part, Random Forests and multilinear regression models were used to predict day-to-day changes of PM$_{10}$ daily concentration. The same database as presented above was used in these studies, including measurements from the Balice station, ERA-5 reanalysis, two circulation types, day of the week, month, day of the year and PM$_{10}$ daily concentration at the previous day. The analysis of Random Forests and multilinear regression models showed similar results, the values of RMSE and MAE were equal on average to 30 µg·m^{-3} and 20 µg·m^{-3}, respectively. For this case hyperparameter tuning and parameter selection for Random Forests models did not significantly improve model accuracy (change of RMSE and MAE did not exceed 3%). It is worth to mentioning that variable selection with the use of the Boruta method for the Random Forests model which uses data with 6-h resolution were reduced from 110 to 27 variables. An example plot presenting comparison observations of day-to-day changes with the Random Forests model forecast is included in Appendix C, Figure A8b). In the case of a multilinear linear regression model with the same data set, stepwise the Akaike method selected 46 variables from 110 available. Verification of four models presented with the use of a Taylor diagram in Figure 6b indicates that the differences between them are negligible; however, comparison of density curves for Random Forests and multilinear regression with observations indicates that also the Random Forests model often predicts day-to-day PM_{10} concentration changes in a range between -10 and $25\ \mu g \cdot m^{-3}$ (Figure A9 in Appendix C). For both models groups the most important parameter was the PM_{10} concentration level on the previous day. Figure 6 presents parameters importance for Random Forests models affecting day-to-day PM_{10} concentration changes.

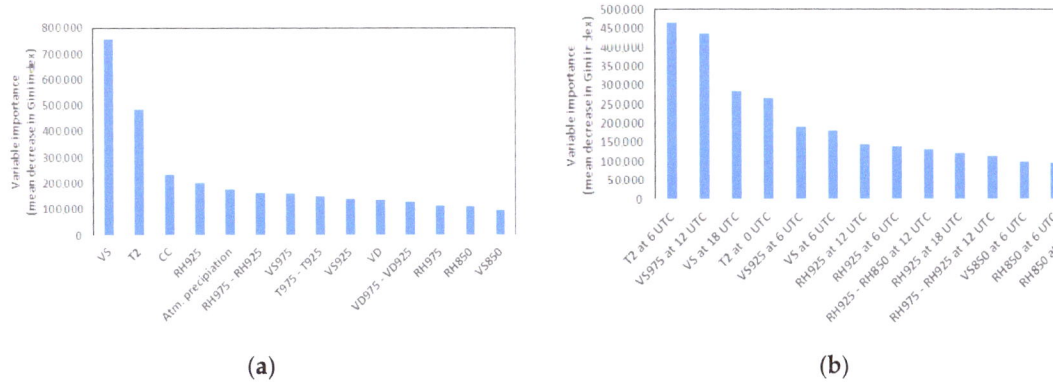

Figure 6. Variable importance plots for Random Forests models trained to predict PM_{10} daily levels in Krasińskiego air quality station with the use of: (**a**) daily averages, and (**b**) 6-h resolution of meteorological parameters in cold half-years.

Analyses of the importance of variables presented at Figure 6 showed that the most important parameters were air temperature and wind speed near the ground and at the closest pressure level 975 hPa. It is worth mentioning that relative air humidity and relative air humidity gradient were more crucial parameters than air temperature gradient in the layer between 975 and 925 hPa concerning the prediction of day-to-day PM_{10} changes (Figure 6a). Two additional sensitivity tests were performed. Firstly, one for the whole period, without dividing the data set into two half-years and another one with training the model with data from 2000 to 2015 and testing it with data from 2016 to 2020 (also without dividing data set into cold and warm half-years). In both cases we achieved a similar order of importance of parameters for both Random Forests models as in Figure 4, while scores were slightly improved, e.g., with a decrease of RMSE around by $7\ \mu g \cdot m^{-3}$ for both models. It can be explained by adding a warm half-year to the data set that is characterized by lower values of PM_{10} concentration level. Results obtained by multilinear linear regression models were slightly worse for both tests, mean differences between Random Forests models were equal to $3\ \mu g \cdot m^{-3}$ for RMSE. As mentioned before, the main motivation for using Random Forests was to determine which meteorological parameters should be considered for further analysis, but a comparison of the accuracy of our forecasts with similar models (both physical and based on machine learning) shows also the good predictive potential of such an approach [60–62].

The optimized Random Forests model built to predict daily PM_{10} concentration based on the daily averages presented above was used to analyze the partial relationship between individual meteorological parameters and the level of the PM_{10} daily concentration.

Figure 7 presents the partial dependence of predicted PM_{10} daily concentration for selected meteorological parameters. The plots show the value of the lower and upper quartiles in the analyzed cold half-years (dashed vertical lines). Detailed analysis of the daily thermal gradient between 925 and 975 hPa, has shown that predicted PM_{10} concentration did not differ significantly for the positive values of vertical gradients (the number of such days in all analyzed cold half-years did not exceed 25%—Figure 7b). For the range of the daily gradient values between $-3\,°C$ and $0\,°C$ in the layer between 925 and 975 hPa, the influence of this parameter on the daily value of the PM_{10} concentration increases significantly. During the days without absence of the elevated inversion in the layer between 850 and 925 hPa, the predicted value of PM_{10} concentration was close to the minimum value (low statistical importance). When the daily temperature gradient in the layer between 850 and 925 hPa decreases below $-3\,°C$, a significant increase in the predicted pollutant concentration can be observed. The plots of the dependence of the air humidity at the ground (Figure 7d) and at the pressure level of 925 hPa (Figure 7e) on predicted PM_{10} concentration has shown a different relationship. For the days when daily relative humidity of 2 m a.g.l. exceeds 80%, there is a linear increase of the predicted daily PM_{10} concentration. On the other hand, with a decrease in relative humidity at the height of 925 hPa the predicted PM_{10} concentration increases. The plots of predicted PM_{10} concentration from the relative humidity gradient in the layer between 925 and 975 hPa have shown gradual deterioration of air quality with the decrease of humidity gradient in the range of from 0 to -25%. The relationship between the average daily wind speed at 10 m a.g.l and predicted air pollution level presented at Figure 7g indicates a strong decrease of PM_{10} concentration for the wind speed in the range from 0 up to 5 $m·s^{-1}$; above this value the increase of wind speed did not significantly improve the air quality in the city. The relationship between the vertical wind shear in the layer between 925 and 975 hPa and the pollution concentration is presented in Figure 7h,i. When wind shear increases wind speed in the vertical profile, the crucial point is the exceedance of 5 $m·s^{-1}$. For such situations the increase in the speed difference between layer 925 and 975 hPa does not significantly improve the air quality. When the wind shear is associated with a significant change in the wind direction between the level of 925 and 975 hPa, an increase in the difference in wind directions negatively affects the predicted air quality in the valley.

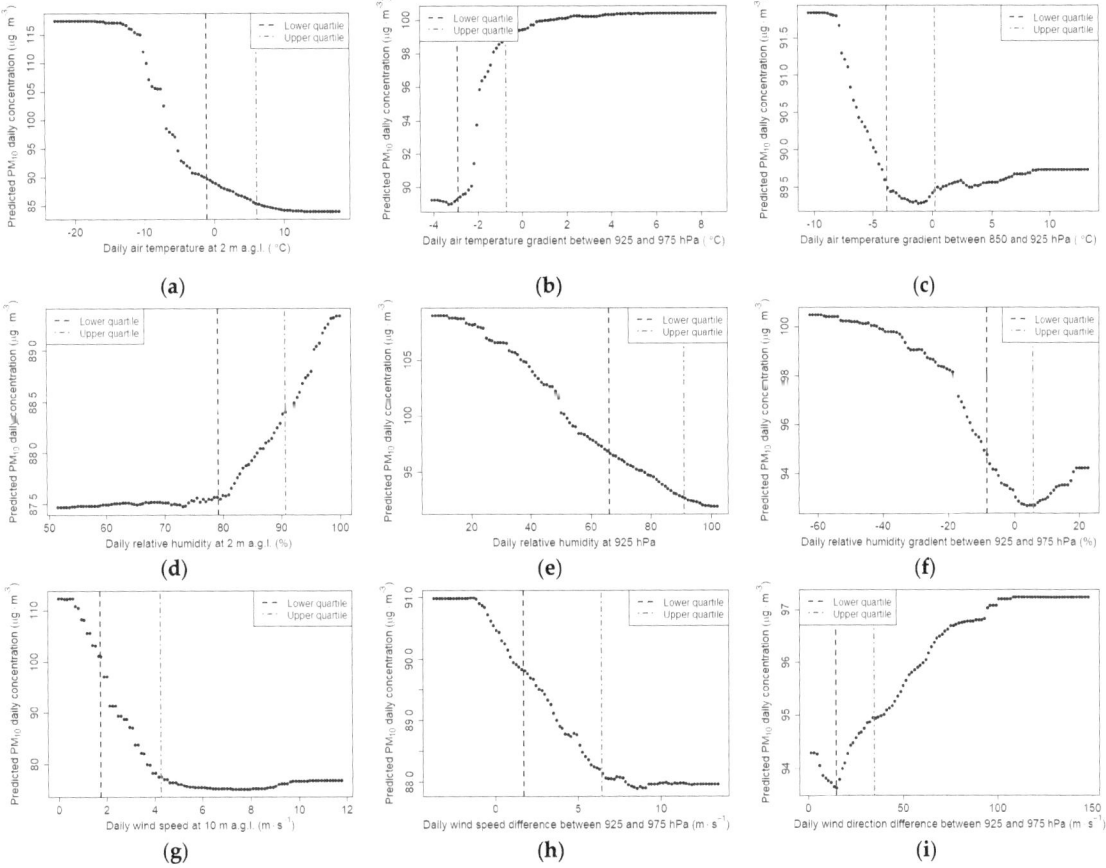

Figure 7. Partial dependence plots of daily (**a**) air temperature at 2 m a.g.l., (**b**) air temperature gradient between 925 and 975 hPa, (**c**) air temperature gradient between 925 and 975 hPa, (**d**) relative humidity at 2 m a.g.l., (**e**) relative humidity at 925 hPa, (**f**) relative humidity gradient between 925 and 975 hPa, (**g**) wind speed at 10 m a.g.l., (**h**) wind speed difference between 925 and 975 hPa and (**i**) wind direction difference between 925 and 975 hPa for the cold half-year obtained from Random Forests model. Subfigures present lower and upper quartiles of selected parameters in the cold half-years.

4. Discussion

The statistical analysis of the impact if meteorological conditions and atmospheric circulation types on air quality in Kraków with the use of machine learning methods made possible an objective selection of crucial parameters influencing the pollutant concentration level. In the studies presented, we have compared results from multilinear regression and the Random Forests method to predict daily PM_{10} concentration and its day-to-day changes for two sets of input data which differed in the temporal resolution of meteorological parameters. The application of 6-h resolution meteorological data in comparison with daily averages to predict daily PM_{10} concentration and its day-to-day changes showed similar results for both methods. This confirms the statement that the use of daily averages of meteorological parameters is sufficient to predict PM_{10} daily concentration a day ahead. Studies of the importance of variables' in predicting PM_{10} day-to-day changes the with the use of 6-h resolution data indicated that the number of crucial parameters was significantly lower than for predicting PM_{10} daily concentration for the Random Forests model (equal to 27 from 111 possible variables). It is also worth mentioning that in the case of predicting

PM_{10} concentration with the use of numerous variables (6-h data resolution), the multilinear regression model significantly reduced the number of variables while the Random Forests model selected more variables as important. Analyses of model performance showed that for this case Random Forests results were better than for multilinear regression models.

Additional tests with the use of measurements from October 2000 to September 2020 were carried out to estimate the impact of changes in meteorology and emissions during the recent cold season (October 2021 to December 2021). Eight different sets of training data were prepared for tests for Random Forests models and multilinear regression models (sets of two temporal resolutions: 6-h and daily averages; data for cold half-years only and for both half-years; training data for period from October 2000 to September 2020 and from January 2015 to September 2020). Analysis of the shorter period for model training was done to answer the question how models' performance is affected. An analysis of Taylor diagram plots (Figure A10 in Appendix C) indicated that the multilinear regression models had a higher standard deviation than the observations, indicating an excessively high variability of predicted PM_{10} concentration. The results obtained from Random Forests models were closer to the observations than the results from linear regression models (lower value of RMSE and standard deviation closer to the observation's). Secondly, using shorter periods for model training showed better results (lower RMSE and lower overestimation of PM_{10} concentration), however in individual smog episodes PM_{10} daily concentration was underestimated in comparison with forecasts using a longer training data set (Figure A11 in Appendix C). Analysis of the time course of predicted PM_{10} concentration showed that the multilinear model in some cases overestimated the PM_{10} decrease, while the Random Forests model performed better.

The conducted experimental studies (based on data from 2021) have shown that such analyses should take into account the circular data with greater resolution than the daily one. Considering only one type of circulation for the whole day does not make it possible to take into account the dynamics of circulation changes, which is particularly high over Central Europe in winter. On the basis of the existing classifications, especially the local one, a method of assessing the atmospheric circulation should be developed, taking into account the daily course (at least with 3 h resolution).

Furthermore, more detailed analysis of the importance of individual parameters on PM_{10} daily concentration level was available with the use of the Random Forests model. The results obtained were used in further analysis to investigate the dispersion of selected parameters for individual types of circulation.

The study on the influence of atmospheric circulation patterns on air quality made it possible to distinguish the dominant circulation types during which the probability of occurrence of poor air quality (Group 1) and a significant improvement in air quality conditions (Group 2) was the highest. Days with the high PM_{10} concentration at cold half-year, occurred mostly during the advection of air masses from the S-SW sector, non-directional anticyclonic situations (Ca + Ka type) and also anticyclonic situations with air advection from the W-NW sector. Such days were characterized by lower wind speed and air temperature at ground level and greater stability of the atmosphere during the day and night periods in comparison with days not assigned to both special groups (remaining days). According to the Mann–Whitney U test, the distribution of daily sums of precipitation was similar for dominant circulation types for days in Group 1 and remaining days, but the frequency of precipitation was lower for days in Group 1. Furthermore, during the daytime for days with high PM_{10} concentration, a local minimum of relative air humidity at level 925 hPa occurred frequently. The partial dependence of meteorological parameters obtained from the Random Forests model has also confirmed the negative effect of strong negative relative humidity gradient in layers between 925 and 975 hPa on air quality. In this case, advection of dry air masses at a height of 925 hPa, especially frequent for the S + SWa and W + NWa types, contributed to the increase in the stability of the atmosphere in the valley and resulted in a longer persistence of humid cold air pool. During the winter, when foehn wind occurs, there is often the advection of warm and dry air masses above the analyzed region [63]. Additionally, previous studies pointed out that circulation types S + SWa, Ca + Ka enhance the occurrence of fog in Kraków which confirms the poor air pollution dispersion conditions linked to those circulation types [64]; the occurrence of haze and fog episodes is studied widely in different parts of the world in the context of air pollution control [65].

The second case studied consisted of days characterized by a significant improvement of air quality (Group 2). A significant reduction in the PM_{10} daily concentration occurred mostly for three circulation patterns (air advection from W-NW sector and cyclonic type with differentiated air advection—Cc + Bc type). These days were characterized by increased wind speed and a greater share of days with precipitation in comparison with remaining days. Atmospheric stratification (relative air humidity and air temperature gradient) was similar for dominant circulation patterns for days assigned to Group 2 and to remaining days. Near-ground temperature inversion for days in Group 2 during daytime almost did not occur, which was confirmed by local measurements and ERA5 reanalysis. It is also worth mentioning that during these days, the local maximum of relative air humidity in the layer 925 hPa during daytime occurred frequently.

Days with anticyclonic conditions with air advection from the W-NW sector are characterized by changeable weather conditions, which contributed to improvement or deterioration of air quality conditions during the cold half-year. The improvement is observed when air masses could penetrate into the valley and remove the cold air pool, while deterioration can be seen when air masses pass over the valley. Studies conducted over the region of the Dead Sea Valley using a high resolution WRF model [66] indicated that foehn wind intrusion into the valley depends on synoptic and mesoscale conditions which affect the vertical structure of the lower troposphere. For the cases with a high stable layer over the Dead Sea Valley, the foehn reached the valley floor, while during a low stable layer, it did not.

Studies of air quality in cold and warm half-years have shown that weak wind speed is one of the most important factors which deteriorates air quality. Owing to this fact circulation patterns which are characterized by weak wind speed, caused by the interaction of local orography (air advection from S-SW and W-NW sectors) and also atmospheric stagnation were the most important (non-advective types with anticyclonic situation, according to Niedźwiedź: type Ca + Ka). The high importance of wind speed on air quality was confirmed by numerous previous studies [3,67,68]; however, studies of the partial dependence of meteorological individual parameters for daily PM_{10} level showed nonlinear dependency [31].

Furthermore, studies of individual meteorological parameters have shown that vertical wind shear can worsen but also improve air quality in the valley. An increase of wind speed difference between the layers 925 and 975 hPa had a positive impact on air quality. On the other hand, strong wind shear associated with a change of the wind direction in vertical profile affects the deterioration of air quality by reducing the height of the mixing layer during the daytime. The study of the PM_{10} concentration vertical profiles in Kraków presented in the work of Sekula et al. 2021 [69] indicate that this phenomenon often occurs at the valley bottom height (approx. 100 m a.g.l.). During the cold half-year, poor dispersive conditions are more frequent than in the warm half-year, which in combination with high rates of emission from the residential sector led to accumulation of pollutants inside PBL. Analysis of the deseasonalized multiyear PM_{10} trend has shown that in the decade 2011–2020 a negative trend was observed which may be linked to the positive trend of air temperature.

According to the Random Forests model, adding a vertical gradient between neighboring pressure fields improved the quality of the PM_{10} level forecast. Other studies concerning application of machine learning methods in air quality forecasting confirmed that meteorological parameters like, wind speed, air temperature, relative air humidity and atmospheric precipitation were important factors affecting air quality [70]; however, studies of the effect of atmospheric precipitation on the concentration of particulate matter showed that it mainly washes out coarse particles while having little effect on fine particles [71]. Attention should also be paid to the representation of the atmospheric stability in the machine learning models; in this research we can distinguish two approaches: the application of planetary boundary layer height [31,72] or a more complex one with the application of meteorological conditions at different atmospheric pressure levels [73]. Owing to the fact that the estimation of planetary boundary layer height in numerical atmospheric models still requires validation and further development [74,75] we would like to suggest applying vertical profiles of atmosphere rather than PBL height in studies.

5. Conclusions

The analysis of air quality conditions in the multiyear period has proved that wind speed, air temperature, atmospheric stability connected to relative humidity gradient and air temperature gradient at lower troposphere and the occurrence of precipitation significantly influences pollutant concentrations. Apart from the non-directional anticyclonic conditions which affect air stagnation, air advection from the S-SW sector, strongly modified by local topography, has usually caused an increase of PM_{10} levels in the study area. Studies have shown that for the region analyzed the direction of air advection and its intensity is of greater importance than the type of pressure system concerning the impact on PM_{10} levels. Certain types of circulation can be indicated as significant both in terms of improving the dispersion of pollution and its deterioration; this is the result of the modification of large-scale processes by orography and near-ground atmospheric conditions. For example, air masses advection from the W-NW sector may strengthen near-ground thermal inversion and reduce wind speed in the valley, but it can also break thermal inversion in the valley and topographically channel the air flow. Research has indicated that particular types of circulation may affect the deterioration of air quality conditions in the cold half-year. During these circulation types lasting for a few or several days, a continuous increase of air pollution can be observed. Sometimes it leads to extremely high values of PM_{10} concentrations (e.g., types S-SW, W + NWa).

The analysis of the number of days with PM_{10} levels exceeding the daily limit in the study period showed that the emission reduction contributed to a significant improvement in the air quality in the city; however, the occurrence of days with poor air quality in the future is very likely due to the strong influence of meteorological conditions on that element. The number of days with low thermal inversion in the 975–925 hPa layer in a cold half-year turned out to be particularly important. Significant factors influencing the improvement of air quality in the cold half-year were the occurrence of longer rainfall (rainfall during the day and night), high daily wind speed in the valley and negative air temperature gradient.

One of the limitations in the studies presented above is the assignment of a single circulation type to the whole day; on days when an atmospheric front or the pressure center passes over a certain area, the meteorological conditions may change significantly during the day. Therefore, for the detailed analysis of atmospheric circulation, the daily fluctuations of circulation conditions should be taken into account. Currently, studies on application of the Convolutional Neural Network in automatic classification of atmospheric circulation according to the Niedźwiedź classification with the use of ERA5 reanalysis are conducted in our research group. The first results obtained are promising, however some model optimizations are still necessary.

Analysis of hourly PM_{10} concentration data and meteorological parameters with the use of cross correlation function have shown the occurrence of delayed time response of PM_{10} concentration level in the city to the change in meteorological conditions. For instance, at Krasińskiego station, the delayed time response obtained for the PM_{10} level for the wind speed, wind gusts, air temperature, as well as the ground thermal gradient was equal to 2 h; however, it should be mentioned here that air quality in the city may vary significantly on spatial and temporal scale as it was presented in other studies [8].

Further research on the intra-city spatial dependency from meteorological parameters and circulation patterns is necessary from the point of view of habitability and health risk.

Previous studies indicated that each technique of circulation patterns classification has some limitations e.g., there is a problem of equally sized classes, separation of different types, seasonal or inter-annual variability of a class frequency. In the case of the subjective classifications, there are high inter-annual variability and larger long-term trends of the frequencies of the types' in comparison to the automated circulation classification methods; however, subjective classification includes important expert knowledge concerning analyzed geographical regions, which is difficult to formulate in precise rules for automated classification methods [52]. In conclusion the authors would suggest using local circulation classification methods in studies for different regions, owing to the effect of topography on modifications of atmosphere dynamics; however, because of the obvious limitations of the use of manual approaches, connected to their subjective nature, it seems that the best solution would be to use a local subjective classification of circulation types, which could be automated. Such approaches are known in the literature, although, they were applied in larger spatial scales [51].

Owing to the fact that machine learning methods create great opportunities in air quality studies, further development works are planned using the Random Forests method to analyze and forecast air quality on a larger spatial scale (e.g., cities in Central Europe) by supplementing the model with additional data such as land cover, topography, and turbulence parameters, as well as the results from operational forecasts of numerical air quality models to improve model accuracy. The further step in air quality studies will be an application of multi-step time series forecasting to model daily cycle of air pollution but also to predict daily pollution levels for three days ahead by using weather forecast and air pollution levels on the current day [75,76]. The next direction of development of the current research focuses on the analysis of spatial and temporal variability of pollution for large cities using data from air quality stations as well as non-governmental air quality systems. The first tests of using convolutional neural networks to determine air pollution level with respect to circulation patterns over larger domains are very promising, and so it is also planned to further investigate those methods.

Author Contributions: Conceptualization, P.S., Z.U., A.B. and M.Z.; methodology, P.S., Z.U. and A.B.; software, P.S. and B.B.; validation, P.S. and B.B.; formal analysis, P.S.; investigation, P.S., A.B., Z.U., B.B. and M.Z.; resources, P.S., Z.U., A.B. and B.B.; data curation, P.S. and B.B.; writing—original draft preparation, P.S.; writing—review and editing, P.S, B.B., Z.U., M.Z. and A.B.; visualization, P.S. and B.B.; supervision, P.S.; project administration, P.S.; funding acquisition, Z.U., M.Z. and B.B. All authors have read and agreed to the published version of the manuscript.

Funding: This research has been partly supported by the EU Project POWR.03.02.00-00-I004/16 (PS). This work was (partially) supported by the AGH UST statutory tasks No. 11.11.220.01/1 within subsidy of the Ministry of Science and Higher Education.

Institutional Review Board Statement: Not applicable.

Informed Consent Statement: Not applicable.

Data Availability Statement: Publicly available datasets were analyzed in this study. This data can be found here: Chief Inspectorate for Environmental Protection (CIEP). Available online: https://powietrze.gios.gov.pl/pjp/archives (accessed on 8 March 2022). Calendar of circulation types. Available online: http://www.kk.wnoz.us.edu.pl/nauka/kalendarz-typow-cyrkulacji/ (accessed on 8 March 2022). Institute of Meteorology and Water Management, National Research Institute. Available online: https://danepubliczne.imgw.pl/datastore (accessed on 8 March 2022).

Conflicts of Interest: The authors declare no conflict of interest. The funders had no role in the design of the study; in the collection, analyses, or interpretation of data; in the writing of the manuscript, or in the decision to publish the results.

Appendix A

Air quality data analysis with measurements from station at Krasińskiego St. has shown significant differences in the distribution of the daily PM_{10} concentration during warm and cold half-years. The multiannual trend of the number of days with exceedance of limit 50, 100, and 200 $\mu g \cdot m^{-3}$ were calculated for cold and warm half-year separately. The period covered from October 2000 to September 2020. For a selected number of days a linear curve was fitted by using the Theil-Sen estimator [77] provided in the RobustLinearReg R package was fitted [78]. The number of days with an exceedance of the daily pollution level (equal to 50 $\mu g \cdot m^{-3}$) is characterized by a decreasing trend equal to -5.33 days/year for the warm half-year (R-squared was equal to 0.51), while for the cold half-year the number of days with exceedance of 50 $\mu g \cdot m^{-3}$ has no positive or negative multiyear trend (Figure A1a,b). During the cold half-years in the period 2000–2020 there was a visible negative trend of the number of days with exceedance of limit 100 $\mu g \cdot m^{-3}$ equal to -2.31 days/year with R-squared equal to 0.23. In the study period, there are visible fluctuations in the number of days with exceedance of the daily PM_{10} limit, during warm and cold half-years (Figure A1c,d), which clearly indicates the impact of weather conditions on the frequency of smog episodes.

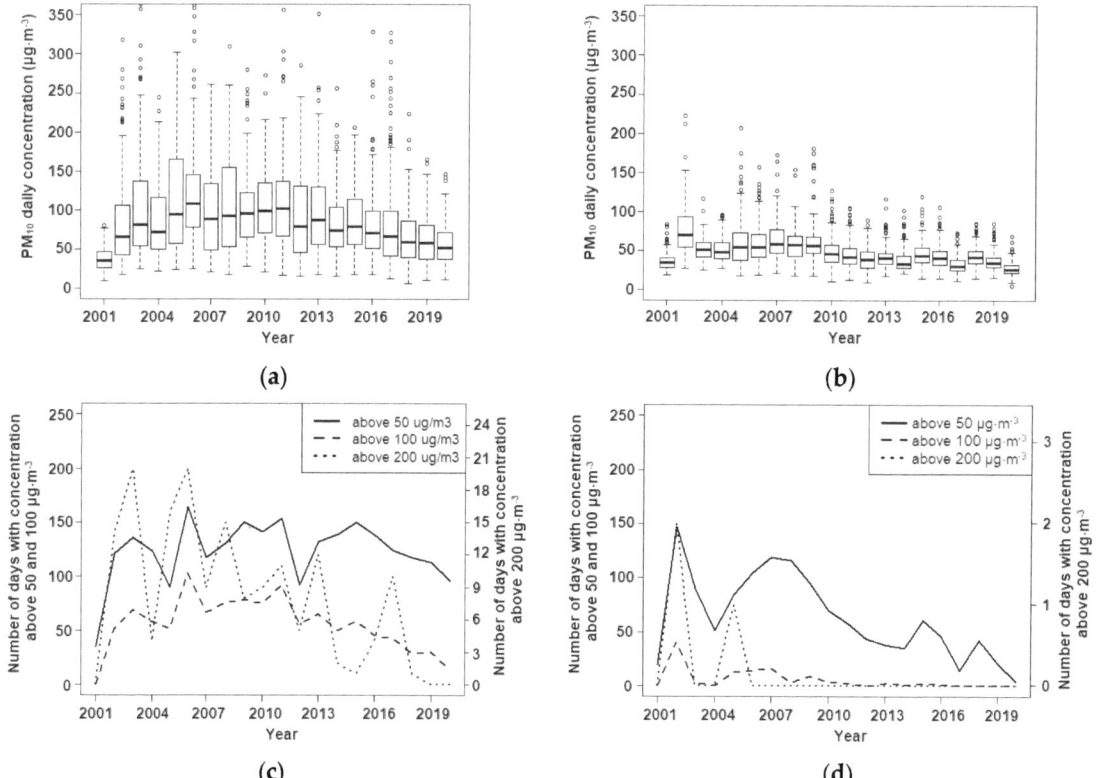

Figure A1. Distribution of daily PM_{10} concentration at the Krasińskiego station in the cold (**a**) and warm half-year (**b**), and the number of days with exceedance of limit 50, 100 and 200 $\mu g \cdot m^{-3}$ in the cold (**c**) and warm (**d**) half-year in the period 2000–2020.

In an effort to better analyze the multiannual trend in the period from October 2000 to September 2020 of the PM_{10} concentration, the Theil-Sen estimator with the switched option of seasonal trend decomposition using loess was used (Figure A2). By default, the values of the averaging period and autocorrelation were used, equal to month and the 95% confidence level. The function used for this calculation was provided by the openair R package [79]. Analysis has shown that during the analyzed period there is a negative trend equal to -1.94 µg·m^{-3}/year. In the study period 2000–2020, the months of January 2001 and 2006 differ significantly from the entire study period. This situation was caused by anomalies of air temperature at 2 m a.g.l. (Figure A3) and atmospheric stability in the layer between 975 and 925 hPa (not shown in the article). Detailed analysis of the multiyear trend of PM_{10} has shown that stronger negative trend of PM_{10} concentration occurred during the period 2011–2019 than for 2002–2010, equal to -2.54 µg·m^{-3}/year and -0.64 µg·m^{-3}/year, respectively. For the same sub-periods, air temperature trends also differ significantly, in the period 2011–2019 the trend was equal to 0.21 °C/year, while for the earlier period amounted to -0.07 °C/year. On the other hand, analysis of the deseasonalized air temperature gradient in the layer between 975 and 925 hPa and daily wind speed at 10 m a.g.l. did not show any significant trends throughout the multiannual period (not shown in the article). The significant positive trend in air temperature in the last decade may be a crucial factor in determining PM_{10} emission in the cold half-years. Studies of warm temperature extremes for Central Europe in the period 1950–2020 have shown a positive trend of intensity and frequency of hot events during winter periods [80].

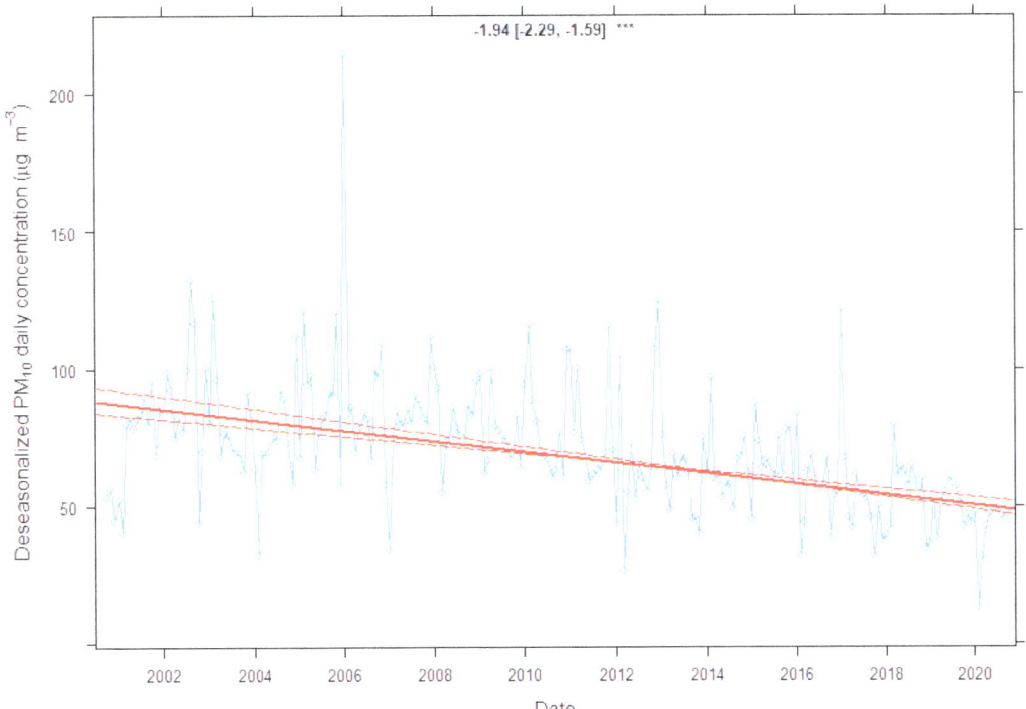

Figure A2. Deseasonalized multiyear trend of PM_{10} daily concentration in period from October 2000 to September 2020. *** indicates that the obtained trends are significant to the levels 0.001.

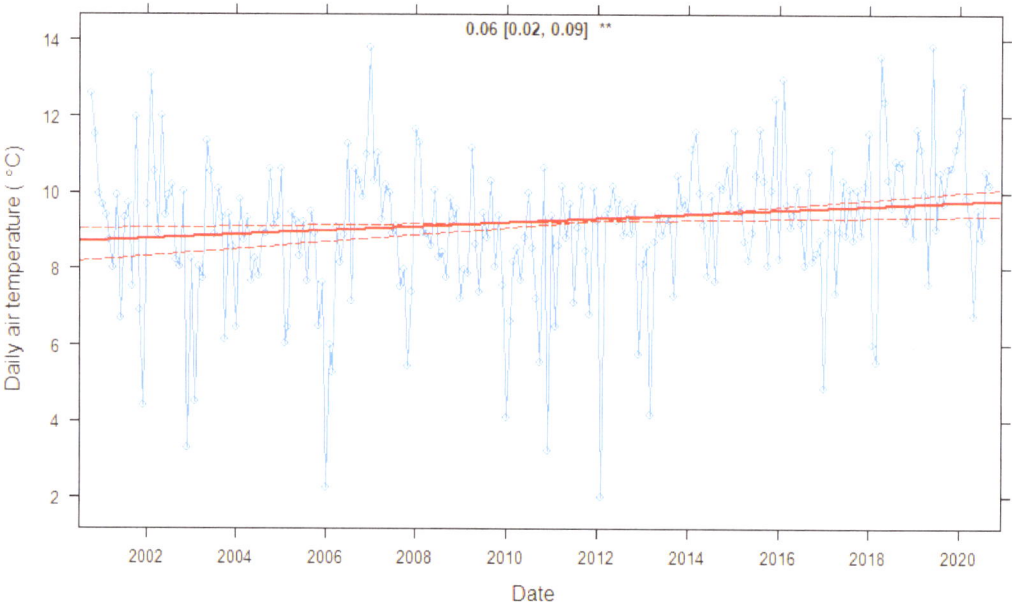

Figure A3. Deseasonalized multiyear trend of air temperature at 2 m a.g.l. in period from October 2000 to September 2020. ** indicates that the obtained trends are significant to the levels 0.01.

Appendix B

Appendix B.1 Niedźwiedź Circulation Classification

The classification of circulation types for Southern Poland [81] is available for the period from September 1873 to the present day. The classification was based on the typology of atmospheric circulation developed by Lamb [82] for the British Isles, with some modifications, especially regarding nonadvection situations. On the basis of synoptic maps of Europe, the direction of air mass movement (N, NE, E, SE, S, SW, W, NW) and the type of baric system (a—anticyclonic situation, c—cyclonic situation) were determined. Finally, 16 types of atmospheric circulation were distinguished. In addition, there are two non-advectional types: Ca—high-pressure center and Ka—anticyclonic wedge or ridge, and two cyclonic types of differentiated air advection: Cc—low pressure center and Bc—cyclonic troughs. The aric col and low-gradient situations, which are difficult to classify, are marked with the letter "x". Thus, the entire classification includes 21 types (10 anticyclonic types, 10 cyclonic types and one indefinite type). By combining adjacent types, a shortened version is also obtained for 11 situations (N + NEa or c; E + SEa or c; S + SWa or c; W + NWa or c; Ca + Ka, Cc + Bc, x), which is useful in studies of periods shorter than 30 years [29].

Due to the fact that the study period covers 20 years, the classification version with 11 types of circulation was used, in order to increase the size of samples and obtain more reliable statistical results. For the analyzed period from October 2000 to September 2020 the multiannual trend was determined for warm and cold half-years, respectively, for each of 11 Niedźwiedź circulation types (Table A1 in Appendix B). During each half-year, the number of days with a specific circulation type were determined. At the next step, linear curves were fitted for each circulation type by using the Theil-Sen estimator from RobustLinearReg R package. Analysis has shown that the strongest negative trend was observed for type Ca + Ka during the warm and cold half-year, and it was equal to −0.73 and −0.60 days/half-year, respectively. Studies of 21 Niedźwiedź circulation types has shown that a strong negative trend was observed for type Ca + Ka was caused by decrease of number of days with anticyclonic wedge or ridge situation (type Ka), for which the trend was equal −0.76 and −0.57 days/half-year, for cold and half-years, respectively (not shown in the article). During the cold half-year multiyear trend of circulation types S + SWa and N + NEc was equal to 0.27 and 0.21 days/half-year, while for circulation type S + SWc the multiyear trend was negative (equal to −0.2 days/half-year). For the warm half-year, the trend of cyclonic conditions with air masses advection from sectors N–NE, S–SW was positive equal to 0.44 and 0.26 days/half-year, respectively. On the other hand, multiyear trend of W + NWc and Cc + Bc types in warm half-years were negative and equal to −0.50 and −0.33 days/half-year, respectively. The total frequency of particular atmospheric circulation types in the period October 2000–September 2020 in warm and cold half-years is presented in Figure A4 in Appendix B.

During the warm half-year, the shares of nonadvection anticyclonic types (Ca + Ka) and cyclonic types with differentiated air advection (Cc + Bc) are the highest and equal to 15% and 16%, respectively. Air advection from the W–NW sector during cyclonic and anticyclonic situations occurs often, and comprises 23% of all cases.

The cold half-year period differs significantly from the warm one; the share of air advection from the SW-NW sector (cyclonic and anticyclonic types) is greater by 15% in cold half-year than in the warm one. Parallel, the share of cyclonic types with differentiated air advection during the cold half-year is lower by 6% compared to the warm half-year.

Table A1. Multiyear trend of 11 atmospheric circulations according to Niedźwiedź classification in warm and cold half-years in period from October 2000 to September 2020.

Circulation Type	Cold Half-Year		Warm Half-Year	
	Trend (Day/Half-Year)	R-Squared	Trend (Day/Half-Year)	R-Squared
N + NEa	−0.09	0.01	0.19	0.06
E + SEa	0.10	0.01	0.13	0.02
S + SWa	0.27	0.07	0.13	0.04
W + NWa	0.00	0.00	0.17	0.02
Ca + Ka	−0.60	0.23	−0.73	0.35
N + NEc	0.21	0.02	0.44	0.12
E + SEc	0.15	0.02	0.00	0.00
S + SWc	−0.20	0.02	0.26	0.12
W + NWc	0.00	0.00	−0.50	0.16
Cc + Bc	0.16	0.04	−0.33	0.17
x	0.12	0.06	0.00	0.00

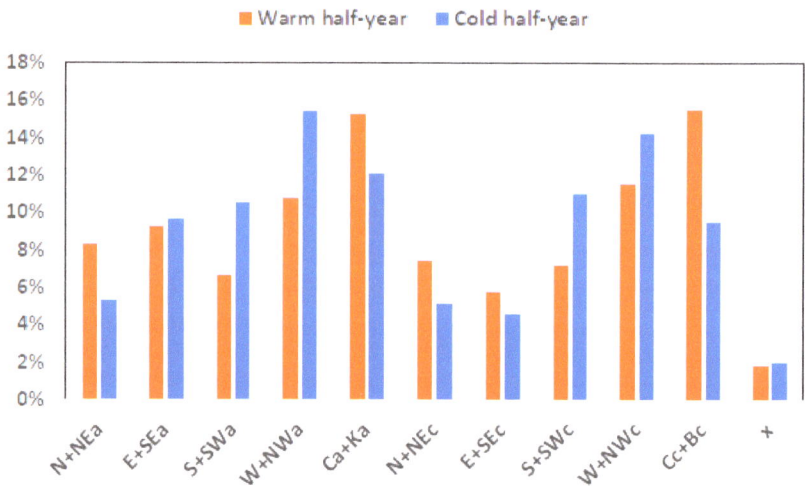

Figure A4. Frequency of Niedźwiedź circulation types during the period October 2000—September 2020 in warm and cold half-year.

Appendix B.2 Lityński Circulation Classification

One of the classifications of atmospheric circulation types widely applied in Poland is the threshold-based method proposed by J. Lityński. Lityński developed his objective classification to be applied to Poland and Central Europe [38,53]. Synoptic types were defined using the following indicators: zonal (Ws), latitudinal (Wp) and Warsaw air pressure (Cp) using sea-level synoptic maps over an area defined as 40–65 °N and 0–53 °E. The Ws indicator was derived using a formula for an average longitudinal component of the geostrophic wind. A conversion of this formula was used to determine the latitudinal circulation indicator [38]. The direction of air advection and the type and strength of the pressure systems were determined on a frequency distribution of the Ws, Wp and Cp indicator values and using a three-class equal-probability system. The thresholds employed to calculate the Wp, Ws and Cp indices change from month to month, which results in flattening the seasonal cycle of the occurrence of circulation types [38]. The resulting air advection type was described by three numeric parameters: Wp, Ws and Cp. The following symbols were used to denote the Ws indicator: E (eastern) for most negative values, 0 for near-zero values and W (western) for most positive values. Similarly, the Wp indicator was denoted by the symbols: N (northern) for most negative values, 0 for near-zero values and S (southern) for most positive values. Cp air pressure classes were marked: C (cyclonic), 0—near-zero and A (anticyclonic). These circulation type symbols were combined with the Wp and Ws indicators class symbols, and finally, one of the three Cp air pressure class symbols were added. Lityński distinguished 27 circulation types, three non-advective types (symbol Oo, Oc, Oa) and 8 directional types (with 3 types, cyclonic, anticyclonic and intermediate type, known as the near-zero type). It is worth to note that Litynski's classification system, as one of the scalable methods, is part of the COST 733 classifications catalogue [83]. In the current artile, the Lityński classification has been used with modifications introduced by Krystyna Pianko-Kluczyńska [63]. Recent studies confirm the high level of comparability in the course of circulation indices according to the classifications of Niedźwiedź and Lityński [29].

Figure A5 in Appendix B presents the frequency of Lityński circulation types during the warm and cold half-year of the study period. For the analyzed period from October 2000 to September 2020, the multiannual trend was determined for warm and cold half-years, respectively, for 27 Lityński atmospheric circulation types. During each half-year, the numbers of days with a specific type of circulation were determined. In the next step, for each circulation type, linear curves were fitted by using the Theil-Sen estimator from the RobustLinearReg R package. Detailed analysis has pointed out that during the warm half-year share of circulation types Ec and Wa have the strongest negative trend equal to −0.25 and −0.21 days/half-year, respectively. F°or the type Oa in the warm half-year, the strongest positive trend equal to 0.25 days/half-year was observed. During the cold half-year share of types NEa and Sc are characterized by the strongest negative trend equal to −0.22 and −0.25 days/half-year, respectively. On the other hand, for SEo, SWc, Wa and NWa positive trend equal on average 0.28 days/half-year was observed. Detailed information on individual circulation types during the cold and warm half-year in the period from October 2000 to September 2020 was included in Appendix B, Table A2. In comparison with warm half-year, during cold half-year there is a visible decrease of air advection from the NE direction, with a significant increase of air advection from the SW direction. It is worth noting that the share of cyclonic types with air advection from the W–NW sector is greater during the cold season, whereas anticyclonic types with air advection from the same direction have a lower frequency in comparison with warm half-year.

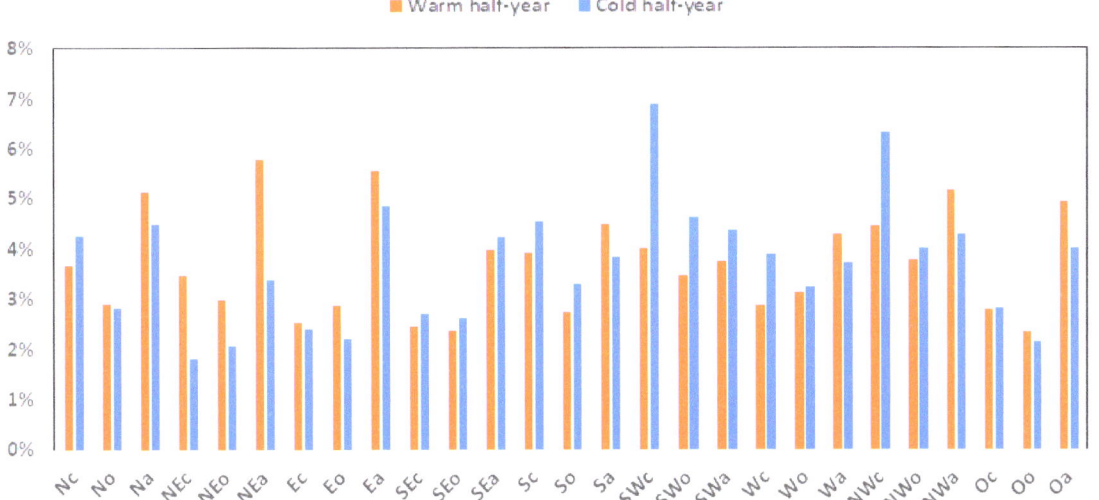

Figure A5. Frequency of Lityński circulation types during the period October 2000—September 2020 in warm and cold half-year.

Table A2. Multiyear trend of 27 atmospheric circulations according to Lityński classification in warm and cold half-years in period from October 2000 to September 2020.

Circulation Type	Cold Half-Year		Warm Half-Year	
	Trend (Day/Half-Year)	R-Squared	Trend (Day/Half-Year)	R-Squared
Nc	−0.13	0.15	0.00	0.00
No	−0.07	0.03	0.20	0.27
Na	−0.04	0.00	−0.09	0.00
NEc	−0.03	0.00	0.00	0.00
NEo	−0.11	0.07	0.08	0.04
NEa	−0.22	0.12	0.00	0.00

Table A2. Cont.

Circulation Type	Cold Half-Year		Warm Half-Year	
	Trend (Day/Half-Year)	R-Squared	Trend (Day/Half-Year)	R-Squared
Ec	0.00	0.00	−0.25	0.18
Eo	0.00	0.00	0.13	0.03
Ea	0.00	0.00	−0.13	0.07
SEc	0.11	0.04	0.00	0.00
SEo	0.25	0.13	0.00	0.00
SEa	0.00	0.00	0.00	0.00
Sc	−0.25	0.06	−0.11	0.02
So	0.00	0.00	−0.08	0.04
Sa	0.00	0.00	0.00	0.00
SWc	0.26	0.08	−0.18	0.10
SWo	0.00	0.00	−0.11	0.13
SWa	0.00	0.00	0.08	0.00
Wc	0.00	0.00	0.13	0.07
Wo	0.00	0.00	−0.10	0.05
Wa	0.27	0.13	−0.21	0.14
NWc	0.00	0.00	−0.13	0.03
NWo	0.14	0.08	0.00	0.02
NWa	0.37	0.17	0.18	0.13
Oc	−0.12	0.04	−0.08	0.00
Oo	0.00	0.00	0.11	0.03
Oa	−0.19	0.11	0.25	0.13

Appendix C

Table A3. Height of PM_{10} daily concentration upper quartile in warm and cold half-years.

Year	Cold Half-Year ($\mu g \cdot m^{-3}$)	Warm Half-Year ($\mu g \cdot m^{-3}$)
2001	47	41
2002	106	94
2003	137	60
2004	116	60
2005	162	73
2006	145	71
2007	134	77
2008	155	69
2009	123	67
2010	135	57
2011	137	53
2012	131	49
2013	130	47
2014	103	43
2015	115	54
2016	99	50
2017	99	38

Table A3. Cont.

Year	Cold Half-Year (µg·m^{-3})	Warm Half-Year (µg·m^{-3})
2018	87	50
2019	82	42
2020	71	32

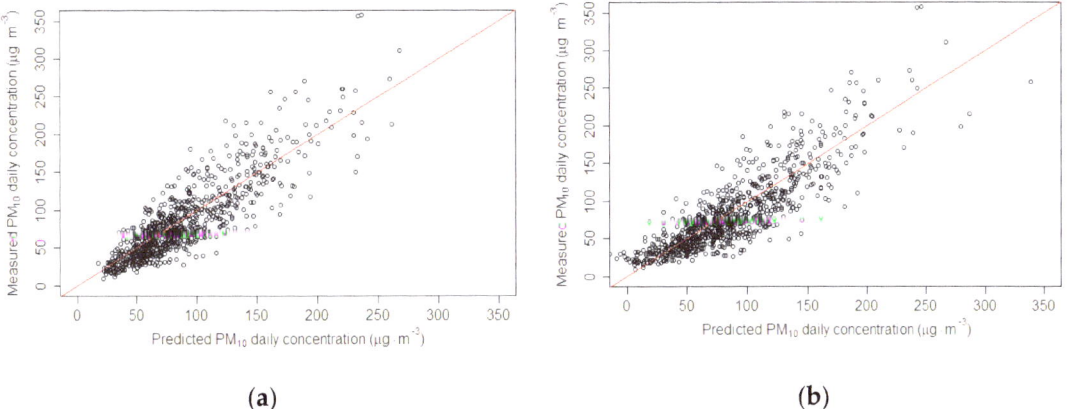

Figure A6. Scatterplot of predicted versus observed PM$_{10}$ daily concentration (**a**) for Random Forests model and (**b**) multilinear regression model with daily meteorological parameters for the cold half-years.

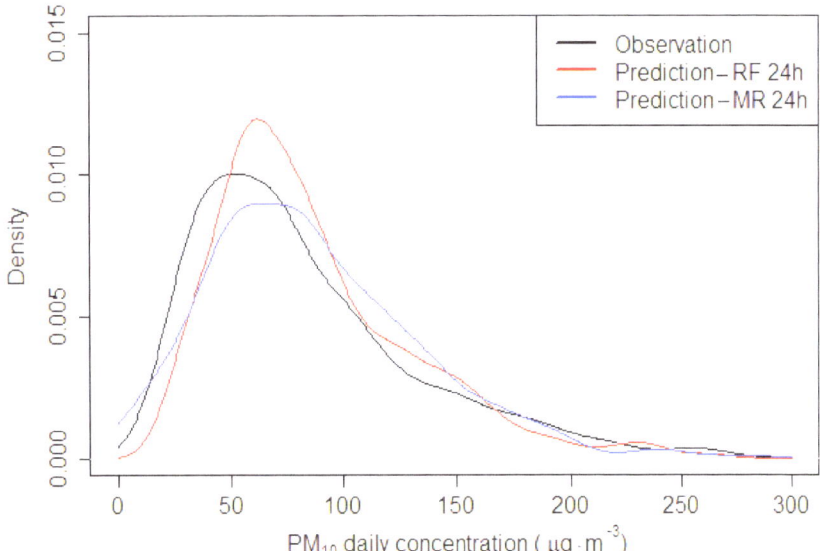

Figure A7. Density plot observed and predicted PM$_{10}$ daily concentration for Random Forests (RF) and multilinear regression (MR) model with daily meteorological parameters for the cold half-years.

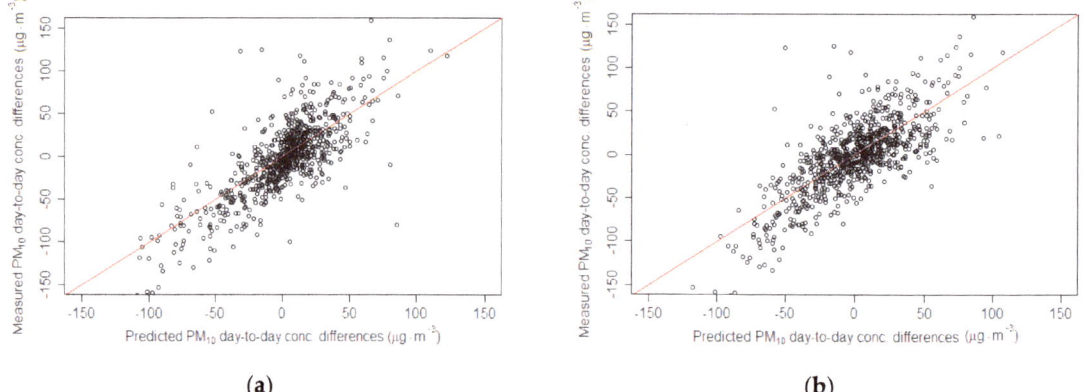

(a) (b)

Figure A8. Scatterplot of predicted versus observed PM_{10} day-to-day concentration changes (**a**) for Random Forests model and (**b**) multilinear regression model with daily meteorological parameters for the cold half-years.

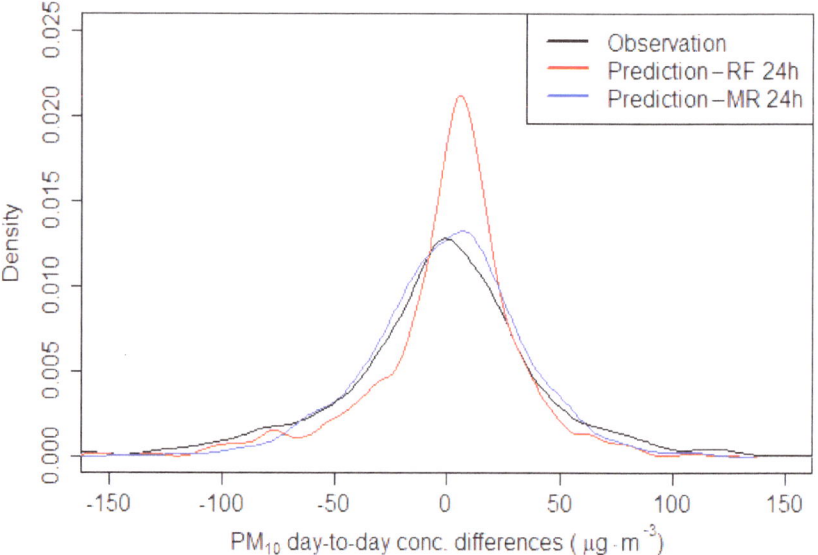

Figure A9. Density plot observed and predicted PM_{10} day-to-day concentration changes for Random Forests (RF) and multilinear regression (MR) models which use daily meteorological parameters for the cold half-years.

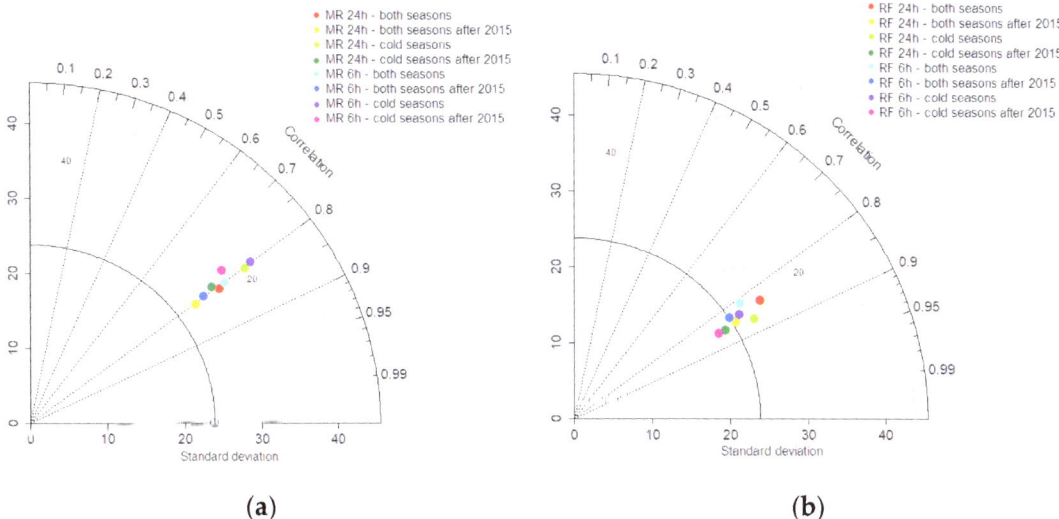

(a) (b)

Figure A10. Taylor diagram plots of predicted PM_{10} daily concentration for the period between October 2021 and December 2021 for (**a**) multilinear regression models and (**b**) Random Forests models for different training data sets.

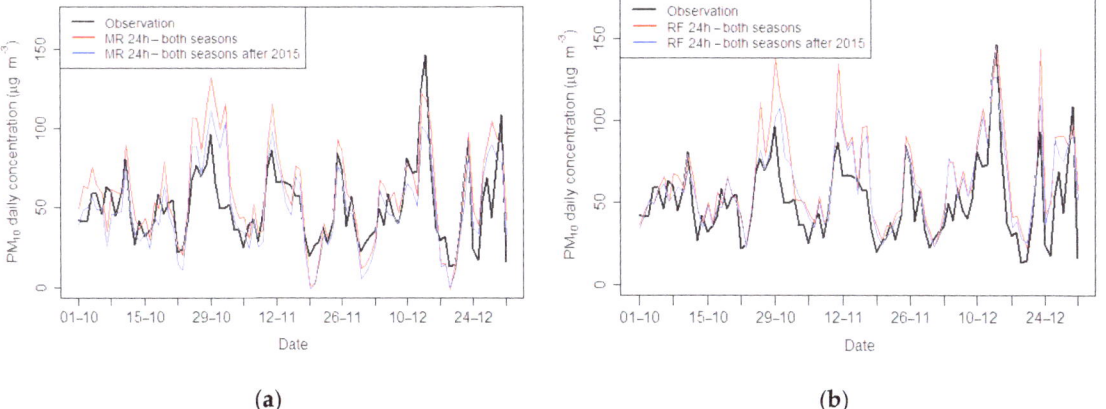

(a) (b)

Figure A11. Time course of observed and predicted PM_{10} daily concentration for period between October 2021 and December 2021 for (**a**) multilinear regression models and (**b**) Random Forests models with use of daily averages of meteorological data for both half-years for two different training periods.

Appendix D

Weather Conditions in Relation to the Circulation Types

The following section describes the distribution of the selected meteorological parameters for 11 Niedźwiedź classification types during the cold half-years in the analyzed period. Studies of daily wind speed for the cold half-year have shown that the lowest wind speed occurs during the advection from sector S–SW which is caused by the local topography (surrounded by highlands from the South, North and West—see Figure 1b). The weak wind in the valley was also frequent at stagnant anticyclonic situations (type Ca + Ka). The conditional probability of the occurrence of a daily wind speed below $1~\text{m·s}^{-1}$ was the highest for the circulation types S + SWa and Ca + Ka, slightly above 30%. For the other types, except for the S + SWc and W + NWa types, the conditional probability did not exceed 5% (for the S + SWc and W + NWa types equal 14% and 10%, respectively). The highest average value of the daily speed was determined for the type of circulation W + NWc, equal to 5 m·s^{-1}.

The analysis of the mean daily cloudiness for the individual circulation types pointed out that the greatest variability of this parameter occurred for the types S + SWa, Ca + Ka and E + SEa (interquartile ranges were greater than 4 oktas). On the other hand, the lowest variability of daily cloudiness was observed for cyclonic types with advection from sectors E–SE and W–NW, as well as for low-pressure center and cyclonic troughs (interquartile range for these types ranged from 0.7 to 1.6 oktas). The conditional probability of a day with daily cloudiness not exceeding 2 oktas was the highest for the circulation types S + SWa and Ca + Ka, equal to 28% and 26%, respectively. The results obtained are consistent with the research of a longer multiyear period [28].

The median of the daily sum of precipitation differs significantly for cyclonic and anticyclonic conditions; for anticyclonic types, the value did not exceed 1 mm/day. The highest value of the median daily precipitation was determined for the types Cc + Bc and E + SEc, equal to 2.5 mm/day. The percentage of days with daily precipitation above 0.1 mm/day was the lowest for anticyclonic types S + SWa and Ca + Ka, equal to 16% and 18%, respectively. For cyclonic types, except for the type S + SWc, the share of days with precipitation in the cold half-years was greater than 70%. A significantly lower share of days with precipitation for the S + SWc type equal to 53% is related to the orographic barrier of the Western Carpathians, which as a result affects the air temperature and the humidity of the air masses and spatial distribution of the atmospheric precipitation [76].

The distribution of the daily relative humidity at the ground level in the cold half-years is similar for most circulation types, except types N + NEc, E + SEc, Cc + Bc and x for which a higher daily humidity was observed. For the selected circulation types, the lower quartile of daily relative humidity ranged from 83% to 87%, while the average value of the lower quartile for the remaining types is equal to 78%.

Analysis of the daily air temperature showed that the largest interquartile range was measured for Ca + Ka and E + SEa types. It should also be noted that for the selected types the value of the lower quartile was the smallest, equal to −4.6 and −3.9 °C, for the types E + SEa and Ca + Ka, respectively. Low values of the daily air temperature for the Ca + Ka type are associated with strong radiative cooling of the surface with the cloudless sky. For days with anticyclonic condition with advection from sector E-SE, in most cases, the analyzed region is under the influence of a strong high-pressure center developed over the area of Eastern Europe and then moved into the West. For this circulation type, advection of polar continental air masses dominates (more than 75% of the cases). On the other hand, the highest values of the median daily air temperature occurred for the types Cc + Bc, W + NWc and S + SWc, equal to 3.9, 4.2 and 4.4 °C, respectively.

The analysis of the vertical profiles obtained from the ERA5 reanalysis indicated that the relative air humidity at pressure levels 925 and 850 hPa strongly depends on the direction of advection, during the day and nighttime periods. The air masses moving from the S-SW sector (for cyclonic and anticyclonic conditions) are characterized by much lower relative humidity than for the other types of circulation (the median value of relative humidity at 925 hPa at 12 UTC for both types were equal to 58 and 67%, respectively, while for the others it was within the range from 77% to 95%). Significant fluctuations in relative air humidity were also observed for stagnant anticyclonic situations (Ca + Ka) and anticyclonic conditions with advection from the W-NW sector at 0 UTC in the pressure level 850 hPa.

It should also be mentioned that for days with advection from the S–SW sector during the daytime and nighttime period, a strong decrease in relative humidity is visible in the layer 975–925 hPa. During the night, at 0 UTC, the median of relative humidity gradient between the levels 925 and 975 hPa for the selected types was lower than −13%, while for the remaining types, the humidity gradient varied between −5% and 4%.

During the day with advection from sector S–SW, a higher relative humidity gradient was observed compared to the remaining circulation types in the layer 925–850 hPa.

Analysis of air temperature at pressure levels 975, 925, and 850 hPa indicated that during the daytime air temperature at the height of 925 hPa is significantly higher for days with advection from the S-SW direction, than for the other types.

The median of the vertical temperature gradient in the layer 925–975 hPa during the night was the highest with the advection from the S–SW sector. During the day, the largest share of days with a positive vertical gradient was observed for the type S + SWa, equal to 30%. The statistically significant share of days with low thermal inversion was also observed for the types E + SEa and Ca + Ka, close to 16%.

Taking into account the distribution of all meteorological parameters for 11 Niedźwiedź classification types, the types Ca + Ka and S + SW (cyclonic and anticyclonic situation) are potentially the most important circulation patterns affecting the deterioration of air quality in the city. For the selected types, weaker wind speed at the ground level, higher frequency of thermal inversions, and stronger negative gradient of relative air humidity were observed in comparison with the remaining circulation patterns. The analysis also showed that the type E + SEa type can have a significant impact on air quality due to the occurrence of low daily air temperatures for this type of circulation.

Group 1: days with the highest PM_{10} concentration.

Table A4 presents the conditional probability of the occurrence of high PM_{10} mean daily concentration for individual types of atmospheric circulation during cold half-year, in the period 2000–2020, according to the Niedźwiedź classification.

Among the 11 types of circulation, only 5 types had the frequency greater than 10% in the cold half-year (S + SWa, W + NWa, Ca + Ka, S + SWc and W + NWc; Figure A4 in Appendix B). The worst air pollution conditions, that is, situations with the highest conditional probability of occurrence of high PM_{10} concentration in Kraków were almost the same: S + SWa (0.52), S + SWc and Ca + Ka (almost 0.4), and W + NWa (0.2), indicating that situations with air advection from the western sector, regardless of the baric center type, have less impact on the deterioration of air quality in the city than other types of circulation most frequent.

The highest number of days with PM_{10} concentration greater than the upper quartile in a cold half-year and exceeding daily limit value of PM_{10} occurred in December and January (162 and 195 days, respectively); the number of cases in November, February and March was similar: 140 days on average. The smallest number of such days was observed in October: 65 days.

Table A4. Conditional probability of the occurrence of high PM_{10} concentration during particular atmospheric circulation types during the cold half-year according to Niedźwiedź classification, number of days with selected circulation type and the number of days with high PM_{10} concentration for individual circulation types in the period 2000–2020.

Circulation Type	Conditional Probability	Number of Days with High PM_{10} Concentration	Total Number of Days in Cold Half-Year
N + NEa	0.06	12	196
E + SEa	0.14	50	354
S + SWa	0.52	202	386
W + NWa	0.20	114	562
Ca + Ka	0.39	171	441

Table A4. Cont.

Circulation Type	Conditional Probability	Number of Days with High PM_{10} Concentration	Total Number of Days in Cold Half-Year
N + NEc	0.14	26	190
E + SEc	0.12	21	170
S + SWc	0.37	149	403
W + NWc	0.04	22	521
Cc + Bc	0.14	49	348
x	0.35	26	74
Total number of days		842	3645

Figures A12 and A13 present the weather conditions from the Balice synoptic station for each circulation type in division into three groups: days with high PM_{10} concentration (Group 1), days with a significant improvement of air quality (Group 2) and remaining days. Comparison of wind conditions for days assigned to Group 1 with the remaining days for four dominant circulation types (boxplots in blue frames in Figure A12a) has shown that these days are characterized by the weakest daily wind speed in the cold half-year. The Mann–Whitney U test calculated for days assigned to Group 1 and the remaining days for wind speed also confirmed that both groups differ statistically significantly for the four circulation types (p-value did not exceed 0.002; Table A5); however, it should be noted that wind speed distribution for circulation patterns S + SWa, S + SWc, and Ca + Ka did not differ significantly between the days from Group 1 and remaining days (the maximum difference of median and upper quartile was equal to 0.2 m·s^{-1} and 0.7 m·s^{-1}, respectively). During the anticyclonic conditions with air advection from W–NW sector (type W + NWa) differences of wind speed distribution between days from Group 1 and remaining days were the highest from all four selected patterns (maximum difference of median and upper quartile was equal to 1.2 m·s^{-1} and 1.6 m·s^{-1}, respectively). Studies of daily air temperature have shown that for days with high PM_{10} concentration for four selected patterns was lower than for the remaining days, the difference of median values ranged from 1.2 °C for type S + SWa to 3.3 °C for type Ca + Ka. It is also worth mentioning that the lowest daily air temperatures occurred for type Ca + Ka, which was partly related to small cloudiness during these days.

Detailed analysis of atmospheric precipitation showed that for anticyclonic conditions (types S + SWa, Ca + Ka, and W + NWa), the share of days with precipitation greater than 0.1 mm/day did not exceed 17% of all days in Group 1, and, moreover, precipitation occurred mostly at nighttime period (more than 90% of all cases). The median and upper quartile values of daily precipitation for these types of circulation were equal on average 0.6 mm and 1.6 mm, respectively. For the remaining days, the share of days with precipitation for types S + SWa and Ca + Ka were similar to days in Group 1 (differences below 10%), while for types W + NWa and S + SWc, the share of days with precipitation was higher compared to days from Group 1 by 25% and 14%, respectively. The Mann–Whitney U test has proven that the distribution of the sum of precipitation for all four types of circulation was similar for days assigned to Group 1 and for the remaining days.

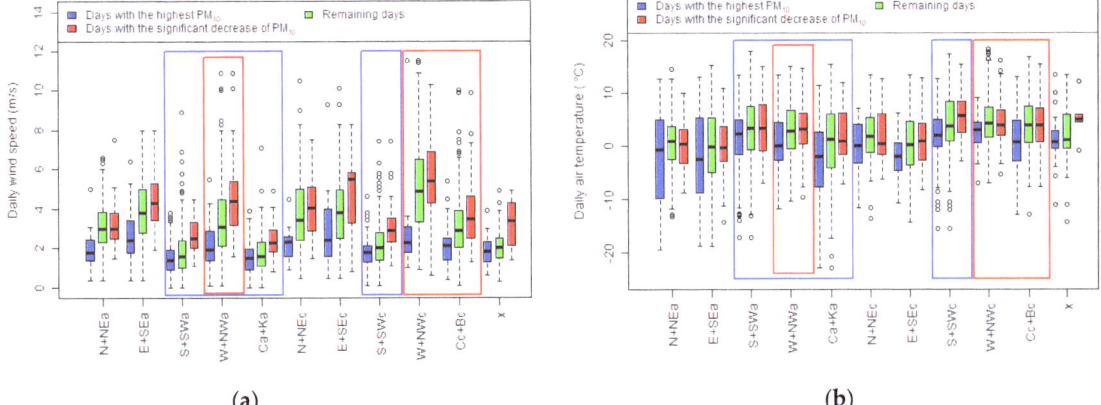

Figure A12. Boxplots of daily (**a**) wind speed and (**b**) air temperature for days with the highest PM$_{10}$ concentration (Group 1), days with a significant decrease of PM$_{10}$ (Group 2) and remaining days at cold half-year for 11 circulation types of Niedźwiedź classification for synoptic station Balice. The red and blue frames at subfigures a and b cover the dominant types of circulation in Group 1 and 2, respectively.

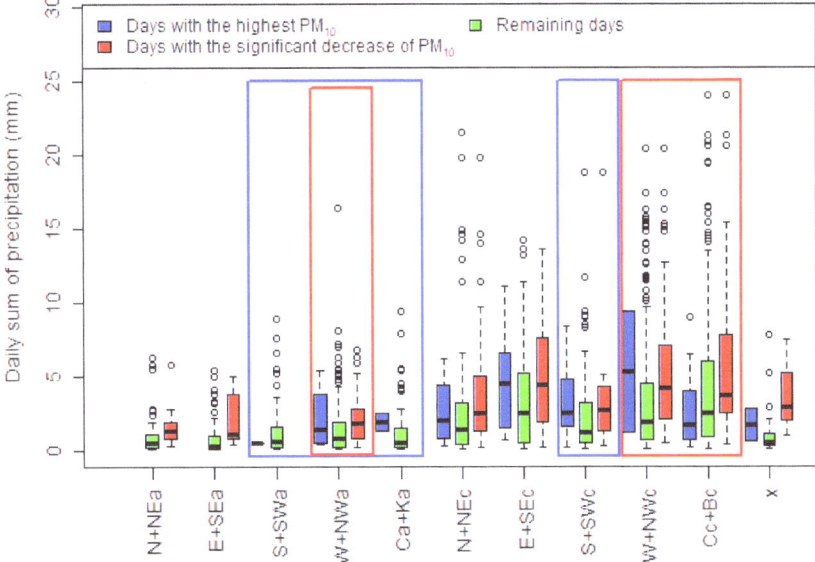

Figure A13. Boxplots of daily atmospheric precipitation for days with the highest PM$_{10}$ concentration, days with a significant decrease of PM$_{10}$ and remaining days during cold half-year for 11 circulation types of Niedźwiedź classification for synoptic station Balice. The red and blue frames cover the dominant types of circulation in both groups of days.

Table A5. A *p*-value calculated with the Mann–Whitney U test between days assigned to Group 1 and remaining days in cold half-year calculated for daily wind speed, air temperature and atmospheric precipitation for four selected Niedźwiedź circulation types.

Circulation Type	Wind Speed	Air Temp.	Precipitation
S + SWa	0.000	0.000	0.736
W + NWa	0.000	0.000	0.813
Ca + Ka	0.002	0.000	0.486
S + SWc	0.000	0.000	0.748

The duration of near-ground thermal inversion for selected circulation patterns for days with the highest PM_{10} concentration, days with significant improvement of air quality and remaining days for day (from 6 to 17 UTC) and nighttime (from 18 to 5 UTC on the next day) were presented at Figure A14. For this purpose, air temperature measurements from the TV mast from two altitudes (2 and 100 m a.g.l.) were used. The lower limit of the occurrence of thermal inversion was set equal to +1 °C. The results presented in Figure A14 show that duration of thermal inversion at day is longer by 3–6 h for days in Group 1 compared to the remaining days for 3 of 4 selected circulation patterns (S + SWa, W + NWa, S + SWc). Furthermore, during days with air advection from sector S-SW (cyclonic and anticyclonic conditions), thermal inversion in special cases persisted for the whole daytime.

The duration of thermal inversion at night for days from Group 1 was the longest for days with circulation type S + SWa. For types S + SWc and W + NWa in most cases, the duration of thermal inversion was greater than 6 h.

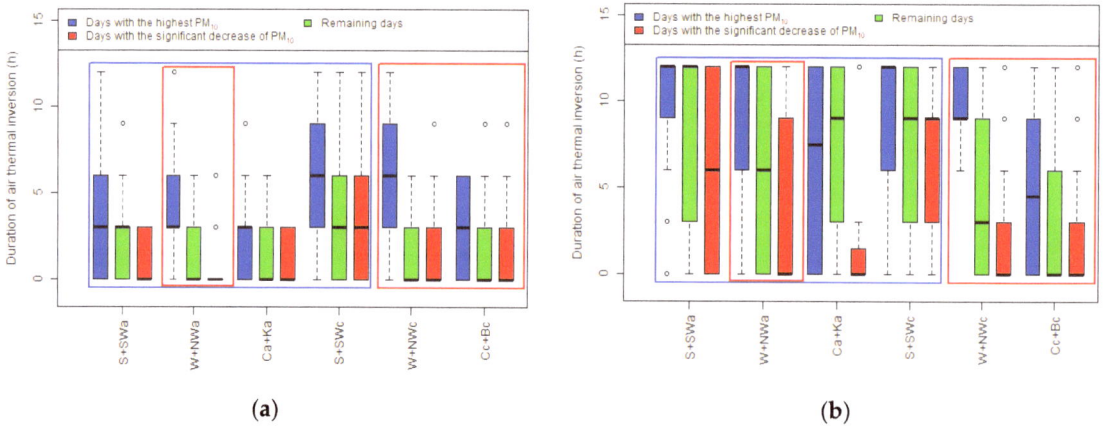

Figure A14. Boxplots of air temperature near-ground inversion duration during daytime (**a**) and nighttime (**b**) according to the data from TV mast for selected Niedźwiedź circulation types for days with the highest PM_{10} concentration (Group 1), days with significant decrease of PM_{10} (Group 2) and remaining days during cold half-year in the period 2010–2020. The red and blue frames at subfigures a and b cover the dominant types of circulation in Group 1 and 2, respectively.

ERA5 reanalysis was used to analyze atmospheric stratification in the lower troposphere, in the first approach only thermal stratification was analyzed; however, the Random Forests machine learning methods presented in the previous subsection have pointed out that air humidity stratification and vertical wind profile were also crucial factors; therefore, the analysis has been extended. Boxplots of the air temperature and relative air humidity gradient for selected circulation types have been presented in Figures A15 and A16. Analysis of the air temperature gradient in the layer 925–975 hPa has shown that atmospheric stability was stronger during night (00:00 UTC current day) and day (12:00 UTC) for all selected circulation types for days in Group 1 than for remaining days (Figure A15a,b, boxplots in blue frames). The differences in the median temperature gradient in the layer 975–925 hPa during the daytime ranged from 1 °C for type S + SWc to 2.5 °C for type S + SWa. On average, for more than 25% of days in Group 1, for four dominant circulation patterns, the temperature gradient in the layer 925–975 hPa was positive during the daytime. In the upper layer (850–925 hPa), the differences in thermal stratification between days assigned to Group 1 and the remaining days were not significant. The highest share of upper thermal inversion during the day for days assigned to Group 1 occurred for Ca + Ka type (56%); for the remaining types of circulation, the frequency of upper inversion was in the range from 39% to 43%.

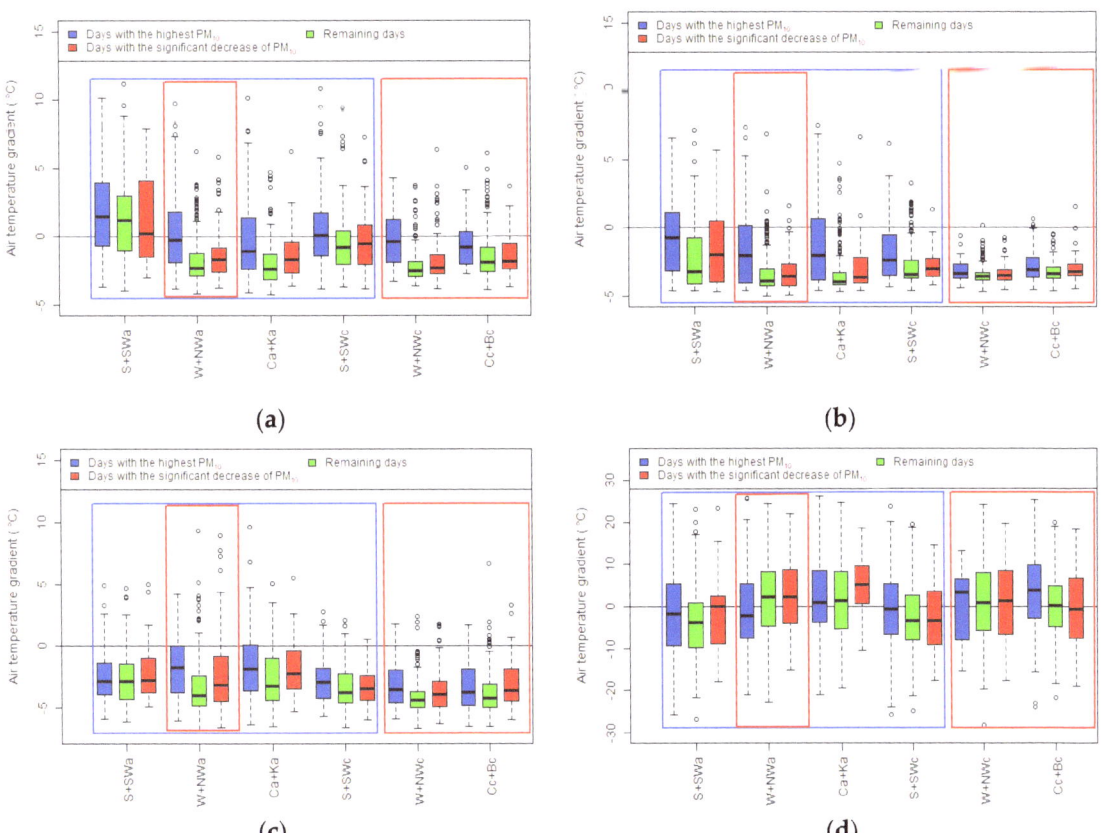

Figure A15. Boxplot of air temperature gradient for selected circulation types for days with high PM_{10} concentration (Group 1), days with significant improvement of air quality (Group 2) and remaining days at cold half-year for layer 925–975 hPa at 00:00 UTC (**a**) and 12:00 UTC (**b**) and for layer 850–925 hPa at 00:00 UTC (**c**) and 12:00 UTC (**d**) from ERA5 reanalysis data for Kraków city.

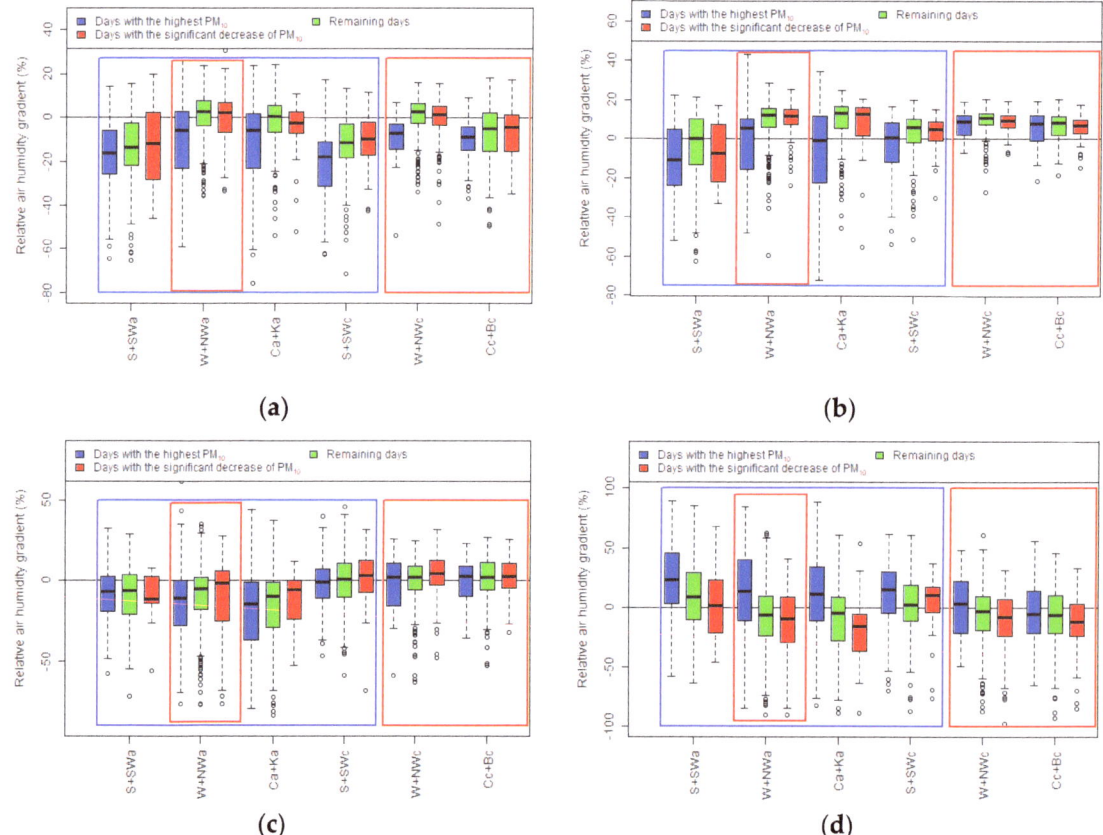

Figure A16. Boxplot of relative air humidity gradient for selected circulation types for days with high PM$_{10}$ concentration (Group 1), days with significant improvement of air quality (Group 2), and remaining days at cold half-year for layer 925–975 hPa at 00:00 UTC (**a**) and 12:00 UTC (**b**) and for layer 850–925 hPa at 00:00 UTC (**c**) and 12:00 UTC (**d**) current day from ERA5 reanalysis data for Kraków city.

Air humidity stratification during days assigned to Group 1 presented for four types of circulation during the nighttime is characterized by a stronger decrease of humidity in layer 925–975 than during the remaining days. Furthermore, during the day, for days with high PM$_{10}$ concentration, local minimum of relative air humidity at the level of 925 hPa occurred frequently.

The number of cases in which relative air humidity in the 925–975 hPa layer was lower than in the neighboring layer (975- 850 hPa) by at least 10%, ranging from 18% for the S + SWc to 43% for S + SWa type. The percentage of such cases during the day and night for days with high levels of PM$_{10}$ and days not assigned to both special groups is presented in Table A6.

Table A6. Frequency of days with a minimum of air relative humidity at 925–975 hPa during daytime and nighttime for days with high PM_{10} concentration and remaining days for selected Niedźwiedź circulation types.

Circulation Type	Days with the Highest PM_{10} Concentration (%)		Remaining Days (%)	
	00:00 UTC	12:00 UTC	00:00 UTC	12:00 UTC
S + SWa	5	44	8	19
W + NWa	4	25	2	4
Ca + Ka	4	33	4	5
S + SWc	15	18	16	10

Analysis of relative air humidity at the level of 925 hPa for days assigned for Group 1 indicated that for the types S + SWa, W + NWa and Ca + Ka median values were significantly lower during the day and night in comparison with the days not assigned for both groups (the greatest difference in medians equal to 25% during day and night was observed for type W + NWa). Furthermore, for the days in Group 1 with circulation type W + NWa and Ca + Ka, the distribution of relative humidity was characterized by a significantly wider interquartile range than for the remaining days. The average daily value of the interquartile range for both selected types was equal 41% and 23% for days in Group 1 and remaining days, respectively. In the case of the circulation type S + SWc, the relative humidity distribution was similar for both groups of days (Group 1 and the remaining days). The relative humidity distribution at the level of 925 hPa for selected types of circulation was presented in Figure A17a). Analysis of the air temperature distribution at the level of 925 hPa for selected circulation types showed no significant differences for days assigned to Group 1 and the remaining days (Figure A17b). The study of the wind speed distribution at the level of 925 hPa for four types of circulation pointed out that the differences of this parameter between days in Group 1 and the remaining days are lower during the day than at night. The greatest differences of the wind speed distribution between days in Group 1 and remaining days at night were observed for W + NWa and S + SWc types. The median values of wind speed at night for days assigned to Group 1 were lower by 3.3 m·s^{-1} and 2.3 m·s^{-1} for W + NWa and S + SWc, respectively (not shown in the article). Furthermore, the analysis of the wind speed difference in layer between 925 and 975 hPa indicated that in the nighttime the wind shear was significantly weaker for the W + NWa and S + SWc types, while in the daytime period the distribution was similar or slightly higher than for the remaining days. For the Ca + Ka type, wind shear was the lowest among the four distinguished circulation types.

Group 2: days with highest decrease of PM_{10} concentration.

The second Group of days selected in the cold half-year consists of days during which a significant decrease in daily PM_{10} concentration occurred in comparison with the previous day. Table A7 presents the conditional probability of the occurrence in a large decrease of PM_{10} concentration for individual types of atmospheric circulation, the number of days with a particular type of circulation and number of days with a significant decrease in PM_{10} concentration for individual types of circulation.

The significant improvement of the dispersion conditions occurred mostly with air advection from the W–NW sector at cyclonic and anticyclonic conditions and nonadvection cyclonic types (Cc + Bc). It is worth mentioning that during circulation type W + NWa the highest levels of PM_{10} concentration occurred, too. The highest conditional probability in significant decrease of PM_{10} concentration was obtained for cyclonic types Cc + Bc and W + NWc (0.29 and 0.25, respectively).

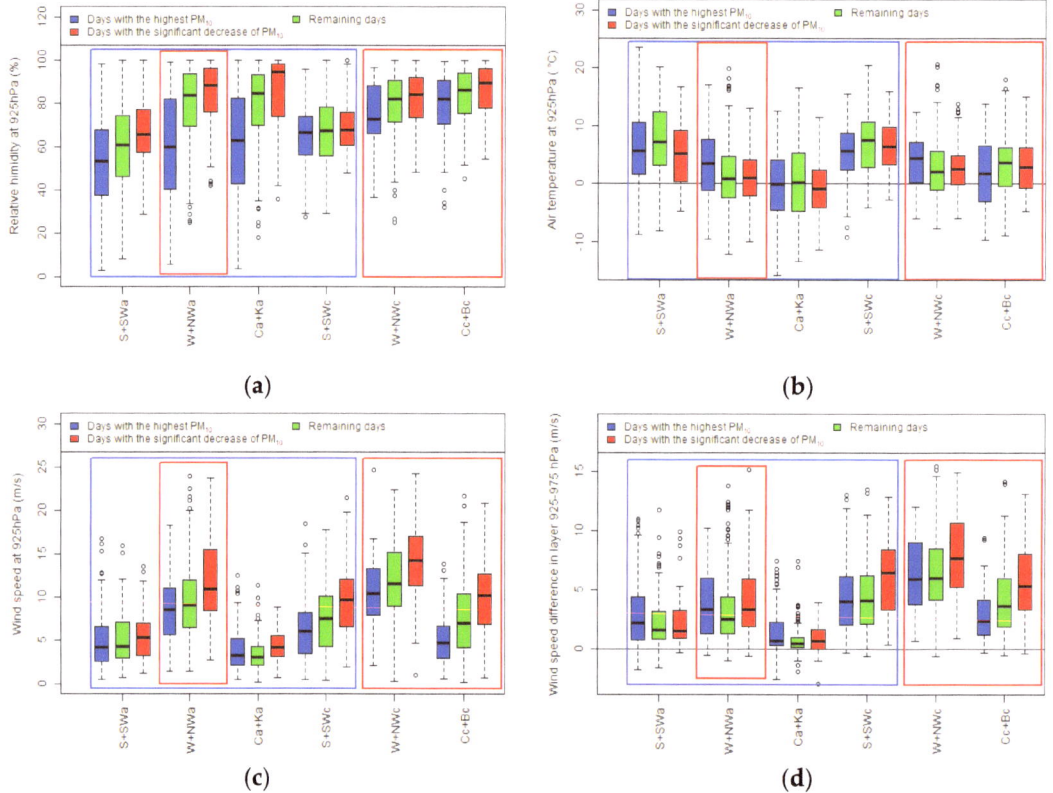

Figure A17. Boxplot of relative air humidity (**a**), air temperature (**b**), wind speed (**c**) at 925 hPa and wind speed differences between 925 and 975 hPa (**d**) at 12:00 UTC for selected circulation types for days with high PM_{10} concentration (Group 1), days with significant improvement of air quality (Group 2) and remaining days at cold half-year for layer 925–975 hPa from ERA5 reanalysis data for Kraków city.

It should be mentioned that the number of days that meet the conditions of significant pollution decrease (more than 25% decrease and at least 20 $\mu g \cdot m^{-3}$) in individual months in the cold half-year is not even; the smallest number of cases meeting the criteria occurred in October (67 days), a similar number of days was selected for the months of November, February, and March (more than 100 days) and the highest number for the period from December to January (more than 120 days).

Analysis of weather conditions from the Balice synoptic station for selected circulation types (Figures A12 and A13; boxplots in red frame) showed that air temperature did not differ significantly for days with a significant decrease of PM_{10} (Group 2) compared to days not assigned to both specific groups (remaining days). The *p*-value calculated with the Mann–Whitney U with significance level α = 0.05 test has also confirmed the similarity of both groups for these meteorological parameters (Table A8). The *p*-values obtained for wind speed and atmospheric precipitation were lower than 0.05, which pointed out that distribution in selected groups is statistically different. For the air temperature groups, the *p*-value were highest. The daily wind speed distribution for three selected circulation patterns were higher for days in Group 2 compared to the remaining days. The highest differences were observed for circulation type W + NWa (the median and upper quartile for this group were higher by 1.3 $m \cdot s^{-1}$ and 0.8 $m \cdot s^{-1}$), while for the remaining types (W + NWc and Cc + Bc), the wind speed was on average higher by 0.5 $m \cdot s^{-1}$.

Table A7. Frequency of days with minimum of air relative humidity at 925–975 hPa during daytime and nighttime for days with high PM_{10} concentration and remaining days for selected Niedźwiedź circulation types.

Circulation Type	Conditional Probability	Number of Days with High PM_{10} Concentration	Total Number of Days in Cold Half-Year
N + NEa	0.18	35	196
E + SEa	0.15	53	354
S + SWa	0.06	25	386
W + NWa	0.17	96	562
Ca + Ka	0.06	28	441
N + NEc	0.31	58	190
E + SEc	0.25	43	170
S + SWc	0.13	54	403
W + NWc	0.29	150	521
Cc + Bc	0.25	86	348
x	0.08	6	74
Total number of days		634	3645

The number of days with precipitation for days in Group 2 compared to days not assigned to both groups for dominant circulation patterns was higher by 18% on average. For the type W + NWa, the number of days with precipitation above 0.1 mm/day was equal to 53% of all days, and for the cyclonic types W + NWc and Cc + Bc it was equal to 84% and 88%, respectively. The sum of daily precipitation for the distinguished types was higher than for the remaining days, the smallest increase was observed for the anticyclonic type W + NWa (greater on average by 0.5 mm/day), while for the cyclonic types, the daily sum of precipitation was greater by more than 1 mm/day. The frequency of rainfall during the day and night was similar for all the distinguished types in Group 2.

Table A8. p-value calculated with the Mann–Whitney U test between days assigned to Group 2 and days not assigned to both groups in the cold half-year calculated for daily wind speed, air temperature and atmospheric precipitation for three selected circulation types from Niedźwiedź classification.

Circulation Type	Wind Speed	Air Temp.	Precipitation
W + NWa	0.000	0.750	0.013
W + NWc	0.005	0.918	0.000
Cc + Bc	0.000	0.871	0.008

Analysis of intra-valley thermal stratification from TV mast data (up to 100 m a.g.l.) indicated that the length of the near-ground inversion persistence did not exceed 3 h for most of the cases. The duration of inversion in the night period was shorter for all the selected circulation types for days with a significant decrease in PM_{10} compared to the reference group that contained the remaining days (Figure A13; boxplots in red frames).

Analysis of the ERA5 data indicated that the vertical gradient of temperature and relative humidity did not differ significantly for the group of days with a significant improvement in air quality compared to days not assigned to both groups (Figures A15 and A16 in Appendix D; data in red frames). Low thermal inversion in the 975–925 hPa layer during the nighttime period did not exceed 20% of all days for the selected types, and during the daytime, low thermal inversion almost did not occur. Upper thermal inversions in the layer 925–850 hPa during daytime accounted for more than 50% of cases, with the highest share equal to 60% for the W + NWa type. The vertical profile of relative humidity during the day was characterized by the local maximum at 925 hPa, the median relative humidity gradient in layers 975–925 hPa and 925–850 hPa was equal on average to +9% and −9%, respectively. The largest number of days, where relative humidity at the level of 925 hPa was higher than neighboring levels by at least 10%, occurred for the W + NWa type, equal to 33%, and the other types accounted for 17% of cases, on average. For the group of remaining days, similar humidity stratification, with local maximum relative humidity at 925 hPa, constituted from 20% for Cc + Bc to 28% for W + NWa type.

Analysis of relative humidity and air temperature at 925 hPa level during the daytime and nighttime period for the three distinguished circulation types did not show significant differences for the days assigned to Group 2 and the remaining days. The distribution wind speed at a pressure level of 925 hPa is for days in Group 2, for all selected types was higher during the day and night in comparison with the remaining days. The wind speed at night and day for all selected circulation types was on average higher by 1.4 m·s^{-1} and 2.6 m·s^{-1} than for the remaining days, respectively (Figure A17c in Appendix D). Furthermore, the wind shear connected to the wind speed change in layer 925–975 hPa was stronger for days with the circulation type W + NWc and Cc + Bc during day and night for days in Group 2 compared to the remaining days.

References

1. Toro, R.; Kvakic, M.; Klaic, Z.B.; Koracin, D.; Morales, R.G.E.; Leiva, M.A. Exploring atmospheric stagnation during a severe particulate matter air pollution episode over complex terrain in Santiago, Chile. *Environ. Pollut.* **2019**, *244*, 705–714. [CrossRef] [PubMed]
2. Xu, Y.W.; Zhu, B.; Shi, S.S.; Huang, Y. Two Inversion Layers and Their Impacts on PM2.5 Concentration over the Yangtze River Delta, China. *J. Appl. Meteorol. Climatol.* **2019**, *58*, 2349–2362. [CrossRef]
3. Ormanova, G.; Karaca, F.; Kononova, N. Analysis of the impacts of atmospheric circulation patterns on the regional air quality over the geographical center of the Eurasian continent. *Atmos. Res.* **2020**, *237*, 104858. [CrossRef]
4. Hadi-Vencheh, A.; Tan, Y.; Wanke, P.; Loghmanian, S.M. Air pollution assessment in China: A novel group multiple criteria decision making model under uncertain information. *Sustainability* **2021**, *13*, 1686. [CrossRef]
5. Zhou, G.; Wu, J.; Yang, M.; Sun, P.; Gong, Y.; Chai, J.; Zhang, J.; Afrim, F.-K.; Dong, W.; Sun, R.; et al. Prenatal exposure to air pollution and the risk of preterm birth in rural population of Henan Province. *Chemosphere* **2022**, *286*, 131833. [CrossRef] [PubMed]
6. Li, G.A.; Wu, H.B.; Zhong, Q.; He, J.L.; Yang, W.J.; Zhu, J.L.; Zhao, H.H.; Zhang, H.S.; Zhu, Z.Y.; Huang, F. Six air pollutants and cause-specific mortality: A multi-area study in nine counties or districts of Anhui Province, China. *Environ. Sci. Pollut. Res.* **2021**, *29*, 468–482. [CrossRef]
7. Jeong, S.J. The Impact of Air Pollution on Human Health in Suwon City. *Asian J. Atmos. Environ.* **2013**, *7*, 227–233. [CrossRef]
8. Vicente, A.B.; Juan, P.; Meseguer, S.; Diaz-Avalos, C.; Serra, L. Variability of PM10 in industrialized-urban areas. New coefficients to establish significant differences between sampling points. *Environ. Pollut.* **2018**, *234*, 969–978. [CrossRef] [PubMed]
9. Penenko, A.; Penenko, V.; Tsvetova, E.; Gochakov, A.; Pyanova, E.; Konopleva, V. Sensitivity Operator Framework for Analyzing Heterogeneous Air Quality Monitoring Systems. *Atmosphere* **2021**, *12*, 1697. [CrossRef]
10. Wang, Y.S.; Yao, L.; Wang, L.L.; Liu, Z.R.; Ji, D.S.; Tang, G.Q.; Zhang, J.K.; Sun, Y.; Hu, B.; Xin, J.Y. Mechanism for the formation of the January 2013 heavy haze pollution episode over central and eastern China. *Sci. China-Earth Sci.* **2014**, *57*, 14–25. [CrossRef]
11. Masiol, M.; Agostinelli, C.; Formenton, G.; Tarabotti, E.; Pavoni, B. Thirteen years of air pollution hourly monitoring in a large city: Potential sources, trends, cycles and effects of car-free days. *Sci. Total Environ.* **2014**, *494*, 84–96. [CrossRef] [PubMed]
12. Tveito, O.E.; Huth, R.; Philipp, A.; Post, P.; Pasqui, M.; Esteban, P.; Beck, C.; Demuzere, M.; Prudhomme, C. *COST Action 733 Harmonization and Application of Weather Type Classifications for European Regions*; Climate & Environment Consulting Potsdam GmbH: Potsdam, Germany, 2016; pp. 243–249.
13. Li, X.; Xia, X.; Wang, L.; Cai, R.; Zhao, L.; Feng, Z.; Ren, Q.; Zhao, K. The role of foehn in the formation of heavy air pollution events in Urumqi, China. *J. Geophys. Res. Atmos.* **2015**, *120*, 5371–5384. [CrossRef]
14. Lesniok, M.; Malarzewski, L.; Niedzwiedz, T. Classification of circulation types for Southern Poland with an application to air pollution concentration in Upper Silesia. *Phys. Chem. Earth* **2010**, *35*, 516–522. [CrossRef]
15. Vautard, R.; Colette, A.; van Meijgaard, E.; Meleux, F.; van Oldenborgh, G.J.; Otto, F.; Tobin, I.; Yiou, P. Attribution of wintertime anticyclonic stagnation contributing to air pollution in western europe. *Bull. Am. Meteorol. Soc.* **2018**, *99*, S70–S75. [CrossRef]
16. Garrido-Perez, J.M.; Ordonez, C.; Garcia-Herrera, R.; Barriopedro, D. Air stagnation in Europe: Spatiotemporal variability and impact on air quality. *Sci. Total Environ.* **2018**, *645*, 1238–1252. [CrossRef] [PubMed]
17. Horton, D.E.; Skinner, C.B.; Singh, D.; Diffenbaugh, N.S. Occurrence and persistence of future atmospheric stagnation events. *Nat. Clim. Chang.* **2014**, *4*, 698–703. [CrossRef] [PubMed]

18. Lee, D.; Wang, S.Y.; Zhao, L.; Kim, H.C.; Kim, K.; Yoon, J.H. Long-term increase in atmospheric stagnant conditions over northeast Asia and the role of greenhouse gases-driven warming. *Atmos. Environ.* **2020**, *241*, 117772. [CrossRef]
19. Flocas, H.; Kelessis, A.; Helmis, C.; Petrakakis, M.; Zoumakis, M.; Pappas, K. Synoptic and local scale atmospheric circulation associated with air pollution episodes in an urban Mediterranean area. *Theor. Appl. Climatol.* **2009**, *95*, 265–277. [CrossRef]
20. Vergeiner, J. *South Foehn Studies and a New Foehn Classification Scheme in the Wipp and Inn Valley*; University of Innsbruck: Innsbruck, Austria, 2004.
21. Sekula, P.; Bokwa, A.; Ustrnul, Z.; Zimnoch, M.; Bochenek, B. The impact of a foehn wind on PM10 concentrations and the urban boundary layer in complex terrain: A case study from Krakow, Poland. *Tellus Ser. B Chem. Phys. Meteorol.* **2021**, *73*, 1–26. [CrossRef]
22. *Air Quality in Europe—2020 Report*; EEA Report No 09/2020; European Environmental Agency: Luxembourg, 2020.
23. Chief Inspectorate for Environmental Protection. *Stan Środowiska w Województwie Małopolskim. Raport 2020 (The State of The Environment in the Lesser Poland Voivodeship. Report 2020)*; National Inspectorate for Environmental Protection: Kraków, Poland, 2020; p. 199.
24. Bokwa, A. Environmental impacts of long-term air pollution changes in Krakow, Poland. *Pol. J. Environ. Stud.* **2008**, *17*, 673–686.
25. Pietras, B. *Meteorologiczne Uwarunkowania Koncentracji Pyłu Zawieszonego w Powietrzu w Krakowie Oraz Próba Określenia Jego Pochodzenia*; Uniwersytet Pedagogiczny: Kraków, Poland, 2018.
26. Wielgosinski, G.; Czerwinska, J. Smog episodes in Poland. *Atmosphere* **2020**, *11*, 277. [CrossRef]
27. Lupikasza, E.; Niedzwiedz, T. Synoptic climatology of fog in selected locations of southern Poland (1966–2015). *Bull. Geogr. Phys. Geogr. Ser.* **2016**, *11*, 5–15. [CrossRef]
28. Matuszko, D.; Weglarczyk, S. Long-term variability of the cloud amount and cloud genera and their relationship with circulation (Krakow, Poland). *Int. J. Climatol.* **2018**, *38*, E1205–E1220. [CrossRef]
29. Niedźwiedź, T.; Ustrnul, Z. Change of Atmospheric Circulation. In *Climate Change in Poland*; Falarz, M., Ed.; Springer: Cham, Switerland, 2021.
30. Ustrnul, Z. Atmospheric circulation conditions. In *Climate of Kraków in the 20th Century*; Matuszko, D., Ed.; Instytut Geografii i Gospodarki Przestrzennej Uniwersytet Jagielloński: Kraków, Poland, 2007; pp. 21–40.
31. Grange, S.K.; Carslaw, D.C.; Lewis, A.C.; Boleti, E.; Hueglin, C. Random forestmeteorological normalisation models for Swiss PM10 trend analysis. *Atmos. Chem. Phys.* **2018**, *18*, 6223–6239. [CrossRef]
32. Vu, T.V.; Shi, Z.B.; Cheng, J.; Zhang, Q.; He, K.B.; Wang, S.X.; Harrison, R.M. Assessing the impact of clean air action on air quality trends in Beijing using a machine learning technique. *Atmos. Chem. Phys.* **2019**, *19*, 11303–11314. [CrossRef]
33. Gariazzo, C.; Carlino, G.; Silibello, C.; Renzi, M.; Finardi, S.; Pepe, N.; Radice, P.; Forastiere, F.; Michelozzi, P.; Viegi, G.; et al. A multi-city air pollution population exposure study: Combined use of chemical-transport and random-Forest models with dynamic population data. *Sci. Total Environ.* **2020**, *724*, 138102. [CrossRef] [PubMed]
34. Hu, X.F.; Belle, J.H.; Meng, X.; Wildani, A.; Waller, L.A.; Strickland, M.J.; Liu, Y. Estimating PM2.5 Concentrations in the conterminous United States using the random forest approach. *Environ. Sci. Technol.* **2017**, *51*, 6936–6944. [CrossRef] [PubMed]
35. Joharestani, M.Z.; Cao, C.X.; Ni, X.L.; Bashir, B.; Talebiesfandarani, S. PM2.5 Prediction based on random forest, XGBoost, and deep learning using multisource remote sensing data. *Atmosphere* **2019**, *10*, 373. [CrossRef]
36. AlThuwaynee, O.F.; Kim, S.W.; Najemaden, M.A.; Aydda, A.; Balogun, A.L.; Fayyadh, M.M.; Park, H.J. Demystifying uncertainty in PM10 susceptibility mapping using variable drop-off in extreme-gradient boosting (XGB) and random forest (RF) algorithms. *Environ. Sci. Pollut. Res.* **2021**, *28*, 43544–43566. [CrossRef]
37. Stafoggia, M.; Johansson, C.; Glantz, P.; Renzi, M.; Shtein, A.; de Hoogh, K.; Kloog, I.; Davoli, M.; Michelozzi, P.; Bellander, T. A random forest approach to estimate daily particulate matter, nitrogen dioxide, and ozone at fine spatial resolution in Sweden. *Atmosphere* **2020**, *11*, 239. [CrossRef]
38. Lityński, J. *Numerical Classification of Circulation Types and Weather Types for Poland*; Pr. PIHM: Kraków, Poland, 1969; Volume 97, pp. 3–14.
39. Ustrnul, Z.; Wypych, A.; Czekierda, D. Composite circulation index of weather extremes (the example for Poland). *Meteorol. Z.* **2013**, *22*, 551–559. [CrossRef]
40. Beck, C.; Philipp, A. Evaluation and comparison of circulation type classifications for the European domain. *Phys. Chem. Earth* **2010**, *35*, 374–387. [CrossRef]
41. Nowosad, M. Variability of the zonal circulation index over Central Europe according to the Lityński method. *Geogr. Pol.* **2017**, *90*, 417–430. [CrossRef]
42. Godłowska, J. *Influence of Meteorological Conditions on Air Quality in Krakow. Comparative Research and an Attempt at a Model Approach*; IMGW-PIB: Warsaw, Poland, 2019; p. 102.
43. Jaagus, J. Climatic changes in Estonia during the second half of the 20th century in relationship with changes in large-scale atmospheric circulation. *Theor. Appl. Climatol.* **2006**, *83*, 77–88. [CrossRef]
44. Hyncica, M.; Huth, R. Long-term changes in precipitation phase in Europe in cold half year. *Atmos. Res.* **2019**, *227*, 79–88. [CrossRef]
45. Statistics Poland. *Area and Population in the Territorial Profile in 2021*; Statistics Poland: Warsaw, Poland, 2021; p. 25.
46. Hess, M. Climate of Kraków. *Folia Geogr. Ser. Geogr.-Phys. Kraków Pol.* **1974**, *8*, 45–102.

47. Oke, T.R. *Initial Guidance to Obtain Representative Meteorological Observations at Urban Sites. Instrument and Observing Methods (IOM)*; Report No. 81, WMO/TD. No. 1250; World Meteorological Organization: Geneva, Switzerland, 2006.
48. Hersbach, H.; Bell, B.; Berrisford, P.; Hirahara, S.; Horanyi, A.; Munoz-Sabater, J.; Nicolas, J.; Peubey, C.; Radu, R.; Schepers, D.; et al. The ERA5 global reanalysis. *Q. J. R. Meteorol. Soc.* **2020**, *146*, 1999–2049. [CrossRef]
49. Chief Inspectorate of Environmental Protection. Available online: https://powietrze.gios.gov.pl/pjp/archives (accessed on 12 March 2022).
50. European Parliament and the Council of the European Union. Directive 2008/50/EC of the European Parliament and of the Council. *J. Eur. Union* **2008**.
51. Huth, R.; Beck, C.; Philipp, A.; Demuzere, M.; Ustrnul, Z.; Cahynova, M.; Kysely, J.; Tveito, O.E. Classifications of atmospheric circulation patterns recent advances and applications. *Trends Dir. Clim. Res.* **2008**, *1146*, 105–152. [CrossRef]
52. Philipp, A.; Bartholy, J.; Beck, C.; Erpicum, M.; Esteban, P.; Fettweis, X.; Huth, R.; James, P.; Jourdain, S.; Kreienkamp, F.; et al. Cost733cat-A database of weather and circulation type classifications. *Phys. Chem. Earth* **2010**, *35*, 360–373. [CrossRef]
53. Ustrnul, Z.; Czekierda, D.; Wypych, A. Extreme values of air temperature in Poland according to different atmospheric circulation classifications. *Phys. Chem. Earth* **2010**, *35*, 429–436. [CrossRef]
54. Pomona. Available online: https://github.com/silkeszy/Pomona (accessed on 12 March 2022).
55. Degenhardt, F.; Seifert, S.; Szymczak, S. Evaluation of variable selection methods for random forests and omics data sets. *Brief. Bioinform.* **2019**, *20*, 492–503. [CrossRef] [PubMed]
56. Akaike, H. Information theory and an extension of the maximum likelihood principle. In *Selected Papers of Hirotugu Akaike*; Parzen, E., Tanabe, K., Kitagawa, G., Eds.; Springer Series in Statistics; Springer: New York, NY, USA, 1998.
57. Cowell, F. *Measurement of Inequality*, 1th ed.; Atkinson, A.B., Bourguignon, F., Eds.; Elsevier: Amsterdam, The Netherland, 2000; p. 938.
58. Tangirala, S. Evaluating the impact of GINI index and information gain on classification using decision tree classifier algorithm. *Int. J. Adv. Comput. Sci. Appl.* **2020**, *11*, 612–619. [CrossRef]
59. Zhang, D.D.; Shen, J.Q.; Liu, P.F.; Zhang, Q.; Sun, F.H. Use of fuzzy analytic hierarchy process and environmental gini coefficient for allocation of regional flood drainage rights. *Int. J. Environ. Res. Public Health* **2020**, *17*, 63. [CrossRef]
60. Wu, C.B.; Li, K.; Bai, K.X. Validation and calibration of CAMS PM2.5 forecasts using in situ PM2.5 measurements in China and United States. *Remote Sens.* **2020**, *12*, 3813. [CrossRef]
61. Pappa, A.; Kioutsioukis, I. Forecasting particulate pollution in an urban area: From copernicus to sub-km scale. *Atmosphere* **2021**, *12*, 881. [CrossRef]
62. Czernecki, B.; Marosz, M.; Jedruszkiewicz, J. Assessment of machine learning algorithms in short-term forecasting of PM10 and PM2.5 concentrations in selected polish agglomerations. *Aerosol Air Qual. Res.* **2021**, *21*, 200586. [CrossRef]
63. Ustrnul, Z. Infulence of foehn winds on air-temperature and humidity in the Polish Carpathians. *Theor. Appl. Climatol.* **1992**, *45*, 43–47. [CrossRef]
64. Bokwa, A.; Wypych, A.; Hajto, M.J. Impact of natural and anthropogenic factors on fog frequency and variability in krakow, Poland in the years 1966–2015. *Aerosol Air Qual. Res.* **2018**, *18*, 165–177. [CrossRef]
65. Han, S.Q.; Hao, T.Y.; Zhang, Y.F.; Liu, J.L.; Li, P.Y.; Cai, Z.Y.; Zhang, M.; Wang, Q.L.; Zhang, H. Vertical observation and analysis on rapid formation and evolutionary mechanisms of a prolonged haze episode over central-eastern China. *Sci. Total Environ.* **2018**, *616*, 135–146. [CrossRef] [PubMed]
66. Kunin, P.; Alpert, P.; Rostkier-Edelstein, D. Investigation of sea-breeze/foehn in the Dead Sea valley employing high resolution WRF and observations. *Atmos. Res.* **2019**, *229*, 240–254. [CrossRef]
67. Stull, R.B. *An Introduction to Boundary Layer Meteorology*; Springer: Dordrecht, The Netherland, 1988.
68. Wang, P.; Cao, J.J.; Tie, X.X.; Wang, G.H.; Li, G.H.; Hu, T.F.; Wu, Y.T.; Xu, Y.S.; Xu, G.D.; Zhao, Y.Z.; et al. Impact of meteorological parameters and gaseous pollutants on PM2.5 and PM10 mass concentrations during 2010 in Xi'an, China. *Aerosol Air Qual. Res.* **2015**, *15*, 1844–1854. [CrossRef]
69. Sekula, P.; Bokwa, A.; Bartyzel, J.; Bochenek, B.; Chmura, L.; Galkowski, M.; Zimnoch, M. Measurement report: Effect of wind shear on PM10 concentration vertical structure in the urban boundary layer in a complex terrain. *Atmos. Chem. Phys.* **2021**, *21*, 12113–12139. [CrossRef]
70. Huang, K.Y.; Xiao, Q.Y.; Meng, X.; Geng, G.N.; Wang, Y.J.; Lyapustin, A.; Gu, D.F.; Liu, Y. Predicting monthly high-resolution PM2.5 concentrations with random forest model in the North China Plain. *Environ. Pollut.* **2018**, *242*, 675–683. [CrossRef] [PubMed]
71. Li, Y.; Chen, Q.L.; Zhao, H.J.; Wang, L.; Tao, R. Variations in PM10, PM2.5 and PM1.0 in an urban area of the sichuan basin and their relation to meteorological factors. *Atmosphere* **2015**, *6*, 150–163. [CrossRef]
72. Stafoggia, M.; Bellander, T.; Bucci, S.; Davoli, M.; de Hoogh, K.; de'Donato, F.; Gariazzo, C.; Lyapustin, A.; Michelozzi, P.; Renzi, M.; et al. Estimation of daily PM10 and PM2.5 concentrations in Italy, 2013–2015, using a spatiotemporal land-use random-forest model. *Environ. Int.* **2019**, *124*, 170–179. [CrossRef] [PubMed]
73. Banks, R.F.; Tiana-Alsina, J.; Rocadenbosch, F.; Baldasano, J.M. Performance evaluation of the boundary-layer height from lidar and the weather research and forecasting model at an urban coastal site in the north-east iberian peninsula. *Bound. Layer Meteorol.* **2015**, *157*, 265–292. [CrossRef]

74. Uzan, L.; Egert, S.; Khain, P.; Levi, Y.; Vadislavsky, E.; Alpert, P. Ceilometers as planetary boundary layer height detectors and a corrective tool for COSMO and IFS models. *Atmos. Chem. Phys.* **2020**, *20*, 12177–12192. [CrossRef]
75. Zhang, K.F.; The, J.; Xie, G.Y.; Yu, H.S. Multi-step ahead forecasting of regional air quality using spatial-temporal deep neural networks: A case study of Huaihai Economic Zone. *J. Clean. Prod.* **2020**, *277*, 123231. [CrossRef]
76. Zhou, Y.L.; Chang, F.J.; Chang, L.C.; Kao, I.F.; Wang, Y.S. Explore a deep learning multi-output neural network for regional multi-step-ahead air quality forecasts. *J. Clean. Prod.* **2019**, *209*, 134–145. [CrossRef]
77. Theil, H. A Rank-invariant method of linear and polynomial regression analysis. In *Henri Theil's Contributions to Economics and Econometrics. Advanced Studies in Theoretical and Applied Econometrics*; Raj, B., Koerts, J., Eds.; Springer: Dordrecht, The Netherland, 1992; Volume 23, pp. 345–381.
78. Hurtado, S. Package 'RobustLinearReg'. Available online: https://cran.r-project.org/web/packages/RobustLinearReg/RobustLinearReg.pdf (accessed on 15 November 2021).
79. Carslaw, D.C.; Ropkins, K. Openair—An R package for air quality data analysis. *Environ. Model. Softw.* **2012**, *27–28*, 52–61. [CrossRef]
80. Sulikowska, A.; Wypych, A. Seasonal variability of trends in regional hot and warm temperature extremes in europe. *Atmosphere* **2021**, *12*, 612. [CrossRef]
81. Niedźwiedź, T. *Synoptic Situations and their Impact on Spatial Differentiation of Selected Climate Elements in the Upper Vistula Basin*; Jagiellonian University: Kraków, Poland, 1981.
82. Lamb, H.H. *British Isles Weather Types and a Register of the Daily Sequence of Circulation Patterns 1861–1971*; Geophysical Memoirs: London, UK, 1972.
83. Pianko-Kluczynska, K. A new calendar of types of atmosphere circulation according to J. Lityński. *Wiadomości Meteorol. Hydrol. Gospod. Wodnej* **2007**, *1*, 65–85.

Article

Operational Data-Driven Intelligent Modelling and Visualization System for Real-World, On-Road Vehicle Emissions—A Case Study in Hangzhou City, China

Lu Wang [1], Xue Chen [1], Yan Xia [1], Linhui Jiang [1], Jianjie Ye [2], Tangyan Hou [1], Liqiang Wang [1], Yibo Zhang [1], Mengying Li [1], Zhen Li [1], Zhe Song [1], Yaping Jiang [1], Weiping Liu [1], Pengfei Li [3,*], Xiaoye Zhang [1,4] and Shaocai Yu [1,*]

[1] Research Center for Air Pollution and Health, Key Laboratory of Environmental Remediation and Ecological Health, Ministry of Education, College of Environment and Resource Sciences, Zhejiang University, Hangzhou 310058, China; 3150105443@zju.edu.cn (L.W.); cxoldsnow@zju.edu.cn (X.C.); xy787933846@163.com (Y.X.); lh_lyn@163.com (L.J.); houtangyan@163.com (T.H.); 11514019@zju.edu.cn (L.W.); ybzhang903@zju.edu.cn (Y.Z.); 11814019@zju.edu.cn (M.L.); lizhen1942@163.com (Z.L.); song_zhe@zju.edu.cn (Z.S.); jiangyaping@zju.edu.cn (Y.J.); wliu@zju.edu.cn (W.L.); xiaoye@cma.gov.cn (X.Z.)
[2] Bytedance Inc., Hangzhou 310058, China; jianjieye@zju.edu.cn
[3] College of Science and Technology, Hebei Agricultural University, Baoding 071000, China
[4] Chinese Academy of Meteorological Sciences, China Meteorological Administration, Beijing 100081, China
* Correspondence: lpf_zju@163.com (P.L.); shaocaiyu@zju.edu.cn (S.Y.)

Abstract: On-road vehicle emissions play a crucial role in affecting air quality and human exposure, particularly in megacities. In the absence of comprehensive traffic monitoring networks with the general lack of intelligent transportation systems (ITSs) and big-data-driven, high-performance-computing (HPC) platforms, it remains challenging to constrain on-road vehicle emissions and capture their hotspots. Here, we established an intelligent modelling and visualization system driven by ITS traffic data for real-world, on-road vehicle emissions. Based on the HPC platform (named "City Brain") and an agile Web Geographic Information System (WebGISs), this system can map real-time (hourly), hyperfine (10~1000 m) vehicle emissions (e.g., $PM_{2.5}$, NO_x, CO, and HC) and associated traffic states (e.g., vehicle-specific categories and traffic fluxes) over the Xiaoshan District in Hangzhou. Our results show sharp variations in on-road vehicle emissions on small scales, which even fluctuated up to 31.2 times within adjacent road links. Frequent and widespread emission hotspots were also exposed. Over custom spatiotemporal scopes, we virtually investigated and visualized the impacts of traffic control policies on the traffic states and on-road vehicle emissions. Such results have important implications for how traffic control policies should be optimized. Integrating this system with chemical transport models and air quality measurements would bridge the technical gap between air pollutant emissions, concentrations, and human exposure.

Keywords: big-data intelligent system; on-road vehicle emissions; traffic monitoring; hyperfine modelling; real-time visualization

1. Introduction

With the simultaneous growth of urban scales and vehicle ownerships, on-road vehicles have the potential to overtake industrial and residential sectors as the dominant emission source in megacities [1–4]. For instance, urban on-road vehicles account for more than 30% of NO_x emissions globally and contribute up to 25% of $PM_{2.5}$ concentrations in China [5–7]. Therefore, the reliable assessment of on-road vehicle emissions is central to air pollution control and human exposure evaluation, which is conducive to the sustainable development of the social environment [8,9]. The vehicle emission inventory can be a valuable tool, as it can well reflect the close link between environmental impact and traffic flow [10]. However, estimating traffic emissions is a very complex process that requires

large amounts of data on emissions-producing activities (e.g., vehicle travelled distance, vehicle type, and operating conditions) and a deep understanding of emission rates [11,12]. With the continuous improvement of the spatiotemporal resolution of road vehicle emissions assessments by current urban pollution control policies, it is essential to accurately quantify real-world road vehicle emissions due to changes in actual traffic characteristics. Therefore, the reliability of activity data and emission factors is a crucial element in the quantification of road vehicle emissions and the quality of emissions inventories. Previous studies have shown that vehicle emissions under actual driving conditions are affected by a variety of factors, including vehicle characteristics (such as vehicle type, age, emission control devices, and operating conditions), urban road types and conditions, fuel type, and environmental conditions (e.g., temperatures and humidity) and traffic conditions [13,14]. Therefore, real-world, on-road vehicle emissions remain largely uncertain in traditional bottom-up emission inventories. The main concern is that road traffic states (e.g., traffic fluxes, road conditions, and vehicle type) can change drastically over a short distance (1~10 km) for a short time (hourly). The traditional inventory of on-road vehicle emissions is established based on historical data on a macro scale. The key concern is that routine frameworks generally rely on spatially coarse proxies (e.g., 1×1~25×25 km^2) and temporally static retrospectives (e.g., a historical year or month) [15–17], focusing on the characteristics or average levels of vehicle emissions, and thus the variations of spatiotemporal vehicle emissions are seldom considered. In addition, due to the limited resolution of vehicle emission calculations [18], they cannot capture on-road vehicle emission hotspots and drivers.

To date, various monitoring systems that can record real-world traffic situations have made significant progress [19–22]. These techniques mainly involve floating cars (e.g., OBD-instrumented diesel trucks and GPS-equipped probe taxis), navigation maps (e.g., Google Map), and on-road video surveillance, each of which has distinct advantages and disadvantages [23–27]. For instance, an individual GPS-instrumented floating car accurately records its speeds and trajectories along with its static information (e.g., its vehicle category). Their fleets enable us to extrapolate surrounding traffic states. Nevertheless, in contrast to real-world fleets, they remain scarce and thus incapable of revealing hyperfine gradients (10 m~1 km) of on-road vehicle emissions [28–30]. Better yet, open-access maps (such as the Google and Baidu Maps) can provide more representative spatiotemporal maps of on-road vehicle emissions. Technically, they treat trajectories of mobile phones as spatiotemporal surrogates of traffic fluxes and, on this basis, establish hierarchical traffic congestion indexes. Despite this, vehicle-specific information remains unavailable, including vehicle-specific speeds and categories. In order to address this issue, a recent study [31] developed a full-sample enumeration approach (with 19 billion trajectories) via the BeiDou Navigation System to construct a big-data-driven vehicle emission inventory, which, however, was only suitable for trucks. Each technology has distinct advantages and disadvantages, and no source alone can achieve the high-resolution demand for quantifying road vehicle emissions. The solution is to use a more comprehensive road traffic system to obtain sufficient real-time traffic data to support hyperfine-resolution emission inventory and the development of a real-time road vehicle emission system.

Real-world traffic monitoring (e.g., on-road video surveillance and radio frequency identifications) can offer a valuable opportunity to recognize instantaneous and heterogeneous vehicle-specific states [32–35]. Through the mutual complementation of different data sources, the specific traffic status information of the vehicle can be obtained in real-time. Nevertheless, the output data come from independent facilities with distinct formats; thus, multi-source data is incompatible mutually. Subsequently, they are incompatible with the existing model frameworks of the on-road vehicle emissions (e.g., the fleet-specific MOVES) [36–38]. More importantly, the resulting database is projected to be of big data, thus leading to huge computational burdens [31,39–41]. For instance, the Data Throughput in a single hour might frequently exceed 200 MB and fluctuate violently. It should be noted that those facilities freshly achieved full coverage only in a few developed regions due to vast expenses [42]. Collectively, the unique path towards real-world on-road, vehicle-specific emissions is given by all-around traffic monitoring coupled with a big-data-driven ITS and HPC platform [38,43,44]. This systematic framework requires various intelligent techniques (e.g., image recognition) to interconnect and transfer those big and incompatible data. It is a unique opportunity to construct a hyperfine-resolution, on-road vehicle emission model. Besides, real-time data analyses would further maximize the benefit. Web Geographic Information Systems (WebGISs) [45] provide the efficient handling, visualization, and manipulation of geographic and geospatial information.

As a leading developed region in China, the Xiaoshan District in Hangzhou is facing serious air pollution, especially with surface O_3 continually exceeding the air quality standard in the summertime; it is mainly caused by the emission of mobile sources [46]. Moreover, the Xiaoshan District has also become a pioneer of digital government reform. A key achievement is that traffic monitoring has seemingly become ubiquitous since 2017. More than this, a breaking-through development is a big-data-driven, intelligent HPC system (named "City Brain") [47], full of ripe artificial intelligence algorithms. Initially, it was designed for tackling the digital reform of government affairs. Here, it is applied to store, fuse, and transfer comprehensive traffic monitoring data, even with incompatible accesses (due to distinct formats and multiple sources). In this study, we used it to build a bottom-up road vehicle emission calculation model to calculate single-vehicle-specific emissions over each fine-scale (10 m–1 km) road link. The objectives of this study are: (1) to conduct application research of high-temporal and spatial resolution and a visualization for urban road vehicle emissions based on comprehensive traffic data and a bottom-up road vehicle emission calculation model; (2) to visualize significant real-time variations in hyperfine on-road vehicle emissions and analyse the corresponding drivers (such as traffic fluxes and vehicle-specific speed) with an agile WebGIS system; and (3) to efficiently validate the benefits of traffic control strategies. Note that such strategies could be precisely designed for specific road segments and vehicle types via our hyperfine system. Therefore, this big-data-driven intelligent modelling and visualization system can serve as an effective and efficient tool for urban on-road vehicle emission management.

2. Materials and Methods

2.1. System Framework

This work aimed to develop and implement a big-data-driven intelligent modelling and visualization system for real-world, on-road vehicle emissions. As illustrated in Figure 1, we built up this system based on a classic Browser/Server (B/S) architecture [48] with four tiers, i.e., the perceptive, data, server, and presentation layers. The last three layers were erected on the "City Brain". The first layer consisted of comprehensive traffic monitoring, which was the foundation of the whole system. After that, the data layer mainly relied on the MySQL database [49] supported by the Relational Database Service (RDS) [50] on-board the "City Brain". Due to its high performance (e.g., large volume and high flexibility), it took responsibility for the storage and transmission of the big spatiotemporal data, including both the input and output data from other layers. On this basis, the server layer implemented an ITS that could interconnect and operate diverse traffic data from incompatible sources. A hyperfine model for on-road vehicle emissions served as the core of this system. This layer also received the user requests to invoke the information in the data layer, accessed the WebGIS application in the presentation layer, and performed corresponding feedbacks. The WebGIS engine can assemble the comprehensive spatiotemporal data analysis when the server receives user requests. In addition, the Elastic Compute Service (ECS) [51] onboard the "City Brain" accounted for basic operations, such as spatiotemporal data analyses, data sharing, and permission settings.

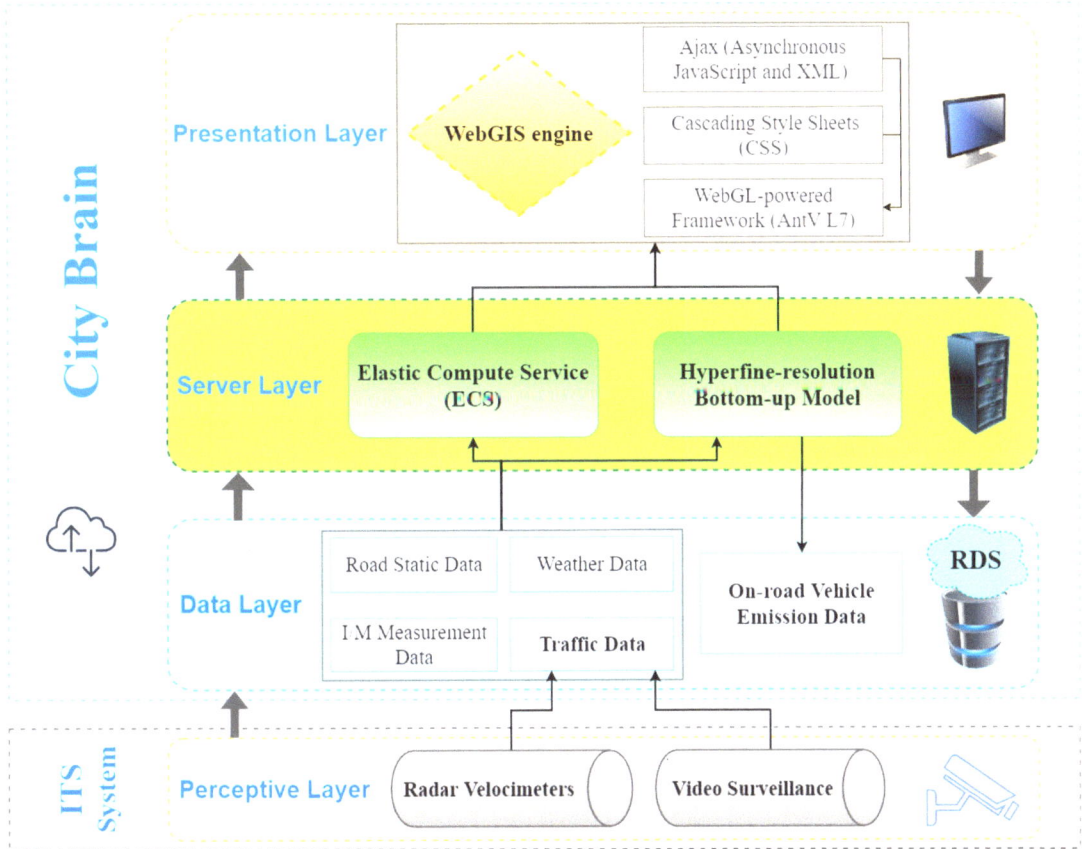

Figure 1. The framework of the big-data-driven intelligent modelling and visualization system for real-world, on-road vehicle emissions. This system based is on a classic Browser/Server (B/S) architecture with four tiers, i.e., the perceptive layer, data layer, server layer, and presentation layer. The last three layers were erected on the "City Brain".

Between the server and presentation layers, the Ajax (Asynchronous JavaScript and XML) technology made asynchronous HTTP requests without reloading client applications [52]. Moreover, cascading style sheets (CSS) technologies were applied to improve user experiences, such as optimizing the interface layout and increasing the response speed [53]. Considering the big data feature of this system, we applied an agile WebGL-powered framework (AntV L7) [54] for large-scale geospatial data visualization and rendering. Hence, the presentation interface can display GIS applications through the B/S architecture and share spatial information resources, thus breaking the limitations of traditional operational methods.

Consequently, this system can map real-time or historical vehicle-specific emissions (i.e., $PM_{2.5}$, NO_x, HC, and CO) and associated traffic states (i.e., traffic fluxes, vehicle-specific images, categories, and speed) from the perspective of spatial (e.g., road links) and temporal (e.g., hourly) dimensions. On this basis, they can be zoomed in and visualized via button selection. Furthermore, the spatiotemporal analysis, such as the top five roads (e.g., in terms of on-road vehicle emissions), was also highlighted. More importantly, relying on this HPC framework, we could virtually investigate and visualize the consequences of traffic control strategies over the custom spatiotemporal scopes.

2.2. Real-World Data Collection

The Xiaoshan District is situated in Hangzhou, Zhejiang Province, China (Figure 2). From its GDP (i.e., CNY 200 billion) perspective, it ranked fifth among districts in Hangzhou in 2019. Within its limited geographical extent (i.e., 1417.8 km^2), there was roughly 1953.7 km of road networks. In this context, the Xiaoshan District emerged as a vital urban transportation hub in Zhejiang Province. This indicates that air quality in urban microenvironment is significantly affected by fine-scale, on-road vehicle emissions.

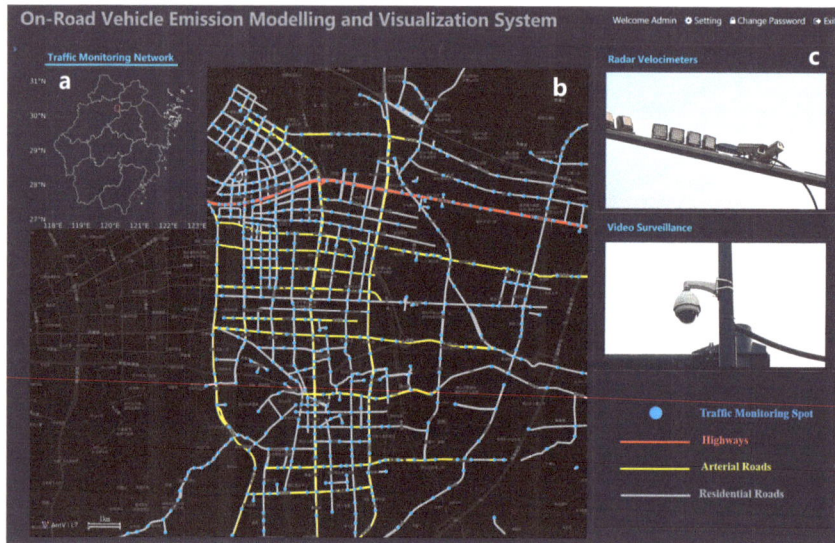

Figure 2. All-around traffic monitoring network. (**a**) The location of the Xiaoshan District in China. (**b**) Spatial distributions of traffic monitoring sites (dot). Correspondingly, all road links are divided into three types, including residential streets, arterial roads, and highways. (**c**) Each site includes radar velocimeters and surveillance cameras. Map data © 2021, AntV L7.

It should be highlighted that all-around traffic monitoring allowed us to collect vehicle-specific data (Figure 2). Specifically, video surveillance and radar velocimeters measure vehicle-specific images and speed, respectively. First, in theory, traffic fluxes and vehicle-specific categories significantly affect the on-road vehicle emission [26,31]. In particular, traffic congestion and high-duty vehicles generally result in emission hotspots. To this end, all-around traffic video surveillances were applied to enable vehicle-specific identifications. From 1 January to 31 December 2021, we established an extensive database of more than 2400 million records. Vehicle-specific categories, licence plates, and fluxes could be identified via intelligent techniques (e.g., image recognition). We defined six vehicle categories, including HDTs (heavy-duty trucks), MDTs (middle-duty trucks), LDTs (light-duty trucks), HDVs (heavy-duty vehicles), MDVs (middle-duty vehicles), and LDVs (light-duty vehicles). It should be noted that these monitoring data were obtained from distinct video facilities, and thus were mutually incompatible. They were further required to be fused spatially and temporally. Detailed information is described in Section 2.3.

Another key driver is vehicle-specific speed, which is of great significance for arranging on-road vehicle emission factors [26]. Along with traffic video surveillance, radar velocimeters were utilized to measure vehicle-specific speed concurrently. All these data were updated in a timely manner (i.e., hourly) and stored in the historical database. Moreover, the vehicle-specific emission factors and road information were relatively static without being updated in real-time. The former was obtained from the vehicle I/M (vehicle Inspect/Maintenance) dataset in the Xiaoshan district [55]. The latter came from the Gaode

Map, divided into 1393 road segments (Figure 2). The spatial resolutions were inconsistent (i.e., 10~1000 m) across road segments, adaptive to the gaps between sets of traffic monitoring. On this basis, all these road links were grouped into three types: residential streets, arterial roads, and highways. Collectively, we achieved a hyperfine map of comprehensive traffic states over the Xiaoshan District, involving vehicle-specific images, speeds, and categories, traffic fluxes, and road segments.

2.3. Real-Time Data Fusion

Comprehensive traffic states, including vehicle-specific categories, speeds, emission factors, and road segments, should be interconnected spatiotemporally (Figure 3). Generally, the standardized data quality controls were completed on respective devices in advance. Yet, vehicle-specific records came from various devices, the links between which should be identified. First, on-road video surveillance offered vehicle-specific images, including vehicle-specific license plates, categories, and fluxes. The I/M dataset would be responsible for double-checking the identification of the vehicle-specific categories according to the license plates. Subsequently, such vehicles would fall into two classes: non-registered vehicles and registered ones. Their emission factors were calculated based on the I/M dataset. Yet, the emission factors of the latter were speed-dependent and vehicle-category-specific, while those of the former were averaged based on vehicle-specific categories (Figure S1). Second, both radar velocimeters and video surveillances were in motion concurrently. We can thus apply time recorders to synchronize the monitoring for fluxes, license plates, vehicle-specific categories, and speed. Third, the static road information was independent of the real-time traffic states. Road links can serve as reliable bridges between them and interconnect vehicle-specific and road-specific information spatially and temporally. The resulting extensive database can be transferred into the subsequent bottom-up emission model and Web GIS interface. The total Data Throughput exceeds 300,000 records and 200 MB per hour.

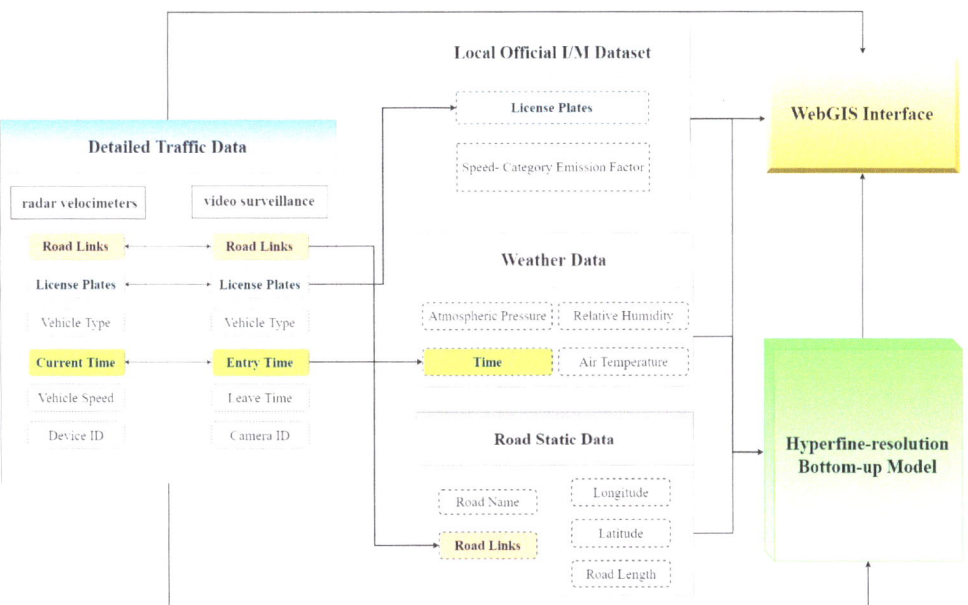

Figure 3. Data fusion. Vehicle-specific records come from diverse devices, the links between which are highlighted.

2.4. Model Framework for On-Road Vehicle Emissions

We applied a hyperfine model framework to estimate real-world, on-road vehicle emissions (i.e., primary PM$_{2.5}$, carbon monoxide (CO), hydrocarbon (HC), and nitrogen oxides (NO$_x$)). Compared to most of the current bottom-up on-road vehicle emission model frameworks, our design was sufficiently elaborate in terms of vehicles, road, space, and time. Figure 1 presents a theoretical flow diagram of the hyperfine on-road vehicle emission model. Overall, the results relied on an ensemble estimate of vehicle-specific speeds, categories, fluxes, emission factors, and road segments [10,25,26]

$$EF_{c,j,l} = \sum_t EF_{c,j}(v) \times TF_{c,h,l} \qquad (1)$$

$$E_{h,j,l} = EF_{c,j,l} \times L_l \qquad (2)$$

where h and l represent the temporal (i.e., hour) and spatial (i.e., road segment) dimensions, respectively. For a given spatiotemporal dimension, $EF_{c,j,l}$ denotes the emission intensity of the pollutant j (g km^{-1} h^{-1}).

Relying on the real-time HPC platform, our outcomes can map significant variations in the on-road vehicle emissions. On this basis, a real-time diagnosis algorithm was generated by comparing the emissions of all road links and vehicles and tracking their spatiotemporal evolution. Consequently, the road links and vehicle categories with high emissions were screened out and identified as key elements. More importantly, with the aid of the vehicle-road links, we can illustrate the contributions of different vehicle categories to different road links. This especially offered precise targets for on-road vehicle emission control strategies.

2.5. "Distance–Decay" Relationship of Hotspot Region

In theory, a comprehensive traffic profile is the basis for estimating hotspot emissions. To analyse the determinants of emission hotspot patterns, we applied an emission–distance relationship $E(d)$ that might be faithfully replicated by the three-parameter unconstrained exponential model as follows:

$$E(d) = \alpha + \beta \exp(-3d/k) \qquad (3)$$

where there are four parameters in this equation: the background emission (α); the length to the hotspot (d, m); the slope boosted by the hotspot emissions (β); and the convergence coefficient (k), which governs the spatial scale over the emission relaxed to background emission (α). In principle, the sum ($\alpha + \beta$) would converge to 1.0, which suggests that the sums of the background emissions and associated increments would reflect the emission levels of the hotspots. We modelled the decay of on-road vehicle emissions, traffic fluxes, and vehicle categories from indicative hotspots outwards on annual weekdays and single hours of weekdays. Among them, annual data were used to prove the rationality of the model. Because annual and single-hour data types are homologous, we used single-hour data fitting to analyse the distance decay characteristics of real-time hotspots emission. The magnitudes of α and β reflected the amplitude of decay from hotspots at the hourly scale.

2.6. Traffic Control Strategies

Over the Xiaoshan district, routine traffic control policies were implemented to mitigate air pollution. A representative measure was on-road vehicle license restrictions, routinely operated during two typical periods, i.e., from 7:00 to 9:00 and from 16:30 to 18:30 on weekdays, so-called the morning and evening rush hours. In theory, such kinds of policies would substantially alter on-road vehicle emissions by affecting traffic states (e.g., traffic fluxes and vehicle-specific speed). Yet, the influences were still elusive.

Here we integrated the hyperfine map of on-road vehicle emissions with an agile WebGIS engine. On this basis, we can picture the impacts of traffic control measures on on-road vehicle emission reductions. We designed four traffic control scenarios with a major focus on traffic fluxes and fleet composition. The main concern was to investigate how to implement traffic control policies spatially and temporally (Table 1). Note that we virtually implemented those policies during a morning rush hour (8:00, Local time) to maximize their influences. First, the weekday scenario (S1) forbade particular vehicles according to the tail numbers of license plates. This scenario focused only on the residential and arterial roads and the morning and evening rush hours. Table 1 summarizes the detailed rules. Second, on the basis of the S1 scenario, the even–odd scenario (S2) applied the even–odd rule and thus halved traffic fluxes. This scenario was obviously more stringent than the weekday scenario (S1). Third, both non-registered and registered trucks were forbidden over the highways (S3). Fourth, we combined the S2 and S3 scenarios to achieve the strictest control on all vehicles (S4). Such a scenario was actually implemented over the Xiaoshan district during the G20 summit in 2016 [56–58].

Table 1. Traffic control policies.

Scenario	Strategy	Vehicle Category	Spatiotemporal Scale
S1	Vehicles with particular tail numbers of license plates are forbidden. Specifically, the prohibited tail numbers were 1 and 9 on Monday, 2 and 8 on Tuesday, 3 and 7 on Wednesday, 4 and 6 on Thursday, and 5 and 0 on Friday.	All	Over residential and arterial roads during morning and evening rush hours from Monday to Friday
S2	Vehicles with even and odd tail numbers of license plates are alternately prohibited.	All	Over residential and arterial roads during morning and evening rush hours from Monday to Friday
S3	HDVs and HDTs are forbidden	HDVs and HDTs	Over highways all day long
S4	All vehicles follow the even–odd rule.	All	Over all roads all day long

3. System Application

3.1. Map of Traffic Characteristics and Hotspots

This system, mainly supported by comprehensive traffic monitoring, ITS, WebGIS, and the bottom-up emission model, provided an unprecedented hyperfine map of urban traffic states in a timely manner (i.e., hourly), including vehicle-specific speed, categories, and traffic fluxes (Figures 4 and 5 and Supplementary File). For instance, regarding the traffic fluxes, the colours of the links evolved from green to red, indicating a gradual increase in traffic fluxes from less than 30/h to more than 100/h. For instance, on 28 December 2021, we found that spatial distributions of traffic states were extraordinarily heterogeneous. First, as expected, the vast majority of traffic fluxes were centred on residential and arterial roads (Figure 4b). Figure 4c presents corresponding hyperfine-resolution variations in a representative 1 km^2 zone. Therein, the hourly traffic flows fluctuated significantly (>25.8 times). Such large fluctuations remained even within individual streets, with an average of more than eight times. An expected finding was the frequent and widespread presence of acute geographical "traffic hotspots" across the traffic monitoring dataset. We treated individual road links or a cluster with traffic flows exceeding the district's average level as hotspots. Through the imaging analysis coupled with all-round video surveillance, Figure 5 presents plausible drivers for some indicative hotspots. We found that traffic congestion played a key role in shaping such hotspots, which, however, were caused by a variety of factors, including high traffic volumes on key arterial or residential roads or road constructions. Further information on the technique for identifying hotspots is given in Supplementary Information.

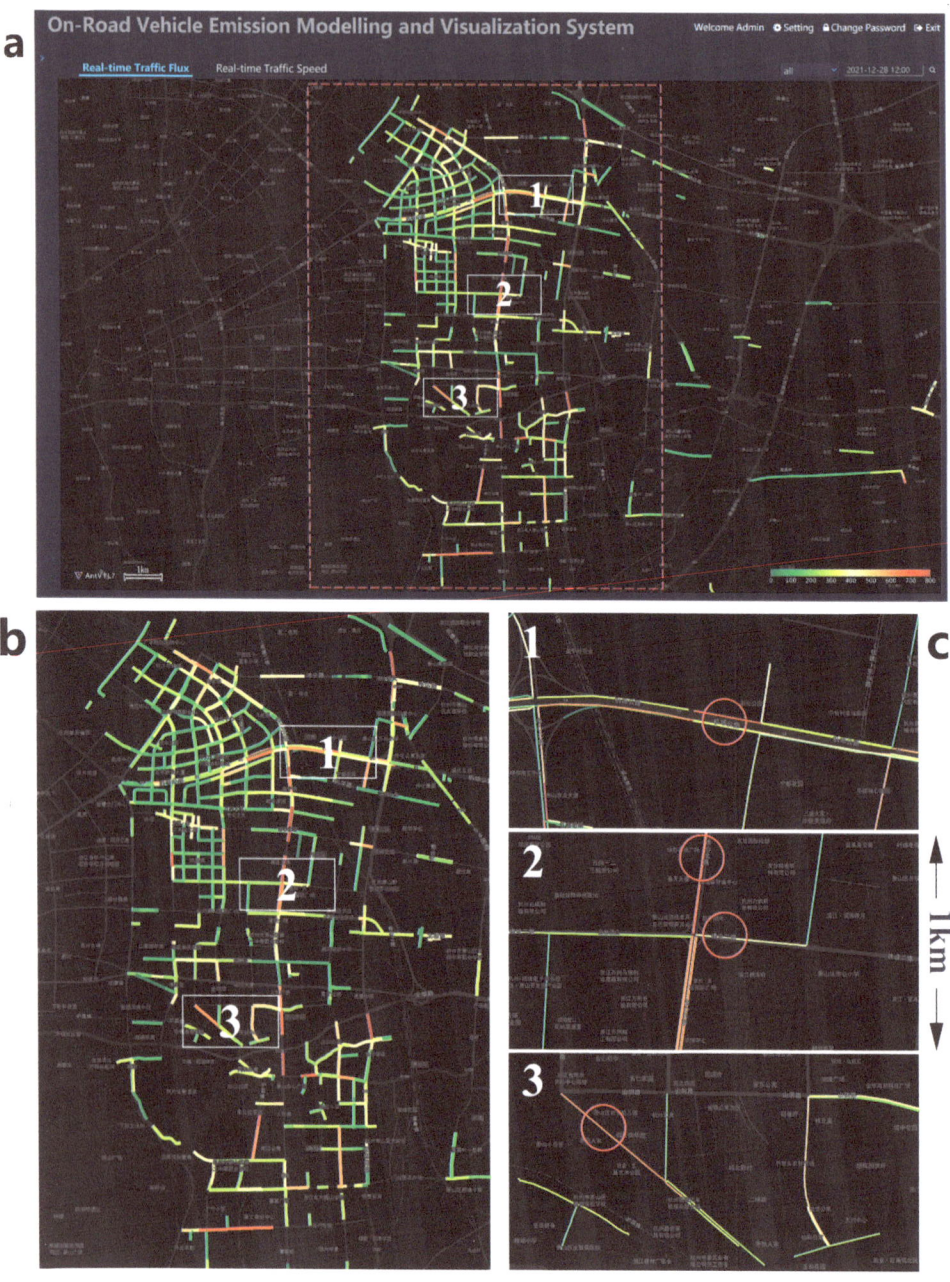

Figure 4. *Cont.*

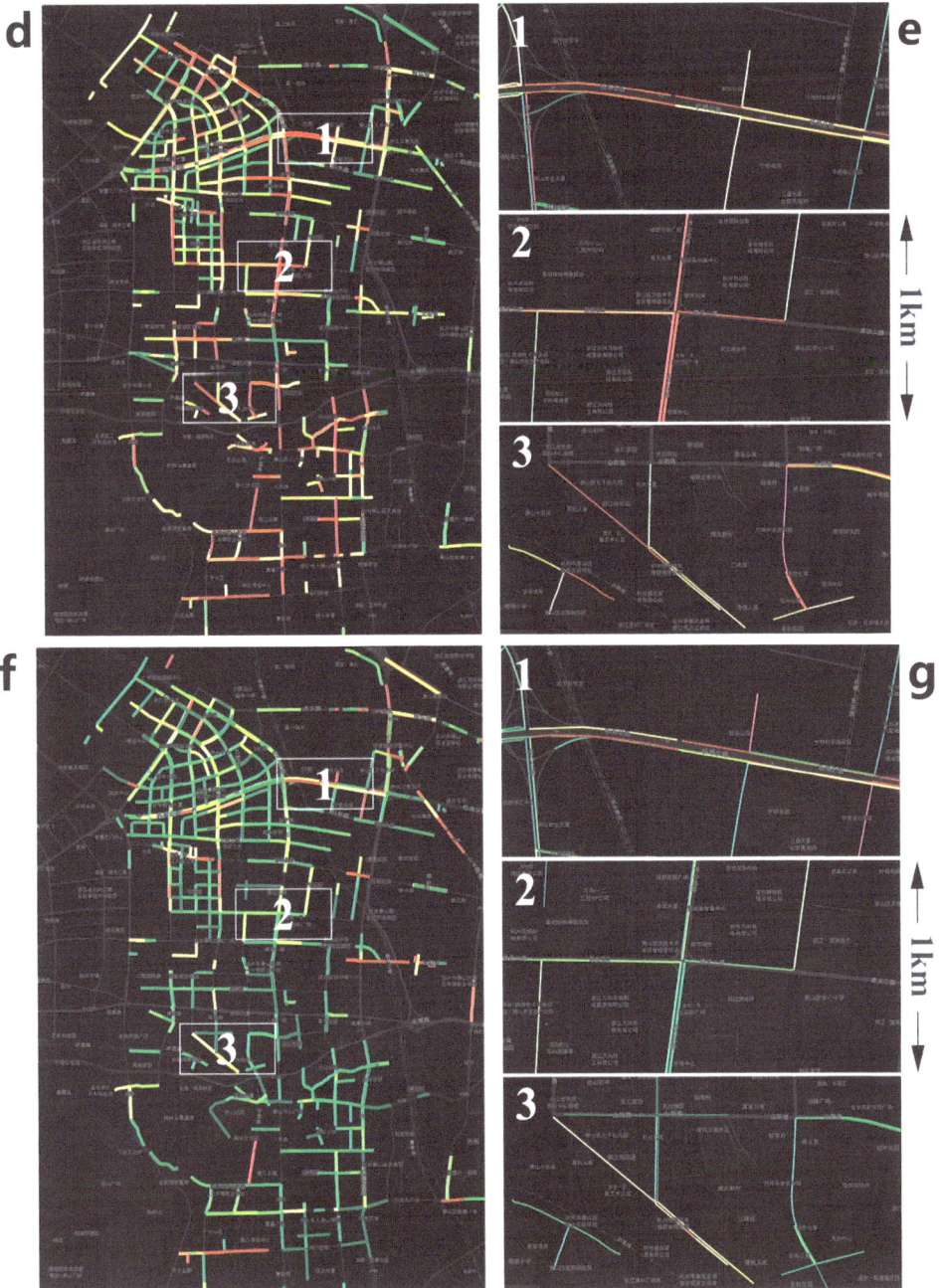

Figure 4. A hyperfine map of traffic fluxes via all-around traffic monitoring on 28 December 2021. (**a**) The whole map is first pictured. (**b**) Hourly (12:00, Local Time) traffic fluxes over the Xiaoshan district and (**c**) three indicative urban zones. The rest subgraphs are similar, but (**d**,**e**) and (**f**,**g**) are for the morning (8:00) rush hour and the hourly traffic fluxes of HDVs and HDTs, respectively. Red circles refer to traffic hotspots in Airport Road, Shixin North Road, Jiansheer Road, and Xiaohang Road. Map data © 2021, AntV L7.

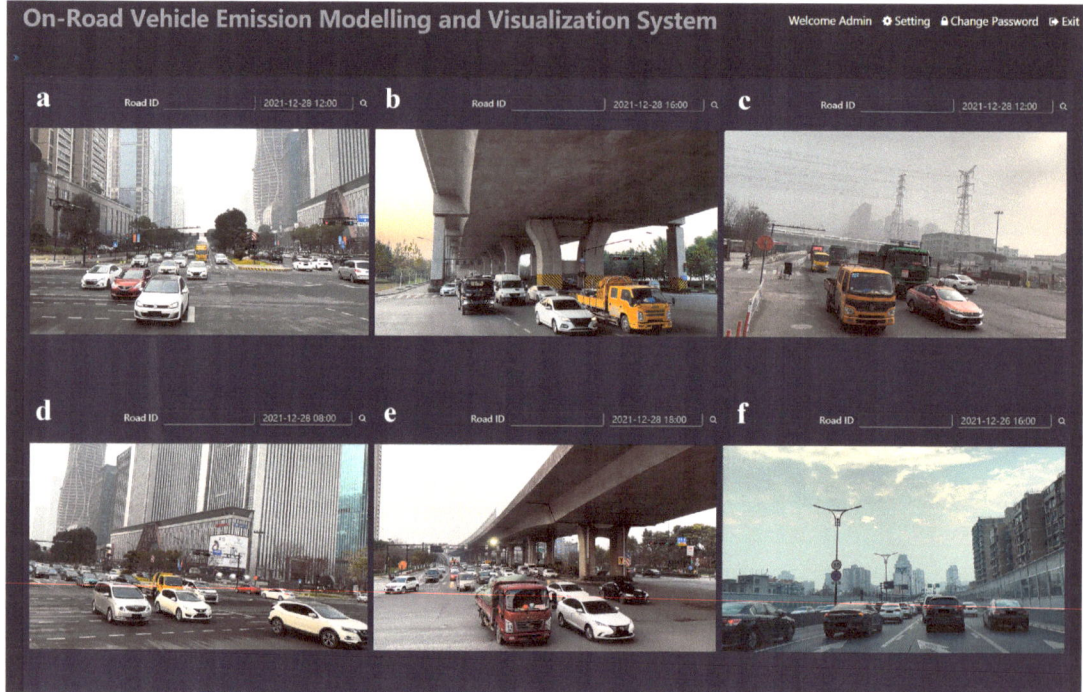

Figure 5. Imagery analyses for illustrative traffic hotspots. The hotspot locations are presented in Figure 3. The common drivers of traffic hotspots are (**a**,**b**) heavy traffic fluxes, (**c**) road constructions, (**d**,**e**) morning and afternoon traffic rushes, and (**f**) heavy traffic fluxes of LDVs and MDVs.

Second, it is worth noting that, on weekdays, the daily averages of traffic fluxes were comparable to those on weekends (Figure S2). Despite this, there was a noticeable difference in hourly variation patterns between weekdays and weekends. It was clear that the morning and evening rush hours had a significant impact on the diurnal traffic fluxes on weekdays, with two maxima at 08:00 and 17:00. Our findings show that, during these periods when the traffic congestion was further exacerbated (Figure 5), the widespread hotspots were geographically stable but quantitatively more conspicuous (Figures 4d,e and S3). By comparison, on weekends, there was a smaller range of traffic fluctuations, and the morning peak arrived two hours later (Figure S2). Specifically, the traffic flux peaks on weekends were roughly 80% of those on weekdays, while their hotspots were also variable, indicating more random trips (Figure S3). Therefore, the hyperfine-resolution patterns of traffic hotspots were significantly heterogeneous, and it was necessary to track them in real-time over the entire district.

Third, the spatial and temporal connections between traffic fluxes and speeds were shown to be substantial. Figure S2 presents that vehicle-specific speeds fluctuated dramatically throughout the day as a result of the varying traffic fluxes. As expected, vehicle-specific speeds were at the lowest level during the peak periods of traffic fluxes. Moreover, those peaks and valleys simultaneously shifted from weekdays to weekends. Spatially, traffic flux hotspots may have determined speed hotspots (Figure S4). Vehicle categories, on the other hand, were unaffected by traffic fluxes. After the morning rush hour, their diurnal changes were stable in regardless of the kind of roads (Figures 4 and S2). Yet the HDVs and HDTs reached their peaks in the early morning hours (i.e., from 1:00 to 5:00). Additionally, the spatial distributions of vehicle categories were especially noteworthy (Figures 4f,g and S5). LDVs, MDVs, LDTs, and MDTs mainly occupied the residential streets and arterial roads, while other kinds of vehicles (i.e., HDTs and HDVs) frequently appeared over the highways. We found the spatial distributions of HDVs and HDTs, the hotspots of which scattered extensively. A unique driver can be related to their large traffic fluxes, which were confirmed via video surveillance. Therefore, the fleet composition can also affect on-road vehicle emission distributions substantially, especially on small scales.

3.2. Real-Time, On-Road Vehicle Emissions

This system produced a real-time map of on-road vehicle emissions, in which widespread emission hotspots were also identified (Figures 6 and S6). Such patterns were distinct from previous maps that can only capture the emissions in downtown areas, which were noticeably higher than those in suburbs. This was mostly related to the spatial distributions of vehicle categories and traffic states (Figures 5 and S4). In particular, high traffic fluxes and low speeds downtown typically led to substantial on-road vehicle emissions hotspots (Figures S14 and S15). It is worth noting that towards the edge of the district, such a phenomenon was not consistent. In contrast, on-road vehicle emissions in residential streets considerably outstripped (>474.2%) those on the neighbouring roads. The spatial patterns of various vehicle categories might explain this discrepancy (Figures 4 and S5). For example, emissions from HDTs and HDVs on a residential street (i.e., the Ningdong Road) contributed 86.2%, far higher than those (8.4%) in its neighbouring arterial roads (the Shixin North–Jianshe Fourth Roads).

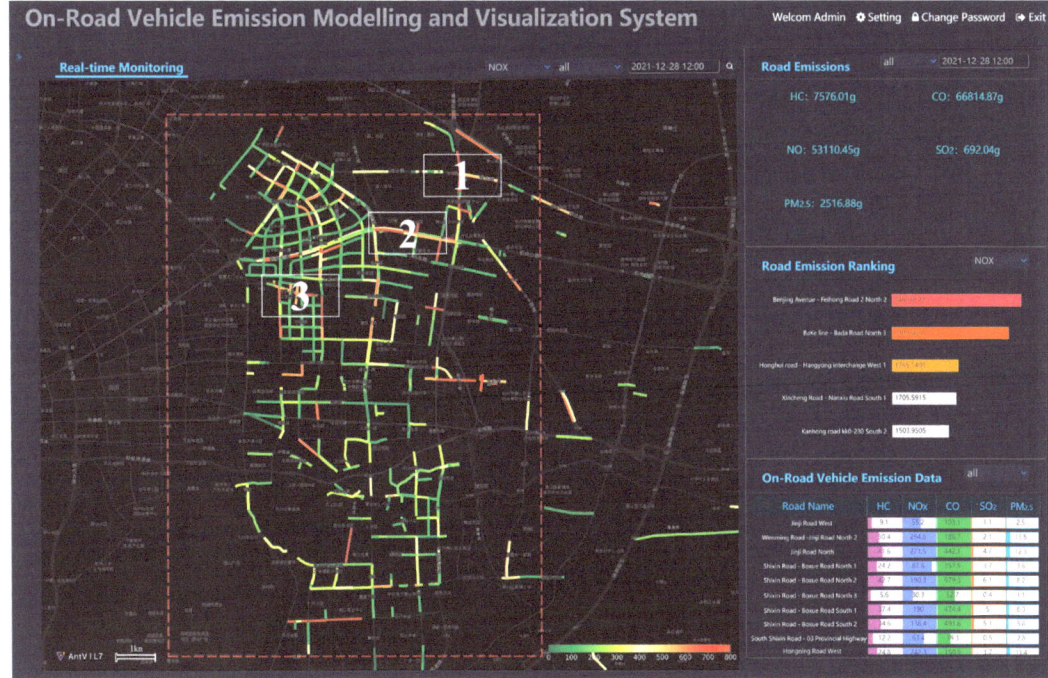

Figure 6. *Cont.*

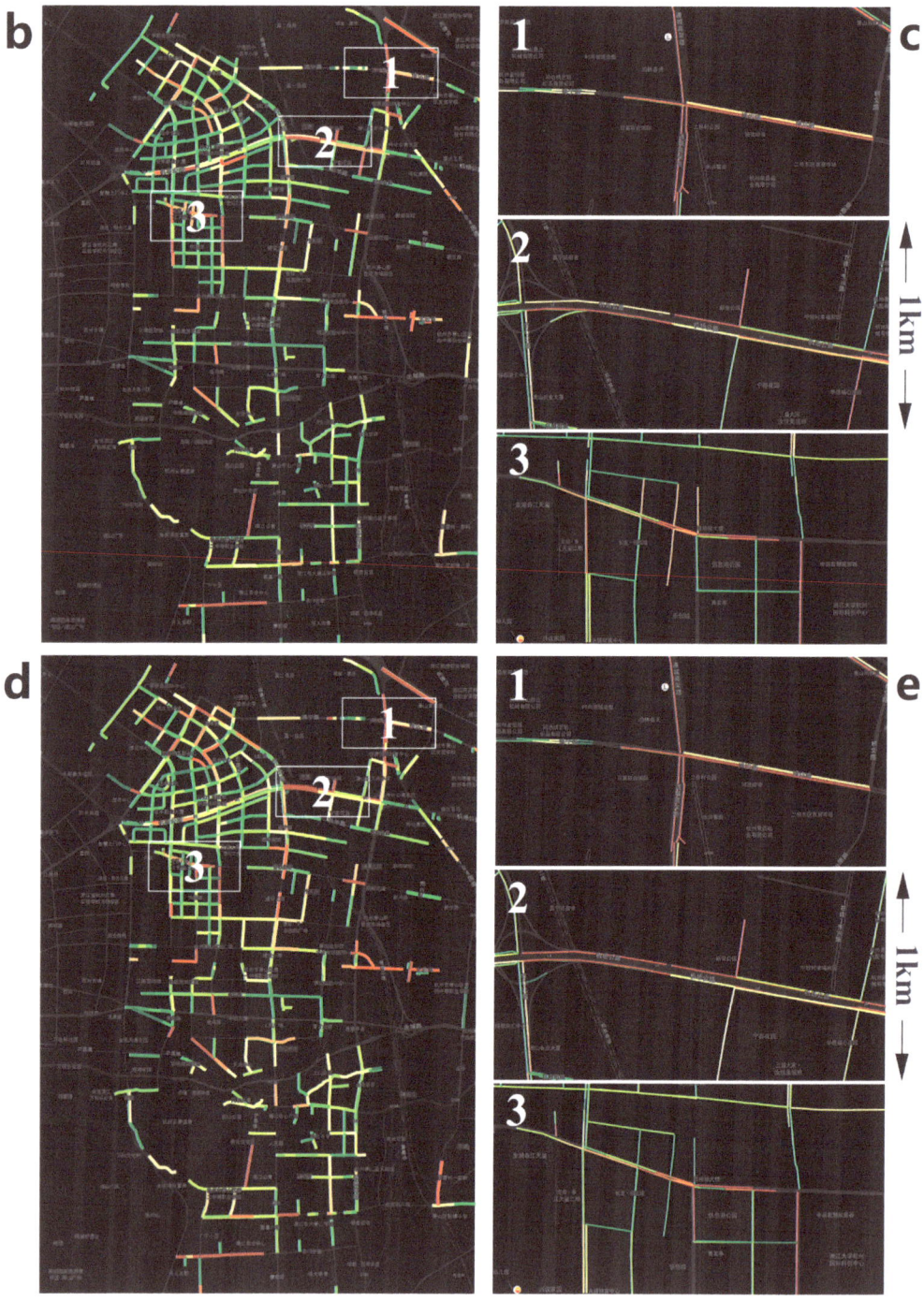

Figure 6. *Cont.*

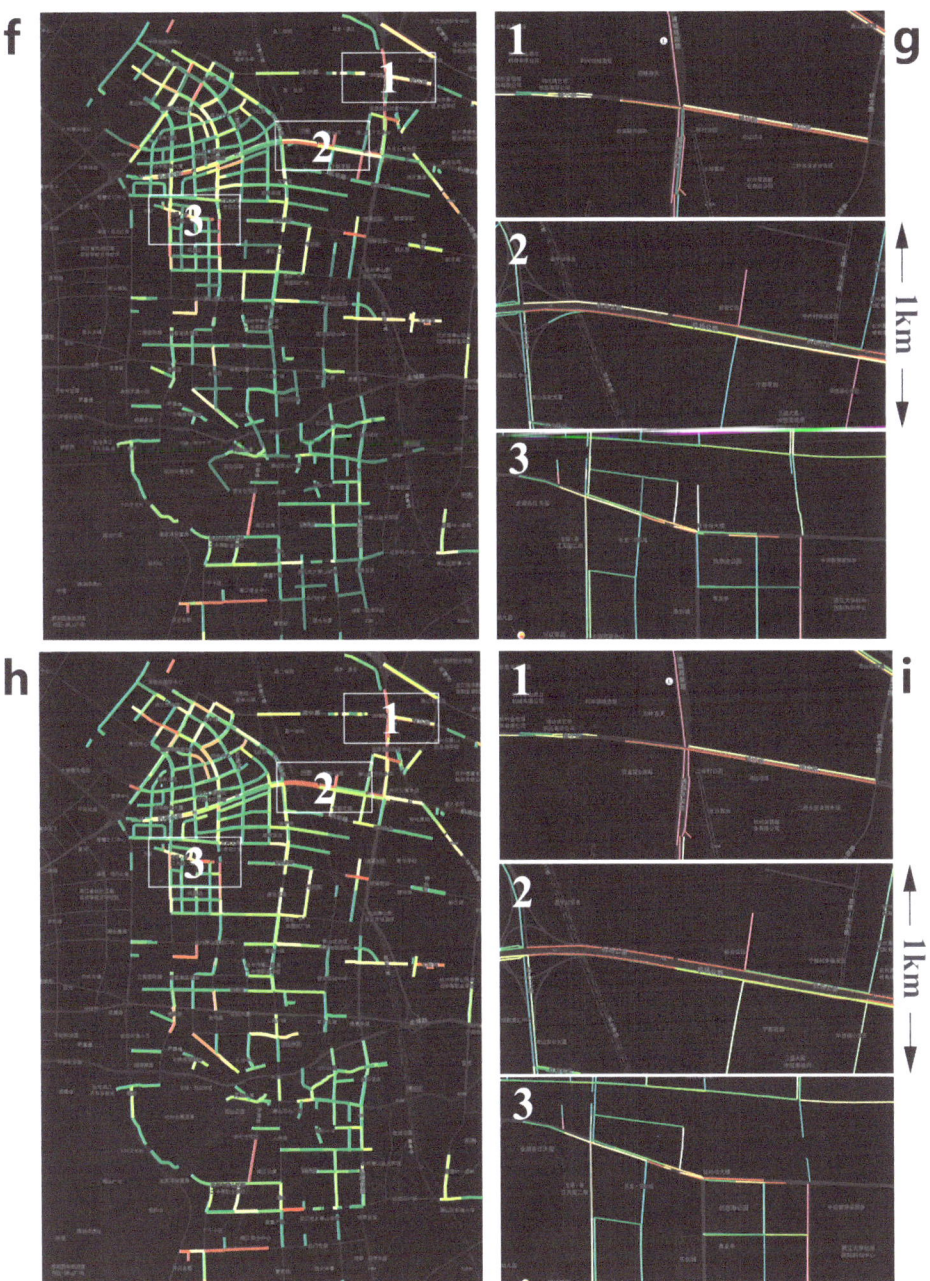

Figure 6. Real-time, on-road vehicle emissions on 28 December 2021. (**a**) The whole system map is first pictured. (**b**) Hourly (12:00, Local time) on-road vehicle NO$_x$ emissions over the whole district and (**c**) three indicative urban zones. The rest subgraphs are similar, but (**d**,**e**), (**f**,**g**), and (**h**,**i**) are for the morning (8:00) rush hour, the emissions of HDTs and HDVs on local time (12:00), and the morning (8:00) rush hours of HDTs and HDVs, respectively. "Road Emission Ranking" displays the five highest road links of specific pollutants in road emissions, and "On-road Vehicle Emission Data" is the specific emission information of each road link. Map data © 2021, AntV L7.

Relying on the classified roads, Table 2 summarizes the hourly emissions of primary $PM_{2.5}$, NO_x, CO, and HC. The top five roads (i.e., Chenhui Road–East of Chaohui Primary School, Tonghui North Road–Hongda Road north, Airport City Avenue–Liqun River bridge west, Airport City Avenue–Minhe Road East, and Tonghui North–Hongda Road east) were also highlighted. We noted that emissions in highways, residential streets, and arterial roads increased in sequence. The primary reason for this sequence was the distinction between vehicle categories on different kinds of roads (Table S1). Taking the hourly NO_x emissions (by Equation (1)) as an example, we found that they are 168.2 g/km, 102.2 g/km, and 126.7 g/km on highways, residential streets, and arterial roads, respectively. It should be noted that highways were of 3.7% of total traffic flows, while contributing 5.6% of total emissions.

Table 2. The summary of on-road vehicle emissions on 28 December 2021.

Road Type	Road Length	Vehicle Category	Emission (g)/Emission Intensity (g/km)			
			CO	HC	NO_x	$PM_{2.5}$
Highways	11.1 km	HDVs and HDTs	113.5/10.2	83.3/7.5	1300.2/116.8	61.1/5.5
		Total	2381.1/213.9	247.4/22.2	1872.2/168.2	77.0/6.9
Arterial roads	63.5 km	HDVs and HDTs	348.7/5.5	257.6/4.1	3988.2/62.8	187.3/3.0
		Total	16,398.9/258.4	1419.1/22.4	8039.3/126.7	299.9/4.7
Residential streets	232.0 km	HDVs and HDTs	1308.0/5.6	978.5/4.2	14,891.9/64.2	698.5/3.0
		Total	36,085.2/155.5	3501.2/15.1	23,703.0/102.2	944.4/4.1

Figure S7 shows that it was roughly consistent throughout the day when it came to on-road vehicle emission patterns of primary $PM_{2.5}$, NO_x, HC, and CO.

From the temporal perspective, it was roughly consistent throughout the day when it came to on-road vehicle emission patterns of CO, HC, NO_x, and $PM_{2.5}$ (Figure S7). Specifically, there was roughly 76.8% of daily NO_x emissions during the daytime. In addition to this, the NO_x emissions fluctuated during the daytime, but were typically stable throughout the various roads. There were, however, noticeable variations in the emissions between weekdays and weekends. As expected, the morning and evening rush hours on weekdays would also lead to peaks of on-road vehicle emissions. On the weekends, though, such trends were difficult to discern.

3.3. Map of Emission Hotpots and Drivers

Figure 6 depicts the hotspots of on-road vehicle emissions on major road intersections. Where two major arterial highways (North Shixin and Jiansheer Roads) intersect, the maximum of hourly average emissions appeared (Figure 4). From the spatial perspective, these emission hotspots varied significantly across various roads. For instance, hotspots in two arterial roads (i.e., the Hongda and Tonghui North Roads) emitted almost the same amount of pollutants as those in two residential streets (i.e., Jinji and Mingxing Roads), respectively (i.e., arterial roads vs. residential streets: 448.6 g/km vs. 251.1 g/km for CO; 41.3 g/km vs. 23.5 g/km for HC; 276.3 g/km vs. 161.3 g/km for NO_x; and 10.6 vs. 6.2 g/km for $PM_{2.5}$). However, residential streets had much lower hotspot emissions than highways and arterial roads. Specifically, hotspot emissions from the arterial roads (highways) outstripped those from residential roads by 1.8 (1.5) times for CO, 1.8 (2.1) times for HC, 1.7 (3.0) times for NOx, and 1.7 (3.2) times for primary $PM_{2.5}$.

Besides, we paid particular attention to highways, in which emission hotspots were widespread and sometimes intensive (Figure 6). For instance, our emission estimates for a highway (i.e., the Airport Road) were consistently higher (1.4~2 times) than those for its neighbouring residential streets (i.e., the Yangfan Road) (Figure S8). The diurnal emission hotspots, on the other hand, were steady, geographically (Figure S9). In contrast, their emission magnitude fluctuated diurnally and between weekdays and weekends

(Figure S10). As expected, the higher emission intensities generally occurred at 08:00 and 17:00.

Generally, such hotspots spanned between 100 and 200 m over the urban zones with varying emissions. Figure 7 shows that the annual hourly average emissions typically followed "distance–decay" relationships outward from the hotspot centres. The results reflected the hourly emission ratios (normalized at the hourly emissions of the hotspots) from hotspots outwards based on the distance (d). In addition, the ratios of the average traffic fluxes and vehicle category proportions were calculated in the same way. Overall, such relationships were the most sensitive for NOx and PM$_{2.5}$. The annual data reflect that the locations of the hotspot areas were relatively fixed in the year. As shown in Figures 3, 5 and 6, we found that the traffic fluxes largely shaped the spatial emission hotspot patterns over the arterial and residential roads. Additionally, the specific vehicle category proportions (i.e., HDVs and HDTs) also played an important role.

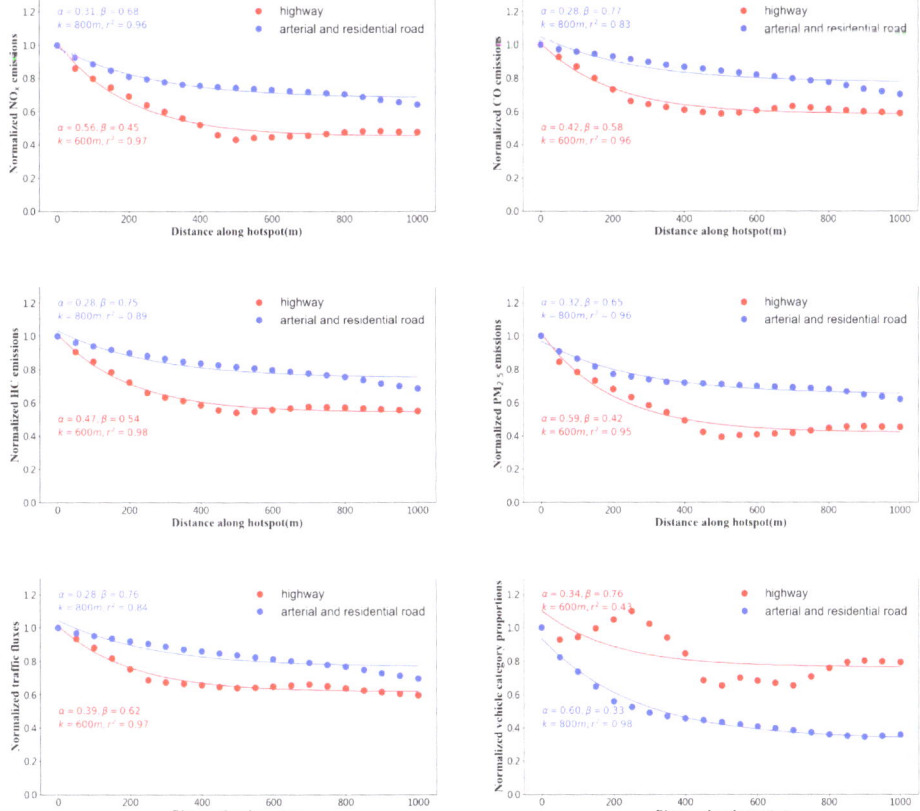

Figure 7. Relationships between the distance to the hotspot cores and normalized values in annual hotspot data. Each dot refers to the annual hourly average emission ratio (normalized to the hourly average emissions of the hotspots).

Moreover, Figure 8 shows the decay patterns of the hourly emissions over urban zones on 28 December 2021. The results reflected a similar pattern to the average data, but we found that there existed variations in some pollutant decay patterns, indicating that road emission hotspots were not fixed consistently, and there existed spatial offsets in the short term. Meanwhile, according to the boxplot analysis of road traffic and pollutant emission data for selected hotspots in the Xiaoshan District in 2021 (Figures S12 and S13), the highway hotspot data are most consistent with the pattern. This is because there is only one highway in the Xiaoshan District, and its traffic activities have a high stability. Arterial roads and residential roads may be affected by various factors and are more sensitive to the disturbance of hotspots, but in general, they also show the corresponding pattern.

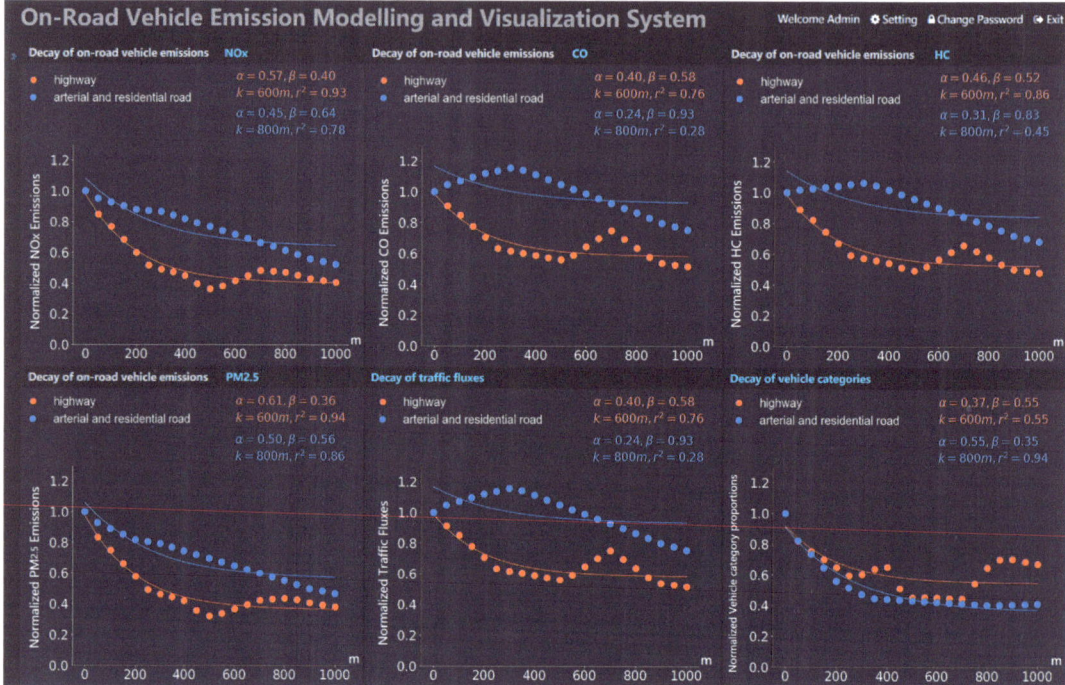

Figure 8. Relationships between the distance to the hotspot cores and normalized values. Each dot refers to the hourly emission ratio (normalized to the hourly emissions of the hotspots). We took 28 December 2021 as an example of the weekdays.

The traffic flows and emissions on highways were likewise relevant to the "distance–decay" functions, although the vehicle-specific categories kept stable therein (Figure 7). This demonstrates that traffic flows were crucial in shaping the spatial patterns of emission hotspots along the highways. Collectively, not only traffic fluxes but also specific particular vehicle categories (i.e., HDTs and HDVs) played a key role in boosting emission hotspots.

3.4. Impacts of Traffic Control Scenarios

As expected, each scenario significantly altered the traffic states spatiotemporally. The first two scenarios aimed to reduce the traffic flows, while the last two ones took into account not only traffic flows but also vehicle categories (Table 1). As a result of Table 3, the first scenario (S1) had no discernible impact on the traffic fluxes, only reducing the traffic fluxes by 3.3%. The second scenario (S2) achieved more reductions of traffic flows (8.3%). In the third scenario (S3), the fleet composition was thoroughly altered. The last scenario (S4) realized the largest reductions of the traffic flows (53.3%).

Table 3. Impacts of traffic control policies in custom spatiotemporal scopes.

Scenario	Traffic Fluxes Reduction	On-Road Vehicle Emissions Reduction			
		CO	NO	HC	$PM_{2.5}$
S1	3.3%	3.4%	2.7%	3.1%	2.3%
S2	8.3%	8.5%	6.8%	7.7%	5.6%
S3	3.7%	4.8%	69.4%	33.7%	79.3%
S4	53.3%	53.3%	54.1%	53.6%	54.3%

As a result of Figure 9 and Figure S11, the S1 scenario, the hourly (8:00) emission levels dropped by the modest portions (2.3% for primary $PM_{2.5}$, 2.7% for NO_x, 3.1% for HC, and 3.4% for CO). By comparison, more decreases were achieved over the urban zones (i.e., the residential streets and arterial roads) (5.6% for primary $PM_{2.5}$, 6.8% for NO_x, 7.7% for HC, and 8.5% for CO) in the S2 scenario. In parallel, the S3 scenario reduced a significant portion (4.8~79.3%) of on-road vehicle emissions over the highways. As a consequence, the S4 scenario realized the largest emission reductions (i.e., 54.3% for primary $PM_{2.5}$, 54.1% for NO_x, 53.6% for HC, and 53.3% for CO). On this basis, as shown in the Figure 9 (S4), the emission hotspots mostly disappeared. It should be noted that, if such scenarios came true, additional traffic states, such as vehicle-specific speeds, would also be altered. Hence, we need to conduct more realistic studies in order to better simulate the feedback associated with traffic conditions.

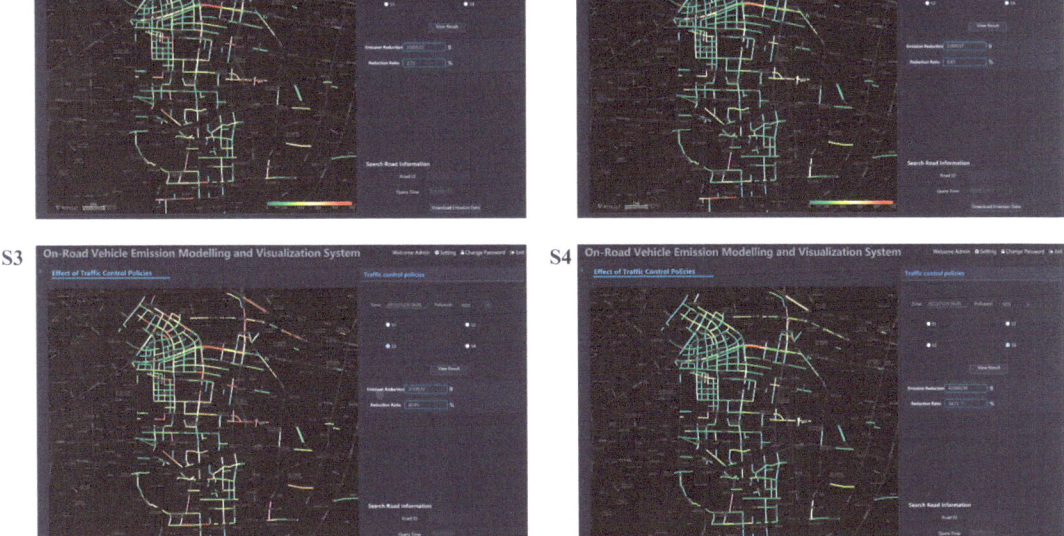

Figure 9. Effects of traffic control measures on on-road vehicle NO_x emissions on 28 December 2021. The traffic control policies were applied during a morning rush hour (8:00, Local time) to maximize their influence. Map data © 2021, AntV L7.

4. Conclusions

This paper described a system that establishes and visualizes real-time, hyper-fine, real-world, on-road vehicle emission distributions. Our results achieved an unprecedented temporal (i.e., hourly) and spatial resolution (i.e., 10 m~1 km, one to three orders higher than ever before). A key technical prerequisite is the comprehensive interconnections between the ITS and ubiquitous traffic monitoring over the Xiaoshan District. As a result, this system reveals frequent and widespread on-road vehicle emission hotspots. Around them, significant variabilities (up to 8~15 times) are exposed and attributed to large traffic fluxes and distinctive vehicle categories. This system also allows us to simulate the benefits of traffic control policies. We confirm that the most serious traffic control policy could achieve far more than 50% of emission reductions.

In this system, the traffic states, including vehicle-specific categories and speeds, are measured in real-time. By comparison, the vehicle-specific emission factors derived from the I/M dataset are of higher uncertainties [59]. Additionally, fuel-dependent differences are not taken into account when determining the emission factors. For instance, HDVs and HDTs are presumed to run on diesel fuel, whereas other vehicle types run on gasoline. In addition, the aging impacts of vehicles were overlooked. These hypotheses are supported by earlier research [26,60]. In the future, near-road emission monitoring might be used to reduce these errors. More than this, low-cost sensors such as those on taxis and mobile phones might drastically reduce the expenses of collecting data and thus widely expand our system.

Overall, the operational application of this system could reform the study of road vehicle emissions. Once our system is linked to a full CTM, real-time, hyper-fine, real-world air quality emulations would also become possible. By combining CTM output and data from near-road air quality managements, a high resolution of air quality response to emissions becomes possible. This could help investigate the complex response of air quality to anthropogenic emissions and even address exposure misclassification. Such results may have additional sociological implications, including for future urban planning and sustainable development.

Supplementary Materials: The following supporting information can be downloaded at: https://www.mdpi.com/article/10.3390/su14095434/s1.

Author Contributions: S.Y. and P.L. conceived and designed the research. L.W. (Lu Wang), X.C., Y.X., L.J. and J.Y. performed system developments. L.W. (Lu Wang), X.C., Y.X., L.J., J.Y., T.H., L.W. (Liqiang Wang), Y.Z., M.L., Z.L., Z.S. and Y.J. conducted data analysis. W.L. and X.Z. contributed to scientific discussions. S.Y. and P.L. wrote and revised the manuscript. All authors have read and agreed to the published version of the manuscript.

Funding: This study was supported by the National Natural Science Foundation of China (No. 42175084, 21577126, and 41561144004), the Department of Science and Technology of China (No. 2018YFC0213506 and 2018YFC0213503), and the National Research Program for Key Issues in Air Pollution Control in China (No. DQGG0107). Pengfei Li was supported by the National Natural Science Foundation of China (No. 22006030), the Initiation Fund for Introducing Talents of Hebei Agricultural University (412201904), and the Hebei Youth Top Fund (BJ2020032).

Data Availability Statement: All measurements and model results are available upon request.

Conflicts of Interest: The authors declare no conflict of interest.

References

1. Liang, X.; Zhang, S.; Wu, Y.; Xing, J.; He, X.; Zhang, K.M.; Wang, S.; Hao, J. Air quality and health benefits from fleet electrification in China. *Nat. Sustain.* **2019**, *2*, 962–971. [CrossRef]
2. Wang, L.; Chen, X.; Zhang, Y.; Li, M.; Li, P.; Jiang, L.; Xia, Y.; Li, Z.; Li, J.; Wang, L.; et al. Switching to electric vehicles can lead to significant reductions of $PM_{2.5}$ and NO_2 across China. *One Earth* **2021**, *4*, 1037–1048. [CrossRef]
3. Zhang, Q.; Zheng, Y.; Tong, D.; Shao, M.; Wang, S.; Zhang, Y.; Xu, X.; Wang, J.; He, H.; Liu, W.; et al. *Drivers of Improved $PM_{2.5}$ Air Quality in China from 2013 to 2017*; National Academy of Sciences: Washington, DC, USA, 2019; Volume 116, pp. 24463–24469.
4. Xue, Y.; Cao, X.; Ai, Y.; Xu, K.; Zhang, Y. Primary Air Pollutants Emissions Variation Characteristics and Future Control Strategies for Transportation Sector in Beijing, China. *Sustainability* **2020**, *12*, 4111. [CrossRef]
5. Gao, J.; Wang, K.; Wang, Y.; Liu, S.; Zhu, C.; Hao, J.; Liu, H.; Hua, S.; Tian, H. Temporal-spatial characteristics and source apportionment of $PM_{2.5}$ as well as its associated chemical species in the Beijing-Tianjin-Hebei region of China. *Environ. Pollut.* **2018**, *233*, 714–724. [CrossRef] [PubMed]

6. Ecology, M.O.; China, E.O.T.P. *China Vehicle Environmental Management Annual Report*; Ministry of Ecology and Environment of the People's Republic of China: Beijing, China, 2020.
7. Song, C.; Wu, L.; Xie, Y.; He, J.; Chen, X.; Wang, T.; Lin, Y.; Jin, T.; Wang, A.; Liu, Y.; et al. Air pollution in China: Status and spatiotemporal variations. *Environ. Pollut.* **2017**, *227*, 334–347. [CrossRef] [PubMed]
8. Anenberg, S.C.; Miller, J.; Minjares, R.; Du, L.; Henze, D.K.; Lacey, F.; Malley, C.S.; Emberson, L.; Franco, V.; Klimont, Z.; et al. Impacts and mitigation of excess diesel-related NOx emissions in 11 major vehicle markets. *Nature* **2017**, *545*, 467–471. [CrossRef]
9. Ogunkunle, O.; Ahmed, N.A. Overview of Biodiesel Combustion in Mitigating the Adverse Impacts of Engine Emissions on the Sustainable Human–Environment Scenario. *Sustainability* **2021**, *13*, 5465. [CrossRef]
10. Zhang, S.; Wu, Y.; Huang, R.; Wang, J.; Yan, H.; Zheng, Y.; Hao, J. High-resolution simulation of link-level vehicle emissions andconcentrations for air pollutants in a traffic-populated eastern Asian city. *Atmos. Chem. Phys.* **2016**, *16*, 9965–9981. [CrossRef]
11. Lyu, P.; Wang, P.S.; Liu, Y.; Wang, Y. Review of the studies on emission evaluation approaches for operating vehicles. *J. Traffic Transp. Eng.* **2021**, *8*, 493–509. [CrossRef]
12. Mangones, S.C.; Jaramillo, P.; Fischbeck, P.; Rojas, N.Y. Development of a high-resolution traffic emission model: Lessons and key insights from the case of Bogotá, Colombia. *Environ. Pollut.* **2019**, *253*, 552–559. [CrossRef]
13. Xue, H.; Jiang, S.; Liang, B. A study on the model of traffic flow and vehicle exhaust emission. *Math. Probl. Eng.* **2013**, *2013*, 736285. [CrossRef]
14. Agarwal, A.K.; Mustafi, N.N. Real-world automotive emissions: Monitoring methodologies, and control measures. *Renew. Sustain. Energy Rev.* **2021**, *137*, 110624. [CrossRef]
15. Janssens-Maenhout, G.; Crippa, M.; Guizzardi, D.; Dentener, F.; Muntean, M.; Pouliot, G.; Keating, T.; Zhang, Q.; Kurokawa, J.; Wankmüller, R.; et al. HTAP_v2.2: A mosaic of regional and global emission grid maps for 2008 and 2010 to study hemispheric transport of air pollution. *Atmos. Chem. Phys.* **2015**, *15*, 11411–11432. [CrossRef]
16. Li, M.; Zhang, Q.; Kurokawa, J.; Woo, J.; He, K.; Lu, Z.; Ohara, T.; Song, Y.; Streets, D.G.; Carmichael, G.R.; et al. MIX: A mosaic Asian anthropogenic emission inventory under the international collaboration framework of the MICS-Asia and HTAP. *Atmos. Chem. Phys.* **2017**, *17*, 935–963. [CrossRef]
17. Zhang, S.; Wu, Y.; Liu, H.; Wu, X.; Zhou, Y.; Yao, Z.; Fu, L.; He, K.; Hao, J. Historical evaluation of vehicle emission control in Guangzhou based on a multi-year emission inventory. *Atmos. Environ.* **2013**, *76*, 32–42. [CrossRef]
18. Lv, W.; Hu, Y.; Li, E.; Liu, H.; Pan, H.; Ji, S.; Hayat, T.; Alsaedi, A.; Ahmad, B. Evaluation of vehicle emission in Yunnan province from 2003 to 2015. *J. Clean. Prod.* **2019**, *207*, 814–825. [CrossRef]
19. Gately, C.K.; Hutyra, L.R.; Peterson, S.; Sue Wing, I. Urban emissions hotspots: Quantifying vehicle congestion and air pollution using mobile phone GPS data. *Environ. Pollut.* **2017**, *229*, 496–504. [CrossRef]
20. Gately, C.K.; Hutyra, L.R. Large Uncertainties in Urban-Scale Carbon Emissions. *J. Geophys. Res. Atmos.* **2017**, *122*, 242–260. [CrossRef]
21. Jiang, L.; Xia, Y.; Wang, L.; Chen, X.; Ye, J.; Hou, T.; Wang, L.; Zhang, Y.; Li, M.; Li, Z.; et al. Hyperfine-resolution mapping of on-road vehicle emissions with comprehensive traffic monitoring and an intelligent transportation system. *Atmos. Chem. Phys.* **2021**, *21*, 16985–17002. [CrossRef]
22. Jing, B.; Wu, L.; Mao, H.; Gong, S.; He, J.; Zou, C.; Song, G.; Li, X.; Wu, Z. Development of a vehicle emission inventory with high temporal–spatial resolution based on NRT traffic data and its impact on air pollution in Beijing–Part 1: Development and evaluation of vehicle emission inventory. *Atmos. Chem. Phys.* **2016**, *16*, 3161–3170. [CrossRef]
23. Liu, Y.; Ma, J.; Li, L.; Lin, X.; Xu, W.; Ding, H. A high temporal-spatial vehicle emission inventory based on detailed hourly traffic data in a medium-sized city of China. *Environ. Pollut.* **2018**, *236*, 324–333. [CrossRef] [PubMed]
24. Wen, Y.; Zhang, S.; Zhang, J.; Bao, S.; Wu, X.; Yang, D.; Wu, Y. Mapping dynamic road emissions for a megacity by using open-access traffic congestion index data. *Appl. Energy* **2020**, *260*, 114357. [CrossRef]
25. Wu, L.; Chang, M.; Wang, X.; Hang, J.; Zhang, J.; Wu, L.; Shao, M. Development of the Real-time On-road Emission (ROE v1.0) model for street-scale air quality modeling based on dynamic traffic big data. *Geosci. Model. Dev.* **2020**, *13*, 23–40. [CrossRef]
26. Yang, D.; Zhang, S.; Niu, T.; Wang, Y.; Xu, H.; Zhang, K.M.; Wu, Y. High-resolution mapping of vehicle emissions of atmospheric pollutants based on large-scale, real-world traffic datasets. *Atmos. Chem. Phys.* **2019**, *19*, 8831–8843. [CrossRef]
27. Yang, W.; Yu, C.; Yuan, W.; Wu, X.; Zhang, W.; Wang, X. High-resolution vehicle emission inventory and emission control policy scenario analysis, a case in the Beijing-Tianjin-Hebei (BTH) region, China. *J. Clean. Prod.* **2018**, *203*, 530–539. [CrossRef]
28. Liu, J.; Han, K.; Chen, X.M.; Ong, G.P. Spatial-temporal inference of urban traffic emissions based on taxi trajectories and multi-source urban data. *Transp. Res. Part C Emerg. Technol.* **2019**, *106*, 145–165. [CrossRef]
29. Song, X.; Guo, R.; Xia, T.; Guo, Z.; Long, Y.; Zhang, H.; Song, X.; Ryosuke, S. Mining urban sustainable performance: Millions of GPS data reveal high-emission travel attraction in Tokyo. *J. Clean. Prod.* **2020**, *242*, 118396. [CrossRef]
30. Xia, C.; Xiang, M.; Fang, K.; Li, Y.; Ye, Y.; Shi, Z.; Liu, J. Spatial-temporal distribution of carbon emissions by daily travel and its response to urban form: A case study of Hangzhou, China. *J. Clean. Prod.* **2020**, *257*, 120797. [CrossRef]
31. Deng, F.; Lv, Z.; Qi, L.; Wang, X.; Shi, M.; Liu, H. A big data approach to improving the vehicle emission inventory in China. *Nat. Commun.* **2020**, *11*, 2801. [CrossRef]
32. Beaton, S.P.; Bishop, G.A.; Zhang, Y.; Stedman, D.H.; Ashbaugh, L.L.; Lawson, D.R. On-Road Vehicle Emissions: Regulations, Costs, and Benefits. *Science* **1995**, *268*, 991–993. [CrossRef]

33. Mcgaughey, G.R.; Desai, N.R.; Allen, D.T.; Seila, R.L.; Lonneman, W.A.; Fraser, M.P.; Harley, R.A.; Pollack, A.K.; Ivy, J.M.; Price, J.H. Analysis of motor vehicle emissions in a Houston tunnel during the Texas Air Quality Study 2000. *Atmos. Environ.* **2004**, *38*, 3363–3372. [CrossRef]
34. Paul, J.; Malhotra, P.; Dale, S.; Qiang, M. RFID based vehicular networks for smart cities. In Proceedings of the 2013 IEEE 29th International Conference on Data Engineering Workshops (ICDEW), Brisbane, Australia, 8–12 April 2013; pp. 120–127.
35. Song, J.; Zhao, C.; Lin, T.; Li, X.; Prishchepov, A.V. Spatio-temporal patterns of traffic-related air pollutant emissions in different urban functional zones estimated by real-time video and deep learning technique. *J. Clean. Prod.* **2019**, *238*, 117881. [CrossRef]
36. Liu, B.; Frey, H.C. Variability in Light-Duty Gasoline Vehicle Emission Factors from Trip-Based Real-World Measurements. *Environ. Sci. Technol.* **2015**, *49*, 12525–12534. [CrossRef] [PubMed]
37. Wu, Y.; Song, G.; Yu, L. Sensitive analysis of emission rates in MOVES for developing site-specific emission database. *Transp. Res. Part D Transp. Environ.* **2014**, *32*, 193–206. [CrossRef]
38. Yang, Z.; Peng, J.; Wu, L.; Ma, C.; Zou, C.; Wei, N.; Zhang, Y.; Liu, Y.; Andre, M.; Li, D.; et al. Speed-guided intelligent transportation system helps achieve low-carbon and green traffic: Evidence from real-world measurements. *J. Clean. Prod.* **2020**, *268*, 122230. [CrossRef]
39. Apte, J.S.; Messier, K.P.; Gani, S.; Brauer, M.; Kirchstetter, T.W.; Lunden, M.M.; Marshall, J.D.; Portier, C.J.; Vermeulen, R.C.H.; Hamburg, S.P. High-Resolution Air Pollution Mapping with Google Street View Cars: Exploiting Big Data. *Environ. Sci. Technol.* **2017**, *51*, 6999–7008. [CrossRef]
40. Guo, L.; Dong, M.; Ota, K.; Li, Q.; Ye, T.; Wu, J.; Li, J. A Secure Mechanism for Big Data Collection in Large Scale Internet of Vehicle. *IEEE Internet Things J.* **2017**, *4*, 601–610. [CrossRef]
41. Louhghalam, A.; Akbarian, M.; Ulm, F. Carbon management of infrastructure performance: Integrated big data analytics and pavement-vehicle-interactions. *J. Clean. Prod.* **2017**, *142*, 956–964. [CrossRef]
42. Gately, C.K.; Hutyra, L.R.; Wing, I.S.; Brondfield, M.N. A Bottom up Approach to on-road CO_2 Emissions Estimates: Improved Spatial Accuracy and Applications for Regional Planning. *Environ. Sci. Technol.* **2013**, *47*, 2423–2430. [CrossRef]
43. Avila, A.M.; Mezić, I. Data-driven analysis and forecasting of highway traffic dynamics. *Nat. Commun.* **2020**, *11*, 1–16. [CrossRef]
44. Zhang, S.; Niu, T.; Wu, Y.; Zhang, K.M.; Wallington, T.J.; Xie, Q.; Wu, X.; Xu, H. Fine-grained vehicle emission management using intelligent transportation system data. *Environ. Pollut.* **2018**, *241*, 1027–1037. [CrossRef] [PubMed]
45. Ding, H.; Cai, M.; Lin, X.; Chen, T.; Li, L.; Liu, Y. RTVEMVS: Real-time modeling and visualization system for vehicle emissions on an urban road network. *J. Clean. Prod.* **2021**, *309*, 127166. [CrossRef]
46. An, J.; Huang, Y.; Huang, C.; Wang, X.; Yan, R.; Wang, Q.; Wang, H.; Jing, S.; Zhang, Y.; Liu, Y.; et al. Emission inventory of air pollutants and chemical speciation for specific anthropogenic sources based on local measurements in the Yangtze River Delta region, China. *Atmos. Chem. Phys.* **2021**, *21*, 2003–2025. [CrossRef]
47. Hua, X. The City Brain: Towards Real-Time Search for the Real-World. In Proceedings of the 41st International ACM SIGIR Conference on Research and Development in Information Retrieval, Ann Arbor, MI, USA, 8–12 July 2018; pp. 1343–1344.
48. Li, Q.; Zhu, H.; He, J. An Inconsistency Free Formalization of B/S Architecture. In Proceedings of the 31st IEEE Software Engineering Workshop (SEW 2007), Columbia, MD, USA, 6–8 March 2007; pp. 75–88.
49. Williams, H.E.; Lane, D. *Web Database Applications with PHP and MySQL: Building Effective Database-Driven Web Sites*; O'Reilly Media, Inc.: Sevastopol, CA, USA, 2004.
50. Carazo, J.M.; Stelzer, E.H.K. The BioImage Database Project: Organizing Multidimensional Biological Images in an Object-Relational Database. *J. Struct. Biol.* **1999**, *125*, 97–102. [CrossRef] [PubMed]
51. Balduzzi, M.; Zaddach, J.; Balzarotti, D.; Kirda, E.; Loureiro, S. A Security Analysis of Amazon's Elastic Compute Cloud Service. In Proceedings of the 27th Annual ACM Symposium on Applied Computing, Trento, Italy, 26–30 March 2012; pp. 1427–1434.
52. Bruno, E.J. Ajax: Asynchronous JavaScript and XML. *Dr. Dobb's J.* **2006**, *31*, 32–35.
53. Waldman, C.G.; Hagel-Sorensen, C. Dynamic Generation of Cascading Style Sheets. U.S. Patent 20,070,220,480, 20 September 2007.
54. Hu, Z.; Zhuge, C.; Ma, W. Towards a Very Large Scale Traffic Simulator for Multi-Agent Reinforcement Learning Testbeds. *arXiv* **2021**, arXiv:2105.13907.
55. Wu, Y.; Wang, R.; Zhou, Y.; Lin, B.; Fu, L.; He, K.; Hao, J. On-Road Vehicle Emission Control in Beijing: Past, Present, and Future. *Environ. Sci. Technol.* **2011**, *45*, 147–153. [CrossRef]
56. Ji, Y.; Qin, X.; Wang, B.; Xu, J.; Shen, J.; Chen, J.; Huang, K.; Deng, C.; Yan, R.; Xu, K.; et al. Counteractive effects of regional transport and emission control on the formation of fine particles: A case study during the Hangzhou G20 summit. *Atmos. Chem. Phys.* **2018**, *18*, 13581–13600. [CrossRef]
57. Wang, L.; Yu, S.; Li, P.; Chen, X.; Li, Z.; Zhang, Y.; Li, M.; Mehmood, K.; Liu, W.; Chai, T.; et al. Significant wintertime $PM_{2.5}$ mitigation in the Yangtze River Delta, China, from 2016 to 2019: Observational constraints on anthropogenic emission controls. *Atmos. Chem. Phys.* **2020**, *20*, 14787–14800. [CrossRef]
58. Zhang, G.; Xu, H.; Wang, H.; Xue, L.; He, J.; Xu, W.; Qi, B.; Du, R.; Liu, C.; Li, Z.; et al. Exploring the inconsistent variations in atmospheric primary and secondary pollutants during the 2016 G20 summit in Hangzhou, China: Implications from observations and models. *Atmos. Chem. Phys.* **2020**, *20*, 5391–5403. [CrossRef]

59. Seo, J.; Park, J.; Park, J.; Park, S. Emission factor development for light-duty vehicles based on real-world emissions using emission map-based simulation. *Environ. Pollut.* **2021**, *270*, 116081. [CrossRef] [PubMed]
60. Zhou, Y.; Zhao, Y.; Mao, P.; Zhang, Q.; Zhang, J.; Qiu, L.; Yang, Y. Development of a high-resolution emission inventory and its evaluation and application through air quality modeling for Jiangsu Province, China. *Atmos. Chem. Phys.* **2017**, *17*, 211–233. [CrossRef]

Article

Can Changes in Urban Form Affect PM$_{2.5}$ Concentration? A Comparative Analysis from 286 Prefecture-Level Cities in China

Chuang Sun [1], Xuegang Chen [2,*], Siyu Zhang [3] and Tianhao Li [1]

[1] College of Resource Environment & Tourism, Capital Normal University, Beijing 100048, China; sun_chuang0331@163.com (C.S.); lthde141@163.com (T.L.)
[2] School of Geographical Science and Tourism, Xinjiang Normal University, Urumqi 830054, China
[3] Beijing National Day School, Beijing 100039, China; zhangsiyu20050420@163.com
* Correspondence: caschxg@126.com; Tel.: +86-138-9991-0696

Abstract: It is crucial to the sustainable development of cities that we understand how urban form affects the concentration of fine particulate matter (PM$_{2.5}$) from a spatial–temporal perspective. This study explored the influence of urban form on PM$_{2.5}$ concentration in 286 prefecture-level Chinese cities and compared them from national and regional perspectives. The analysis, which explored the influence of urban form on PM$_{2.5}$ concentration, was based on two types of urban form indicators (socioeconomic urban index and urban landscape index). The results revealed that cities with high PM$_{2.5}$ concentrations tended to be clustered. From the national perspective, urban built-up area (UA) and road density (RD) have a significant correlation with PM$_{2.5}$ concentration for all cities. There was a significant negative correlation between the number of patches (NP) and the average concentration of PM$_{2.5}$ in small and medium-sized cities. Moreover, urban fragmentation had a stronger impact on PM$_{2.5}$ concentrations in small cities. From a sub-regional perspective, there was no significant correlation between urban form and PM$_{2.5}$ concentration in the eastern and central regions. On the other hand, the influence of population density on PM$_{2.5}$ concentration in northeastern China and northwestern China showed a significant positive correlation. In large- and medium-sized cities, the number of patches (NP), the largest patch index (LPI), and the contagion index (CONTAG) were also positively correlated with PM$_{2.5}$ concentration, while the LPI in small cities was significantly negatively correlated with PM$_{2.5}$ concentration. This shows that, for more developed areas, planning agencies should encourage moderately decentralized and polycentric urban development. For underdeveloped cities and shrinking cities, the development of a single center should be encouraged.

Keywords: urban form; urbanization; PM$_{2.5}$; spatiotemporal characteristics; spatial autocorrelation

Citation: Sun, C.; Chen, X.; Zhang, S.; Li, T. Can Changes in Urban Form Affect PM$_{2.5}$ Concentration? A Comparative Analysis from 286 Prefecture-Level Cities in China. *Sustainability* **2022**, *14*, 2187. https://doi.org/10.3390/su14042187

Academic Editors: Weixin Yang, Guanghui Yuan and Yunpeng Yang

Received: 17 December 2021
Accepted: 11 February 2022
Published: 15 February 2022

Publisher's Note: MDPI stays neutral with regard to jurisdictional claims in published maps and institutional affiliations.

Copyright: © 2022 by the authors. Licensee MDPI, Basel, Switzerland. This article is an open access article distributed under the terms and conditions of the Creative Commons Attribution (CC BY) license (https://creativecommons.org/licenses/by/4.0/).

1. Introduction

In the 21st century, China has undergone rapid development in the terms of urbanization. However, at the same time there has been a sharp rise in PM$_{2.5}$ concentrations [1–3]. In particular, extensive economic growth has led to the aggravation of this situation, affecting the sustainable development of cities [4,5]. In the meantime, PM$_{2.5}$ pollution-induced issues wreak great damage to natural ecosystems and have a deleterious effect on the physical and mental health of people [6–8]. In addition, PM$_{2.5}$ pollution can cause other negative outcomes such as crises of government trust and social instability [9–12]. Therefore, it is of great significance to PM$_{2.5}$ pollution mitigation that we determine the distribution of PM$_{2.5}$ concentrations and distinguish the determinants of PM$_{2.5}$ pollution in China.

The relationship between urban form and air quality (especially PM$_{2.5}$) has been given more and more attention by urban planners and environmentalists [13]. Different social and economic conditions and different geographical characteristics affect the development and characteristics of cities in different regions. In some developed countries, some evidence suggests that cities with fairly low levels of urban fragmentation and spread have less

PM$_{2.5}$ pollution than fragmented, dispersed, and complex cities [14,15]. The higher the degree of urban fragmentation, the denser the urban population, and the worse the urban air quality [16]. It seems that compact, low sprawling, and highly contiguous urban forms provide better air quality in developed countries. Some research results also point to the negative effects of a scattered population and inconvenient transportation on air quality [17,18].

China's environmental and socioeconomic conditions are different from those of developed countries. Different socioeconomic factors and geographical and climatic conditions cause great differences in PM$_{2.5}$ concentrations between China and developed countries [19]. Some researchers have studied the relationship between urban form and PM$_{2.5}$ concentration in China. For instance, based on 288 prefecture-level cities, Li [20] pointed out that small-scale, decentralized, and polycentric urban forms improve air quality in China. She et al. [21], through the study of the Yangtze River Delta, discovered that urban expansion accelerates energy consumption, resulting in a positive correlation with PM$_{2.5}$ concentration. Moreover, Zhang and Zhang [22] revealed that high population densities and numbers of cars might contribute to air pollution in urban agglomerations in China. Du et al. [12] came to a similar conclusion in the Pearl River Delta. However, most of these studies focus on the relationship between individual cities or urban agglomerations. In reality, each regional or spatial scale has a specific socioeconomic background and geographical and climatic conditions, which may lead to different research results. A discovery in one region cannot be used for another. The regional difference standpoint has been proven valid in several positive research studies [23–25]. In China, a few regions (such as the BTH region, Yangtze River Delta, and Pearl River Delta) have relatively concentrated populations and economies. Correspondingly, the PM$_{2.5}$ pollution level in these regions is higher than that in other regions. Consequently, it is necessary to study the influence of urban form on PM$_{2.5}$ concentration from both national and regional perspectives.

To correct the deviation in space, a spatial econometric model was used to analyze the impact of urban form on air quality in 286 cities in China. Moreover, spatial autocorrelation and spatial regression were conducted to distinguish the correlations between urban form and air quality in different regions of China. The index of urban form can be divided into two categories: a socioeconomic index and an urban landscape index. The urban landscape index was based on land-use data derived from satellites and calculated by FRAGSTATS software. According to their economic situations, the 286 prefecture level cities are divided into five regions, namely the eastern region, the central region, the northeastern region, the northwestern region, and the southwestern region. Then, the cities are divided into large cities, medium-sized cities, and small cities according to their populations. Considering China's national conditions and the distribution of data samples, cities with a population of 3 million or below are defined as small cities. Cities with a population of 3 million to 5 million are defined as medium-sized cities. Cities with a population of more than 5 million are defined as large cities.

This study differs from existing studies in the following aspects: (1) it conducted long-term spatiotemporal analyses of PM$_{2.5}$ concentrations annually in 286 prefecture-level cities in China. (2) The urban landscape index and urban socioeconomic index were used to characterize urban form. (3) On the one hand, when the road density low, the development of road traffic is conducive to reducing PM$_{2.5}$ concentration. On the other hand, when road traffic develops to a very high level, increasingly crowded roads will increase PM$_{2.5}$ concentration. (4) Excluding the effects of meteorological and geographic conditions, most of the more-developed cities or areas, which have a higher degree of urban development (except for fragmentation, other urban-form indicators have less of an impact on PM$_{2.5}$), should exhibit moderately decentralized and polycentric urban development. (5) For less-developed cities and shrinking cities, the single-center development model can better mitigate PM$_{2.5}$ pollution than the multicenter development model. The above points define the specificity of this study. The research results will further analyze the relationship between urban morphology and PM$_{2.5}$ concentration in combination with existing relevant research so as to provide a reliable reference for urban planning and urban air quality improvement. This paper is separated into five parts: the first part contains the introduction and research objectives; the second part contains the research methods and data sources; the third part contains the data sources and variable calculations; the fourth part contains the analysis and discussion; and the fifth part contains the conclusions and research prospects with a detailed discussion of the study's limitations. The flow chart of the article is shown in Figure 1.

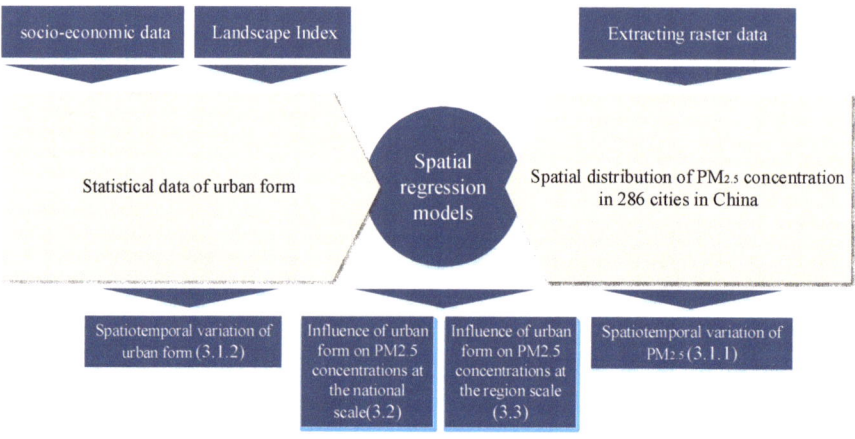

Figure 1. Flow chart of the article.

2. Materials and Methods
2.1. Data Source and Variable Calculations
2.1.1. Study Area

Prefecture-level cities, with large urban area and population scale, have advantages in terms of their economic structure and geographical location. They are not only present in all parts of China but reflect the development trend of urbanization and regional economic characteristics [24]. In this study, 286 prefecture-level cities were selected as our sample. According to the degree of urban development in different regions and the different climatic and geographical conditions, Chinese cities are divided into the eastern region, central region, northeastern region, northwestern region, and southwestern region. Development of the above areas was uneven. The eastern regions have large cities, strong economic strength, and high population density [26]. Being the most developed area of China, the eastern region consumes large amounts of natural resources to maintain urban development and socioeconomic growth. The situation is similar in the eastern and central regions. Relative to the eastern and central regions, on the one hand development of the northeastern and northwestern regions is unbalanced, on the other hand problem of population loss is serious in those area. Most cities in northeastern China are resource-based or resource-exhausted. Urban development has stagnated, and cities have contracted. The degree of urban development in northwestern China is low; most cities are in a stage of rapid or embryonic development, and the urban compactness and the stability of their spatial structure are low. It is worth thinking about how the $PM_{2.5}$ concentration increased between 2000 and 2015 in two regions. To cut down $PM_{2.5}$ concentration, it is essential to determine the relationship between urban form and $PM_{2.5}$ concentration in five different regions. Furthermore, through the comparison between countries and regions, we can better understand the impact of different urban forms on $PM_{2.5}$ concentration and provide better suggestions for decision makers to reduce $PM_{2.5}$ concentration. The classification of the study area is shown in Figure 2.

Figure 2. Study area.

2.1.2. PM$_{2.5}$ Concentration Data

The PM$_{2.5}$ concentrations in prefecture-level cities between 2000 and 2015 were taken from the dataset provided by Donkelaar's team [27]. This dataset combines the AOD inversion results from multiple satellite instruments. The geochemical transport model was used to correlate the total column measurements of aerosols with the near-surface PM$_{2.5}$ concentration. The geographically weighted regression model (GWR) was combined with the global ground survey to adjust the residual PM$_{2.5}$ bias [27]. The gridded data was set at 0.01 degrees. Compared with 210 ground-monitoring data from North America and Europe, this dataset showed a high degree of consistency (R2 = 0.81) [28]. The detection of PM$_{2.5}$ concentration by the China Environmental Monitoring Station depends on the ground detection of air monitoring stations in each city. There are three methods for measuring PM$_{2.5}$ concentration in China's air monitoring stations. The methods include the β Ray plus dynamic heating system method, the β Ray plus dynamic heating system combined with light scattering method, and the micro oscillating balance plus film dynamic measurement system method [29]. Compared with ground monitoring data (data from air monitoring stations in different cities), satellite measurement has both advantages and disadvantages. Since some of China's cities lacked air monitoring stations between 2000 and 2015, one of the advantages of this method is that it can measure PM$_{2.5}$ concentrations in more cities. Its disadvantage is that it has low accuracy in areas with high reflectivity, such as snow-covered areas and is prone to extreme values. This will cause the PM$_{2.5}$ concentrations in snow-covered areas (e.g., some cities in northern China, particularly in Xinjiang and Heilongjiang Provinces) to be lower than they should be.

2.1.3. Urban Form Metrics

The urban form indicators were divided into two categories: urban index and landscape metric. The urban index is composed of urban built-up area (UA), population density (PD), and road density (RD). Urban built-up area (UA) reflects the degree of urban development. Population density (PD) represents the intensity of human social activities in a city. Road density (RD) reflects the development of a city's transportation infrastructure and can also represent the horizontal development of a city. The above three factors were obtained from the Chinese City Statistical Yearbook from 2000 to 2015 [30]. Landscape metric is extremely useful for describing urban forms. It has several advantages for characterizing the heterogeneity of urban landscapes and the gap between urban land use patterns and governance processes, as well as analyzing urban development [31–34]. Three landscape metrics were ultimately selected to indicate urban forms in this study. These metrics were number of patches (NP), largest patch index (LPI), and contagion index (CONTAG). NP describes the heterogeneity of the whole urban landscape and represents the degree of fragmentation of urban patches. LPI represents the proportion of the largest urban patch to the whole urban landscape area. It reflects the direction and strength of human activities. CONTAG describes the agglomeration degree or extension trend of different urban patches in the landscape. The high value of CONTAG indicates that some urban patches in the landscape show good connectivity; otherwise, it indicates that the city has a dense pattern. Urban landscape data were derived from land-use datasets (30 m × 30 m) for China from 1998 to 2015 that were produced by the Institute of Remote Sensing and Digital Earth at the Chinese Academy of Sciences through the interpretation of Landsat TM or ETM images. The overall accuracy of classification for these datasets is more than 85% [35]. We used Fragstats 4.2 to calculate three urban landscape indicators for each city.

2.2. Methods

2.2.1. Spatial Autocorrelation Test

Moran's I statistic is one of the most commonly adopted measures for spatial autocorrelation. It has been used to test spatiotemporal characteristics by identifying spatial correlations and spatial heterogeneity [36]. When researching spatial correlations and spatial distribution patterns, respectively, we usually divide Moran's I into global Moran's I and local Moran's I [37]. The global Moran index describes the average correlation degree of all spatial units with the entire surrounding region, allowing us to explore whether there is spatial correlation at the regional level. The equation for calculating global Moran's I is as follows:

$$I_G = \frac{n}{\sum_{i=1}^{n}\sum_{j=1}^{n} W_{ij}} \times \frac{\sum_{i=1}^{n}\sum_{j=1}^{n} W_{ij}(y_i - \bar{y})(y_j - \bar{y})}{\sum_{i=1}^{n}(y_i - \bar{y})^2} \quad (1)$$

where y_i and y_j represent the attribute values of the ith spatial element and the jth spatial element, respectively; $\bar{y}$ denotes the mean value of y; n is the total number of spatial elements; and W_{ij} is the spatial weight value.

The local Moran index can be used to observe the spatial aggregation in the local areas. The equation for calculating local Moran's I is as follows:

$$I_L = (y_i - \bar{y}) \sum_{i \neq j}^{n} W_{ij}(y_j - \bar{y}) \quad (2)$$

The values of global Moran index and local Moran index both range from −1 to 1. For the global Moran index, when $I_G > 0$, it means that the attribute values of all regions are positively correlated; otherwise, the attribute values of all regions are negatively correlated. For the local Moran index, an I_L value above zero means that the indicators of a city are similar to those of surrounding cities. An I_L value below zero means that a city is surrounded by cities with different indicators. To clarify the spatial aggregation of each urban form index, we generated a LISA cluster map in GeoDa on the basis of the Moran scatter diagram and the Moran index.

2.2.2. Spatial Regression Models

Regression analysis has been used in most studies to explore the relationship between urban form and $PM_{2.5}$ concentration [27,35,38,39]. Some regression analyses ignore the influence of spatial heterogeneity, such as least-squares regression analysis and ridge regression analysis [24]. A multitude of studies has proven that the spatial regression model can effectively solve the spatial dependence issue [40]. There are two commonly used spatial regression models: the spatial lag model (SLM) and the spatial error model (SEM). The calculation formula is as follows:

$$y = \rho Wy + X\beta + \varepsilon \quad (SLM) \qquad (3)$$

$$y = \gamma W\varepsilon + X\beta + \delta \quad (SEM) \qquad (4)$$

where y represents the $PM_{2.5}$ concentration of each prefecture-level city; ρ is the spatial lag coefficient of urban forms; W represents the spatial weight; X represents the urban form index; β represents the regression coefficient vector; ε denotes the random error vector; γ is the residual correlation parameter, and δ represents a vector of the error terms. To determine which model to use, mainly relied on the Lagrange multiplier (LM) test and the robust LM test.

3. Results
3.1. Spatiotemporal Characteristics Analysis
3.1.1. Spatiotemporal Variation of $PM_{2.5}$

As shown in Figure 3, the average annual concentration of $PM_{2.5}$ surged from 32.34 µg/m^3 in 2000 to 47.33 µg/m^3 in 2015; a growth rate of 46.7%. The mean value as a whole showed a trend of first increasing and then decreasing, and the median also showed the same trend. The median values for 2000, 2005, 2010, and 2015 were 29.85 µg/m^3, 47.3 µg/m^3, 47.1 µg/m^3, and 45.4 µg/m^3, respectively. It is worth noting that the mean is always greater than the median. This indicates that there are some cities with more serious $PM_{2.5}$ pollution, making the mean value larger. In addition, $PM_{2.5}$ concentrations increased sharply in 2000–2005. The annual growth rate of the average concentration of $PM_{2.5}$ was more than three times that of the entire 15 years. The change in $PM_{2.5}$ concentration standard-reaching rate between cities also illustrates this point. In China's ambient air quality standards, the concentration limit of fine particulate matter ($PM_{2.5}$) is divided into level I and level II, of which the level I standard is 15 µg/m^3. This standard applies to areas such as nature reserves. The secondary standard is 35 µg/m^3. This standard applies to residential areas, commercial areas, and other areas. In 2000, only 97 of 286 prefecture-level cities had $PM_{2.5}$ concentrations exceeding 35 µg/m^3, with a standard-reaching rate of 66%. However, 225 cities had $PM_{2.5}$ concentrations above 35 µg/m^3 in 2005, and the standard-reaching rate was only 21.1%. Among the 286 prefecture-level cities, the number of cities that met the Grade II Standards was 189, 61, 69, and 78 in 2000, 2005, 2010, and 2015, respectively. For detailed data on PM2.5 concentration, see Table S1 in the Supplementary Materials.

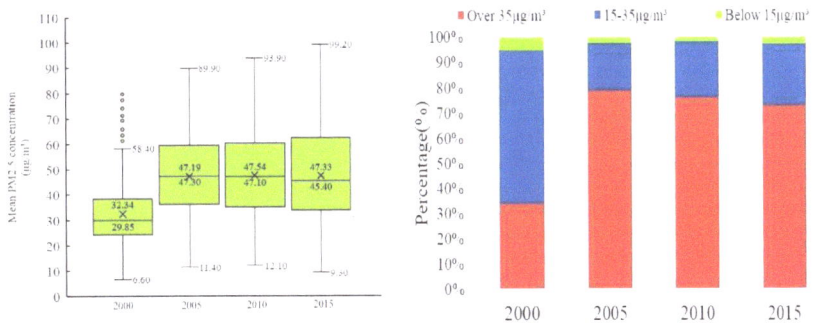

Figure 3. Mean $PM_{2.5}$ concentration and annual statistics of the proportion of days with different $PM_{2.5}$ levels from 2000 to 2015.

As shown in Figure 4, the distribution of $PM_{2.5}$ gradually concentrated in some areas (the northeastern region, the central region, and the eastern region). The northeastern region exhibited a high concentration of $PM_{2.5}$ in 2015. This scenario could be attributed to the coal consumption and winter heating in that region. A similar problem is also present in developed European countries. In Poland, for example, solid fuel heating in Krakow causes air pollution in surrounding cities [41]. $PM_{2.5}$ levels were also extraordinarily high in some cities in Central China, especially in the BTH region. This may be due to most cities in the central region being dependent upon secondary industries for their economic activities [42]. Additionally, heavy vehicle emissions also lead to an increase in $PM_{2.5}$ concentrations [43]. Apart from the above factors, the $PM_{2.5}$ pollution level in cities can be significantly affected by the pollution levels of their neighboring areas, especially in cities with high concentrations [44,45].

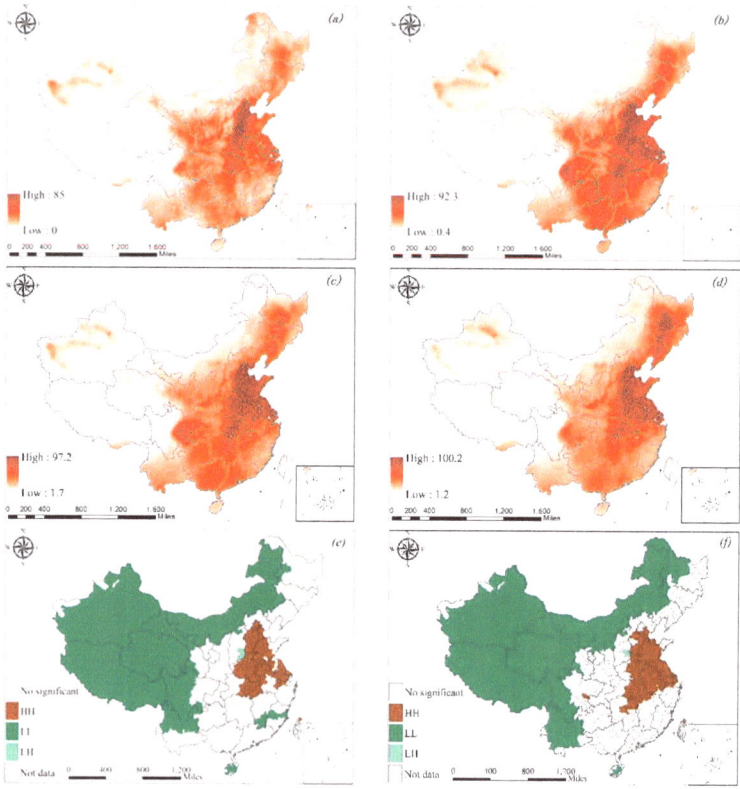

Figure 4. *Cont.*

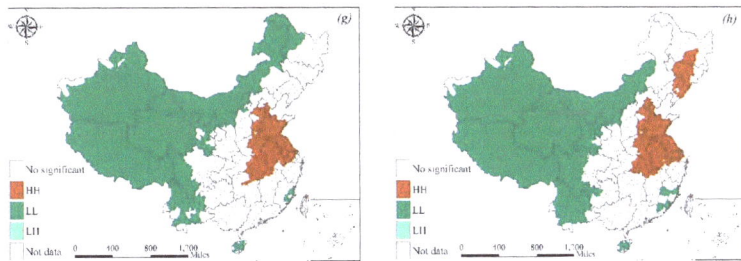

Figure 4. Spatial variation of PM$_{2.5}$ concentration from 2000 to 2015. (**a**–**d**) show the changes of PM concentration from 2000 to 2015. (**e**–**h**) show the spatial clusters of PM concentration from 2000 to 2015. Not significant indicates that the region is not significantly clustered. No data indicates that there is no data in this area.

The spatial autocorrelation was weaker in 2000 (Moran's I = 0.734) than in 2005 (Moran's I = 0.777), 2010 (Moran's I = 0.78), and 2015 (Moran's I = 0.779). The trend of the change in spatial autocorrelation was the same as that of PM$_{2.5}$ concentration. As shown in Table 1.

Table 1. Results of the spatial autocorrelation analysis.

Years	PM$_{2.5}$	
	Moran's I	Z-score
2000	0.734	17.8813
2005	0.777	18.9055
2010	0.78	19.0748
2015	0.779	19.039

3.1.2. Spatiotemporal Variation of Urban Form

As is shown in Figure 5, the population density of all cities increased slightly from 423.4 in 2000 to 433.9 in 2015, indicating that the urban population has become more concentrated over the past two decades. Urban built-up areas also became larger during this period, as is indicated by the urban area index, which increased from 61.69 in 2000 to 137.21 in 2015. Road density increased from 0.1 in 2000 to 0.14 in 2015, reflecting the concentration of traffic and the degree of urban development. The number of patches showed a trend of increasing first and then decreasing, from 2206.1 in 2000 to 2278.7 in 2010 to 2209.3 in 2015, indicating that the urban form experienced dispersion and aggregation as cities developed. The trend of the largest patch index and contagion also reflects this situation. The largest patch index initially decreased from 32.17 in 2000 to 31.37 in 2010, followed by an increase to 33.14 in 2015. Contagion decreased initially from 47.1 in 2000 to 46.4 in 2010 but then increased to 47.2 in 2015. The trends of these three types of indicators indicate that the development of prefecture-level cities changed from decentralization to centralization. The contiguity and compactness of urban areas as a whole have increased.

The LISA cluster map of urban form in Chinese cities from 2000 to 2015 is shown in Figure 6. As can be seen, the spatial distribution of population density had hardly changed. From the graph, high–high clusters of population density mainly exist in the eastern region, while low–low clusters of population density exist in the northeast, the northwest, and Guangxi. The main changes were concentrated in Shaanxi Province and Henan Province. From 2000 to 2015, the spatial aggregation of population density in Shaanxi Province became stronger, while that of Henan province became weaker. On the whole, urban built-up areas can be divided into four categories of spatial pattern.

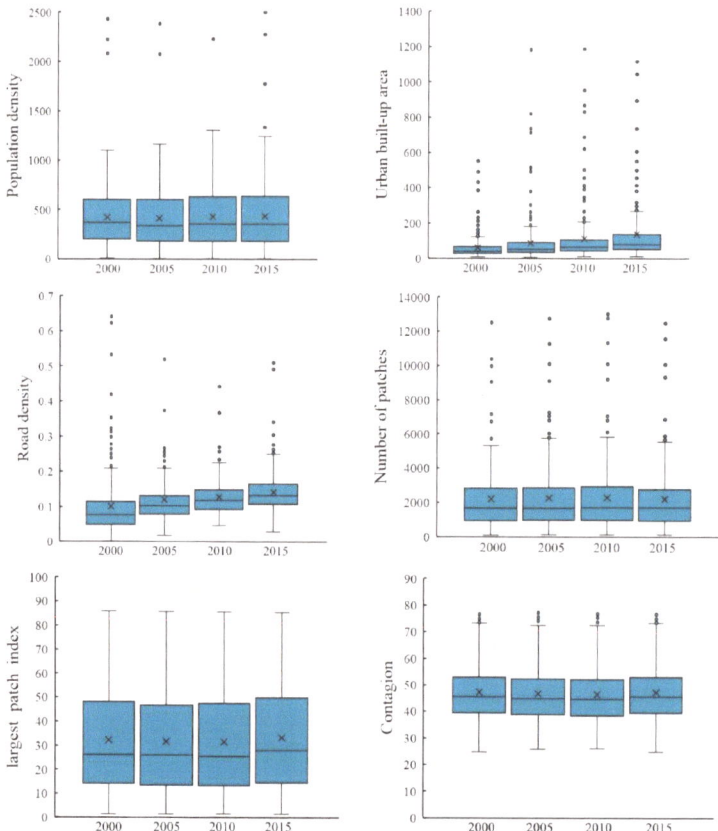

Figure 5. Urban form metrics for 286 different-sized prefecture-level cities between 2000 and 2015.

High–high clusters of urban built-up areas exist in the BTH (Beijing, Tianjin, and Hebei) region and Yangtze River Delta region (Shanghai, Zhejiang, and Jiangsu); low–low clusters of urban built-up areas exist in Shanxi, Ningxia, and central Gansu; high–low clusters of urban built-up area exist in the southeast coastal regions, including Guangxi and Guangdong; and low–high clusters of urban built-up area exist in central Liaoning and Jilin and some areas of Sichuan, Guizhou, and Hebei. The spatial aggregation changes of urban construction land are mainly concentrated in Hebei, Guangxi, and Chongqing. From the figure, we can see that the spatial distribution of traffic density in Shandong, Zhejiang, and Jiangsu is relatively concentrated. The spatial distribution of road density varies greatly from 2000 to 2015. The spatial aggregation of traffic density in Heilongjiang Province and Jilin province changed from insignificant to low–low clusters between 2000 and 2005. Shaanxi Province exhibited a high concentration in 2015. As time has gone on, the low–low clusters in Sichuan, Shanxi, Hunan, and Hubei have disappeared and some new clusters and outliers have emerged.

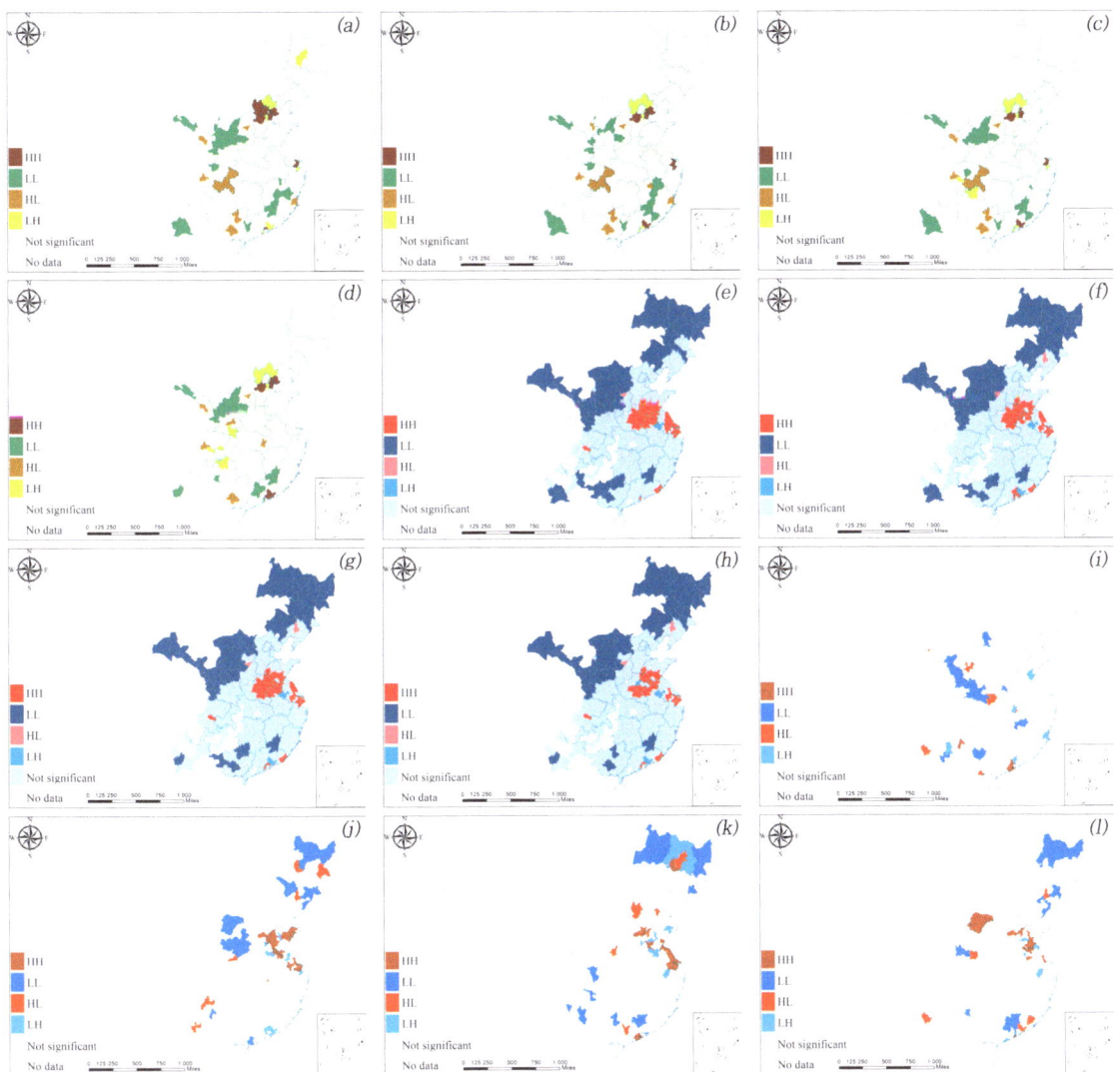

Figure 6. LISA cluster map of urban index in Chinese cities from 2000 to 2015. Panels (**a–d**) show the spatial clusters of population density between 2000 and 2015; panels (**e–h**) show the spatial clusters of urban areas (built-up areas) from 2000 to 2015. Panels (**i–l**) show the spatial clusters of road density between 2000 and 2015. Specifically, HH indicates a city with a high population density (urban area, road density) surrounded by cities with high population density (urban area, road density); LL indicates a city with a low population density (urban area, road density) surrounded by cities with low population density (urban area, road density); HL indicates a city with a high population density (urban area, road density) surrounded by cities with low population density (urban area, road density); and LH indicates a city with a low population density (urban area, road density) surrounded by cities with high population density (urban area, road density).

3.2. Influence of Urban Form on $PM_{2.5}$ Concentrations at the National Scale

It can be seen from Table 2 that the test results of the correlation coefficient (R2), bass information content criterion (SC), log likelihood (log likelihood), and Akaike info criterion (AIC) are significant, which proves that SEM has relatively high goodness of fit and can accurately evaluate the influence of urban form on the long-term variations in $PM_{2.5}$ concentration. From the perspective of all cities in China, not all urban form indicators are significantly correlated with $PM_{2.5}$ concentration. Specifically, UA (0.015 in 2000 and 0.009 in 2015) was significantly positively correlated with $PM_{2.5}$ concentrations. RD showed a negative correlation in 2000 (-7.516) and a significant positive correlation in 2015 (22.432) with $PM_{2.5}$ concentrations. This phenomenon shows that when the urban road density is at a low level, increasing road construction will reduce road congestion and lead to a decrease in $PM_{2.5}$ concentration, while when the urban road density is at a higher level, increasing road construction will increase the concentration of $PM_{2.5}$. In general, urban expansion and urban road network density have become the major factors impacting $PM_{2.5}$ concentrations in all cities in China.

3.3. Influence of Urban Form on $PM_{2.5}$ Concentrations at the Region Scale

The spatial regression results of five regions by city size are shown in Tables 3–7. Some coefficients of the influence of urban form on $PM_{2.5}$ concentration in eastern and central China are basically consistent with the national results. For instance, the increase in urban built-up area promoted the increase in $PM_{2.5}$ concentrations in the eastern cities, especially in 2000. In addition, for the eastern cities, road density showed a negative correlation with $PM_{2.5}$ concentration in 2000, but the results were reversed in 2015; there was a positive correlation (Table 3). The reasons for this are manifold. On the one hand, the increase in road density is closely related to economic development. On the other hand, as roads become more developed, the compactness of cities will also increase correspondingly, further affecting $PM_{2.5}$ concentration.

In the northeastern region, only PD had a significant correlation with $PM_{2.5}$ concentration between 2000 and 2015 (0.022 in 2000 and 0.065 in 2015). However, when the analysis was refined according to the size of the cities, the results were different. For large and small cities in northeastern China, in addition to PD, the impact of NP, LPI, and CONTAG on $PM_{2.5}$ was also significant. It is worth noting that NP and LPI were negatively correlated with $PM_{2.5}$ concentration in large cities, while CONTAG was positively correlated with $PM_{2.5}$. The opposite was true of small cities (Table 5). Cities in the northwest faced the same situation as those in the northeast. PD was still the most important factor affecting $PM_{2.5}$ concentration in northwestern China. For small- and medium-sized cities in northwestern China, UA also played a positive role in $PM_{2.5}$ concentration, especially in 2015 (medium 0.02 and small 0.064). UA was negatively correlated with $PM_{2.5}$ concentration in large cities. Otherwise, LPI (0.048) had the greatest positive impact on $PM_{2.5}$ in 2015. This state of affairs also confirms that large cities, which have become diverse, continuous, and uncompact, have been able to reduce $PM_{2.5}$ pollution.

Table 2. The results for cities grouped by urban size (nation-wide).

	All Cities				Large				Medium				Small			
Variables	2000	2005	2010	2015	2000	2005	2010	2015	2000	2005	2010	2015	2000	2005	2010	2015
UA	0.015 **	0.01 **	0.008 **	0.009 ***	−0.007	−0.001	−0.003	0.002	−0.011	0.044 **	3.033 **	0.011	−0.003	−0.017	−0.018	−0.002
PD	0.002	0.006 **	0.003	0.006 ***	0.002	0.003	0.002	0.004	0.001	−0.004	3.014 **	0.001	0.006 **	0.023 ***	0.02 ***	0.022 ***
RD	−7.516 *	−2.018	14.886 **	22.432 ***	−4.982	−8.602	6.667	39.71 **	−5.628	1.614	7.743	−1.257	1.53	75.114 ***	33.068	29.285 **
NP	−0.001	−0.001	−0.001	−0.001	0.001	−0.001	0.001	0.001	−0.001 *	−0.002 ***	−0.001	−0.002 ***	−0.001 *	−0.001 **	−0.001 **	−0.001 *
LPI	0.056	0.085 *	0.101 **	0.191 **	0.145 *	0.349 ***	0.202 **	0.063	0.069	0.012	0.038	−0.033	0.094	0.074	0.179 *	0.098
CONTAG	−0.049	−0.119	−0.186 *	0.067	−0.215	−0.642 ***	−0.431 **	−0.082	0.094	0.18	0.302	0.498 **	0.175	0.09	−0.093	0.175
R2	0.739	0.787	0.804	0.797	0.767	0.673	0.752	0.772	0.519	0.546	0.612	0.602	0.63	0.603	0.595	0.58
S	6.566	7.554	7.648	8.4302	6.854	8.348	7.789	8.263	8.015	8.09	8.337	9.935	8.104	9.7	11.29	11.486
LogL	−972.14	−1018.141	−1027.645	−1049.404	−286.54	−316.188	−356.822	−382.965	−276.332	−303.457	−286.601	−289.524	−448.026	−423.148	−414.448	−408.91
AIC	1958.39	2050.28	2069.29	2112.81	587.095	646.378	727.644	779.93	566.665	620.915	587.202	593.048	894.053	860.296	842.897	831.833
SC	1983.91	2075.83	2094.86	2138.35	603.857	663.558	745.739	798.441	583.162	638.013	603.877	609.363	913.794	879.388	861.607	850.344
Lag coeff	0.883	0.9	0.915	0.901	0.806	0.696	0.779	0.776	0.423	0.509	0.408	0.596	0.638	0.59	0.5	0.578

Notes: *** represents a significance level of 1%, ** represents a significance level of 5%, and * represents a significance level of 10%. The urban population of large cities is more than 5 million, the urban population of medium cities is more than 3 million and less than 5 million, and the urban population of small cities is less than 3 million.

Table 3. The results for eastern cities grouped by urban size.

	All Cities				Large				Medium				Small			
Variables	2000	2005	2010	2015	2000	2005	2010	2015	2000	2005	2010	2015	2000	2005	2010	2015
UA	0.013	0.009	0.013 *	0.002	0.014	0.005	0.001	0.004	0.004	0.086	0.201 **	0.099 ***	0.001	−0.002	−0.102 **	−0.095 **
PD	−0.006 *	−0.004	−0.007 *	0.001	−0.009	−0.008	−0.005	0.003	−0.003	−0.021 *	−0.042 **	−0.012	−0.003	0.001	0.015 *	0.011
RD	−21.21 ***	−1.5	−2.827	21.072	−29.64 **	−4.61	−16.525	20.256	−13.07	−1.369	−74.321 *	82.627 **	20.083	143.258	273.229 ***	133.773 ***
CONTAG	0.182	−0.026	−0.006	0.033	−0.021	−0.549 **	−0.227	−0.058	−0.248 **	−0.123	−0.411	−0.085	0.174	0.039	0.119	−0.733 **
R2	0.673	0.746	0.802	0.836	0.618	0.695	0.685	0.764	0.81	0.801	0.854	0.861	0.246	0.607	0.734	0.627
S	4.982	5.256	6.071	5.753	6.042	5.501	7.236	6.459	3.095	4.498	5.396	5.307	6.543	6.781	6.867	8.825
LogL	−245.994	−253.78	−262.354	−259.667	−117.522	−107.107	−119.776	−133.269	−56.653	−61.522	−49.675	−53.991	−82.527	−80.103	−78.166	−80.139
AIC	505.988	521.567	538.708	533.335	249.041	228.201	253.552	280.538	127.315	137.052	113.352	121.981	181.057	176.206	170.334	174.278
SC	522.481	538.067	555.205	549.832	244.255	238.674	264.236	292.183	134.274	144.021	118.76	127.813	190.812	185.631	178.282	181.915
Lag coeff	0.871	0.943	0.893	0.925	0.653	0.674	0.652	0.701	0.803	0.667	0.003	0.508	0.180	0.197	−0.789	0.668
NP	−0.001	−0.001	−0.001	0.001	−0.003 **	−0.004 ***	−0.003 **	−0.002 **	0.001 *	−0.002 **	−0.007 ***	−0.003 **	−0.001	−0.005	−0.004 ***	−0.004 **

Table 3. Cont.

Variables	All Cities				Large				Medium				Small			
	2000	2005	2010	2015	2000	2005	2010	2015	2000	2005	2010	2015	2000	2005	2010	2015
LPI	−0.025	0.008	0.022	0.018	0.062	0.246 **	0.106	0.048	0.052	0.027	0.351	0.146	−0.05	0.081	0.264	0.524 ***

Notes: *** represents a significance level of 1%, ** represents a significance level of 5%, and * represents a significance level of 10%. The urban population of large cities is more than 5 million, the urban population of medium cities is more than 3 million and less than 5 million, and the urban population of small cities is less than 3 million.

Table 4. The results for central cities grouped by urban size.

Variables	All Cities				Large				Medium				Small			
	2000	2005	2010	2015	2000	2005	2010	2015	2000	2005	2010	2015	2000	2005	2010	2015
								LnPM$_{2.5}$								
UA	−0.018	0.021	0.009	0.008	−0.061	−0.032 *	−0.011	−0.005	−0.105 **	−0.038	−0.003	0.006	−0.058	−0.033	−0.012	−0.059
PD	0.012 *	0.014 **	0.014 ***	0.013 ***	0.008	0.007	0.006	0.012 ***	0.034 **	0.007	0.021	0.01	0.009	0.015 *	0.029 ***	0.027
RD	−27.656 *	18.172	16.434	40.269 **	4.139	6.533	36.294	16.302	36.126	130.853 **	158.854 *	11.315	17.782	37.505	−19.837	−2.604
NP	−0.001	−0.001	−0.001	−0.001	−0.004 ***	−0.004 ***	−0.002 **	−0.001	−0.001	−0.004 ***	−0.002	−0.006 **	−0.001	−0.004 **	0.003	0.001
LPI	0.037	0.009	0.013	−0.042	0.177	0.211 ***	−0.032	−0.037	−0.167	0.359 **	0.518	−0.008	−0.194	−0.229	−0.226	−0.284
CONTAG	−0.012	−0.065	−0.015	0.071	−0.699 **	−0.687 ***	−0.188	−0.03	0.226	−0.487	−0.665	0.203	0.654	0.488 *	0.548 *	0.733
R2	0.625	0.545	0.603	0.615	0.804	0.908	0.813	0.816	0.636	0.728	0.691	0.703	0.289	0.511	0.48	0.604
S	7.516	7.259	7.329	7.339	5.706	2.92	4.655	4.599	7.191	5.915	7.395	7.347	9.728	7.119	7.023	6.827
LogL	−274.394	−267.741	−268.577	−269.07	−89.998	−75.57	−105.614	−113.3	−78.233	−75.343	−68.877	−62.164	−103.48	−91.346	−81.016	−76.902
AIC	562.788	551.481	553.154	554.139	193.998	165.141	225.229	242.601	170.468	164.687	151.755	138.329	222.962	198.693	176.032	169.805
SC	579.285	570.335	572.007	572.993	203.069	174.467	235.913	255.488	178.416	172.636	158.726	144.562	233.62	209.06	184.279	178.889
Lag coeff	0.742	0.501	0.507	0.54	0.785	0.839	0.784	0.754	0.193	−0.534	−0.332	−0.459	0.11	0.095	0.285	0.162

Notes: *** represents a significance level of 1%, ** represents a significance level of 5%, and * represents a significance level of 10%. The urban population of large cities is more than 5 million, the urban population of medium cities is more than 3 million and less than 5 million, and the urban population of small cities is less than 3 million.

Table 5. The results for northeastern cities grouped by urban size.

Variables	All Cities				Large				Medium				Small			
	2000	2005	2010	2015	2000	2005	2010	2015	2000	2005	2010	2015	2000	2005	2010	2015
								LnPM$_{2.5}$								
UA	−0.008	−0.004	−0.003	−0.013	0.067 *	0.074 ***	0.091 ***	0.012	0.137 ***	0.042	0.021	0.046	0.019	0.018	0.021	0.002
R2	0.529	0.633	0.688	0.586	0.844	0.881	0.904	0.897	0.872	0.489	0.616	0.942	0.656	0.705	0.715	0.686

Table 5. Cont.

Variables	All Cities				Large				LnPM$_{2.5}$ Medium				Small			
	2000	2005	2010	2015	2000	2005	2010	2015	2000	2005	2010	2015	2000	2005	2010	2015
S	3.929	5.572	6.035	10.989	2.248	3.296	2.963	5.575	2.354	5.566	5.209	3.461	3.184	4.861	5.927	7.534
LogL	−105.224	−118.119	−121.041	−142.757	−14.772	−17.063	−15.493	−20.756	−18.887	−28.266	−28.548	−26.299	−58.782	−66.497	−71.079	−76.559
AIC	224.449	250.238	256.082	299.515	37.543	42.127	38.987	49.512	45.774	66.538	65.095	60.598	125.564	140.996	150.159	161.119
SC	235.726	261.514	267.358	310.791	36.711	41.294	38.154	48.679	46.092	67.518	65.885	61.387	129.928	145.36	154.523	165.483
Lag coeff	0.626	0.623	0.618	0.555	0.836	0.836	−0.562	0.904	0.583	0.333	−0.678	−0.984	0.572	0.303	0.358	0.401
NP	−0.001	−0.001	−0.001 *	−0.001	−0.006 ***	−0.011 ***	−0.008 ***	−0.027 ***	−0.001 **	−0.002 ***	−0.002 **	−0.002 *	0.001 **	0.001 **	0.001 ***	0.002 ***
LPI	0.109 *	0.098	0.012	−0.078	−0.877 **	−1.713 **	−1.451 **	−4.233 **	0.015	−0.096	−0.169	−0.723 **	0.246 ***	0.361 ***	0.391 **	0.436 **
CONTAG	−0.292 *	−0.307	−0.249	0.007	0.593 **	2.065 **	1.979 ***	4.994 ***	−0.463	−0.102	0.153	3.161 ***	−0.654 ***	−0.944 ***	−1.188 **	−1.242 ***
PD	0.022 **	0.036 **	0.039 ***	0.065 **	−0.031 *	−0.021	−0.003	−0.092 ***	−0.004	0.037 *	0.040 ***	0.061 *	0.032 ***	0.065 ***	0.079 ***	0.084 ***
RD	−5.019	−47.899	7.051	−3.039	93.732	111.675	39.886 **	622.913	510.859 ***	−61.046	−61.169	−82.537 **	20.319	−10.087	20.509	114.183 **

Notes: *** represents a significance level of 1%, ** represents a significance level of 5%, and * represents a significance level of 10%. The urban population of large cities is more than 5 million, the urban population of medium cities is more than 3 million and less than 5 million, and the urban population of small cities is less than 3 million.

Table 6. The results for northwestern cities grouped by urban size.

Variables	All Cities				Large				LnPM$_{2.5}$ Medium				Small			
	2000	2005	2010	2015	2000	2005	2010	2015	2000	2005	2010	2015	2000	2005	2010	2015
UA	0.014	0.024	0.016	0.025 *	−0.001	−0.012 ***	−0.004 ***	−0.005 ***	−0.081 ***	0.009	0.017 ***	0.02 ***	0.042	0.109 ***	0.057 ***	0.064 ***
PD	0.036 ***	0.047 ***	0.041 ***	0.034 ***	−0.001	−0.008	−0.003 ***	−0.005 ***	0.063 ***	0.081	0.01	0.001	0.001	0.056 **	0.046 ***	0.03 *
RD	−23.437 ***	−29.944	1.601	15.619	2.042	−10.758	−2.625	2.827	−130.714 ***	−5.359	4.309	45.063 ***	3.974	−24.226	17.923	0.546
NP	0.001 **	0.001	0.001	0.001	−0.001 ***	0.004	−0.003 **	−0.003	0.001 *	−0.005 *	0.001	−0.002	−0.001	0.001	0.001	0.001
LPI	0.023	0.166	0.222	0.146 *	−0.003	0.102 ***	0.025	0.048 ***	0.199 **	0.262	0.185	0.049	−0.108	0.145	0.067	0.112 *
CONTAG	0.484 *	0.543 *	0.235	0.536 ***	−0.001	0.204 **	0.06 *	0.142 **	1.011 **	−0.227	−2.498 ***	−0.315	0.079	0.044	−0.06	0.002
R2	0.783	0.818	0.799	0.767	0.767	0.854	0.895	0.851	0.944	0.847	0.924	0.971	0.582	0.639	0.683	0.745
S	5.164	6.43	5.611	6.19	0.604	0.701	0.675	0.572	1.168	4.482	1.09	0.848	4.176	6.11	4.368	4.206

Table 6. Cont.

Variables	All Cities				Large				LnPM$_{2.5}$ Medium				Small			
	2000	2005	2010	2015	2000	2005	2010	2015	2000	2005	2010	2015	2000	2005	2010	2015
LogL	−99.016	−108.175	−102.893	−105.546	1.832	−3.885	−2.952	−1.98	−12.1	−21.478	−14.001	−12.889	−63.912	−76.4	−65.835	−63.878
AIC	214.033	230.349	219.786	225.093	0.334	11.77	3.46	7.978	34.2	50.957	36.003	33.779	141.824	166.801	145.672	141.757
SC	225.759	240.609	230.046	235.354	−1.467	9.967	1.657	6.175	33.93	50.741	35.787	33.562	149.461	174.749	153.309	149.394
Lag coeff	0.483	0.752	0.664	0.603	−0.33	−0.33	−0.33	−0.33	0.693	0.752	0.956	0.975	0.571	0.669	0.713	0.533

Notes: *** represents a significance level of 1%, ** represents a significance level of 5%, and * represents a significance level of 10%. The urban population of large cities is more than 5 million, the urban population of medium cities is more than 3 million and less than 5 million, and the urban population of small cities is less than 3 million.

Table 7. The results for southwestern cities grouped by urban size.

Variables	All Cities				Large				LnPM$_{2.5}$ Medium				Small			
	2000	2005	2010	2015	2000	2005	2010	2015	2000	2005	2010	2015	2000	2005	2010	2015
UA	0.032	0.008	0.013	−0.005	0.048 ***	0.007	0.023	0.01	−0.016	−0.073 *	0.113 **	0.023	0.054	−0.309 **	−0.686 ***	0.038
PD	0.006	0.013	0.016 **	0.013 *	0.01 *	0.008	0.009	0.011	−0.007	−0.017 *	−0.035	−0.009	−0.005	−0.024	0.008	−0.031 ***
RD	−2.569	−0.514	−11.826	21.076	17.775 **	−92.773 **	−36.323	68.443 *	−41.895 ***	50.484 *	−118.929	−14.477	−6.993	45.942	115.688 **	34.418 ***
NP	0.001	0.001	0.001	0.002 **	0.002 **	0.001	0.007	0.002 *	−0.003	0.001	−0.006 *	−0.003	−0.001	−0.002	−0.001	−0.001
LPI	−0.002	0.121	0.14	0.153	0.411	1.115 ***	0.48	0.723 ***	−0.199	0.063	−0.103	−0.06	−0.071	−0.078	0.336 *	0.269 ***
CONTAG	0.069	−0.273	−0.116	−0.3	−0.387	−1.335 **	−0.354	−1.163 ***	0.308	−0.533 *	1.096 *	0.339	0.248	0.325	−0.651	−0.52 ***
R2	0.523	0.709	0.78	0.664	0.809	0.734	0.525	0.664	0.653	0.797	0.717	0.301	0.159	0.817	0.699	0.952
S	5.009	7.26	6.242	5.113	2.462	6.108	7.911	4.173	2.133	3.752	5.823	5.655	6.436	4.719	5.567	1.878
LogL	−138.356	−157.129	−149.6	−141.425	−20.904	−39.531	−55.821	−46.427	−36.308	−57.09	−55.285	−50.523	−66.051	−46.312	−41.318	−32.658
AIC	292.713	330.257	315.2	296.851	51.808	93.062	127.643	106.854	86.617	128.18	124.571	117.048	146.103	106.624	98.637	79.316
SC	307.166	344.711	329.654	309.497	52.794	96.457	134.791	112.262	92.025	134.791	129.979	123.228	153.073	111.097	103.157	83.271
Lag coeff	0.492	0.663	0.613	0.669	0.125	0.503	0.82	0.449	−0.608	0.82	0.82	0.161	0.29	0.823	−0.396	0.924

Notes: *** represents a significance level of 1%, ** represents a significance level of 5%, and * represents a significance level of 10%. The urban population of large cities is more than 5 million, the urban population of medium cities is more than 3 million and less than 5 million, and the urban population of small cities is less than 3 million.

4. Discussion

4.1. The Relationship between $PM_{2.5}$ and Urban Area from a National Perspective

According to the estimated results in Table 2, several important conclusions can be drawn at the national scale. First, the correlation coefficients of urban areas (built-up areas) were significantly positive in 286 prefecture-level cities, implying that the expansion of urban areas aggravates the pollution of $PM_{2.5}$ at the national scale and especially in large cities. Liu [46] and She [21] also found a positive relationship between the urban area and urban air pollution. The emergence of this situation may be due to the expansion of the urban area, which leads to population growth and increased road traffic. These factors aggravate energy consumption and increase $PM_{2.5}$ pollution.

Second, the correlation coefficients of population density were significantly positive in 2000, 2005, and 2010 but not significant in 2015, implying that increased population density leads to more $PM_{2.5}$ pollution. An increase in population density not only increases the demand for consumption and work resources but also aggravates housing congestion and traffic jams [15].

Third, the correlation coefficients of road density were significantly negative in 2000 but positive in 2015. It is possible that the road density has a positive correlation with $PM_{2.5}$ when the road density reaches a certain limit, and the correlation increases gradually. It was found that some emerging cities that are developing show opposite results in terms of road density compared to some other large cities. For example, in Yangquan, a city located in the northwestern region of China, the road density increased from 0.095 to 0.1429 between 2005 and 2015, but the annual average concentration of $PM_{2.5}$ decreased from 63.6 µg/m^3 to 57.2 µg/m^3. However, further research is needed by the authors to ascertain exactly what this limit is and whether it is more relevant to industrial development or to urban planning.

Fourth, the correlation coefficient of NP in small- and medium-sized cities is significantly and negative, but not for large cities. This shows that the impact of urban fragmentation on $PM_{2.5}$ concentration is only reflected in cities on a general scale. The correlation coefficients of the large patches index were significant and positive in 2005 and 2010. This indicates that small, dispersed, and polycentric cities exhibited less $PM_{2.5}$ pollution than compact and larger cities. A similar finding was observed by Wu et al. [32] and She et al. [21], who both concluded that a more uniform distribution of urban patches might be better for mitigating particulate matter in large urban agglomerations (for example, the YRD region). The more complex the urban form is, the greater the average distance between urban patches and the smaller the concentration of $PM_{2.5}$. This shows that the multicenter urban form can improve air quality.

Lastly, unlike previous studies, we found that urban compactness (CONTAG) does not promote $PM_{2.5}$ concentrations at the national scale. Urban connectivity, or connectivity between centers, had little effect on $PM_{2.5}$ concentrations. With an increase in the population and a change in policy, cities, especially larger cities, begin to change from a single center to a double- or multicenter model.

4.2. The Relationship between $PM_{2.5}$ and Urban Form from the Sub-Regional Perspective

From the overall situation of each region, the impact of the urban area on $PM_{2.5}$ concentration varies with city size. Compared with small- and medium-sized cities, the change in $PM_{2.5}$ concentration in big cities is more easily affected by the urban area. An increased urban built-up area corresponds to greater traffic demand and energy consumption, causing comparatively worse air quality [14,47]. The impact of the urban built-up area on $PM_{2.5}$ concentration in northwestern China is more significant. The reason for this situation may be that the cities in northwestern China are in the initial stage of development, and the rapid increase in the urban area makes $PM_{2.5}$ pollution more serious.

Additionally, the population density of the cities in the northeast and northwest has a great influence on the concentration of $PM_{2.5}$, but the population density in the east has little influence. This kind of regional difference may be caused by differences in the speed of land urbanization and population urbanization. On the one hand, the reason is that the population density of the cities in the northeast and northwest is smaller than that in the east; on the other hand, in recent years, some cities in northeastern China have been facing resource depletion contraction, which makes population density more sensitive to $PM_{2.5}$ concentration. The differences between northwestern China and northeastern China are as follows: first of all, most cities in northwestern China are in a period of rapid development; the built-up area is increasing and population growth is stagnant. The speed of land urbanization and population urbanization in northwestern cities was in a serious decoupling state: the urban land expansion speed was faster than that of the population expansion [48]. Secondly, the urban contraction in northeastern China is more serious than that in northwestern China, which has resulted in northeastern China becoming a single-core area, increasing the influence of population and road density on $PM_{2.5}$. Lastly, the northwestern region is at a high altitude and has more mountains than the northeastern region; the less compact urban form helps disperse pollutants over the mountainous terrain, which results in less urban air pollution [21,49].

The influence of patch number and maximum patch index on $PM_{2.5}$ concentration was significant in northeastern China. It is worth noting that the results of large cities are the opposite of those of small cities. The results of large cities are similar to those of other developed regions. When cities tend to be polycentric, $PM_{2.5}$ pollution is reduced. However, the results obtained by small cities are just the opposite. When cities develop intensively and reduce the degree of urban fragmentation, it is easier to reduce the concentration of $PM_{2.5}$. This result is not consistent with those obtained by other researchers. For instance, Namdeo et al. [50] found that a more compact urban layout helps to reduce urban traffic and improve industrial efficiency, thereby improving air quality. Lu and Liu indicated a negative correlation between compact urban form and air pollution in most cities of China. Bechle et al. [47] demonstrated that urban compactness was not a significant predictor of air pollution in 83 global cities. Fan et al. [39] found that a more compact urban form leads to less $PM_{2.5}$ pollution in China, especially in the northern region.

In southwestern China, the influence of urban form on $PM_{2.5}$ concentration was not significant. This may be caused by topography, weather, or industrial conditions. Most of the cities in southwestern China are concentrated in intermountain basins, river valleys, and alluvial fans, and the annual rainfall is substantial [26]. These conditions result in lower levels of air pollution in these places than in other places [42]. In addition, in southwestern China, the lack of coal industry was also an important reason for this result.

4.3. Limitations and Future Directions

There are three limitations to this study. The first is that the concentration of $PM_{2.5}$ is affected by many factors, and while it is certain that urban form is one of the key factors, other factors may play a more important role in influencing $PM_{2.5}$ concentration in some areas. For example, many of the southwestern cities are located in the mountainous valley zones, and their terrains are narrow. At the same time, rainfall is more concentrated in these areas. The terrain and meteorological factors are two of the main reasons for the changes in $PM_{2.5}$ concentration. Not all $PM_{2.5}$ pollution can be attributed to urban form. The second point is that the mechanism of influence of urban form on $PM_{2.5}$ concentration is not completely distinct, and it needs to be further studied in the future. The main reason for this is that the resolution of land-use data is low, and it is difficult to accurately determine the spatial distribution of roads, commercial areas, and residential areas through land-use data with a 1 km resolution. The calculation results of urban forms, such as the road density and patch number, may be biased. The last point is that based on data availability, our study focuses on the average annual variation in $PM_{2.5}$ concentration from 2000 to 2015. Future studies should analyze the relationship between urban forms, $PM_{2.5}$ concentrations, and seasonal variations on a spatiotemporal scale.

5. Conclusions

This study, using 286 prefecture-level Chinese cities as its sample data, examined the spatial patterns and temporal trends of PM$_{2.5}$ concentration from 2000 to 2015 and further explored the influence of urban form on PM$_{2.5}$ concentration. The results show that the PM$_{2.5}$ concentration significantly increased during the period from 2000 to 2005. The cities with heavy PM$_{2.5}$ pollution were mainly concentrated in the eastern and central regions of China, especially in the large cities and their surrounding areas. The cities with large changes in PM$_{2.5}$ concentration were mainly distributed in northeastern China. Specifically, many cities in the BTH and Yangtze River Delta regions, as well as central Liaoning and Shandong provinces, had more serious PM$_{2.5}$ pollution. Moreover, cities with high PM$_{2.5}$ concentrations be located close to one another, which indicates that PM$_{2.5}$ concentration is regional. From the national point of view, urban area and road density are related to higher PM$_{2.5}$ concentrations. On the other hand, there is little correlation between urban fragmentation and PM$_{2.5}$ concentration.

In the Northeast and Northwest China, the urban form and population density have more influence on PM$_{2.5}$ concentration. The reason for this is that the speed of land urbanization and population urbanization in the northeastern and northwestern regions of China was in a seriously decoupled state, and there was a serious phenomenon of urban contraction. Therefore, moderately compact and single-center urban development is conducive to the air quality of small- and medium-sized cities in northeastern and northwestern China. For the cities in the eastern and central regions, and most large-scale cities in China, it is more important to control unplanned urban and road expansion to encourage cities to develop in a decentralized and multicenter manner.

Supplementary Materials: The following supporting information can be downloaded at: https://www.mdpi.com/article/10.3390/su14042187/s1, Table S1: PM2.5 concentration data for 286 cities in China.

Author Contributions: Conceptualization, C.S. and X.C.; methodology, C.S. and X.C.; formal analysis, C.S. and T.L.; data curation, C.S., S.Z. and T.L.; writing—original draft preparation, C.S.; funding acquisition, X.C. All authors have read and agreed to the published version of the manuscript.

Funding: This research was funded by the National Natural Science Foundation of China (No. 41861033) and the Innovative Approaches Special Project of the Ministry of Science and Technology of China (No. 2020IM020300).

Institutional Review Board Statement: Not applicable.

Informed Consent Statement: Not applicable.

Data Availability Statement: The data presented in the study is available in the Supplementary Materials.

Acknowledgments: We are grateful to the China Environmental Monitoring Center, the NASA Data Center, and the ECMWF for providing us with the foundational data.

Conflicts of Interest: The authors declare no conflict of interest.

References

1. Song, C.; Wu, L.; Xie, Y.; He, J.; Chen, X.; Wang, T.; Lin, Y.; Jin, T.; Wang, A.; Liu, Y.; et al. Air pollution in China: Status and spatiotemporal variations. *Environ. Pollut.* **2017**, *227*, 334–347. [CrossRef]
2. Cohen, A.J.; Brauer, M.; Burnett, R.; Anderson, H.R.; Frostad, J.; Estep, K.; Balakrishnan, K.; Brunekreef, B.; Dandona, L.; Dandona, R.; et al. Estimates and 25-year trends of the global burden of disease attributable to ambient air pollution: An analysis of data from the global burden of diseases study 2015. *Lancet* **2017**, *389*, 1907–1918. [CrossRef]
3. Liu, Y.; Wu, J.; Yu, D. Characterizing spatiotemporal patterns of air pollution in China: A multiscale landscape approach. *Ecol. Indic.* **2017**, *76*, 344–356. [CrossRef]
4. Cheng, Z.; Jiang, J.; Fajardo, O.; Wang, S.; Hao, J. Characteristics and health impacts of particulate matter pollution in China (2001–2011). *Atmos. Environ.* **2013**, *65*, 186–194. [CrossRef]

5. Health Effects Institute. *State of Global Air 2017*; Special Report; Health Effects Institute: Boston, MA, USA, 2017; Available online: https://www.stateofglobalair.org/resources?resource_category=archives#block-exposedformresources-all (accessed on 16 December 2021).
6. Mansfield, T.A.; Freer-Smith, P.H. Effects of urban air pollution on plant growth. *Biol. Rev.* **1981**, *56*, 343–368. [CrossRef]
7. Lovett, G.M.; Tear, T.H.; Evers, D.C.; Findlay, S.E.G.; Cosby, B.J.; Dunscomb, J.K.; Driscoll, C.T.; Weathers, K.C. Effects of air pollution on ecosystems and biological diversity in the eastern United States. *Ann. N. Y. Acad. Sci.* **2009**, *1162*, 99–135. [CrossRef] [PubMed]
8. Smith, P.; Ashmore, M.R.; Black, H.I.J.; Burgess, P.J.; Evans, C.D.; Quine, T.A.; Thomson, A.M.; Hicks, K.; Orr, H.G. Review: The role of ecosystems and their management in regulating climate, and soil, water and air quality. *J. Appl. Ecol.* **2013**, *50*, 812–829. [CrossRef]
9. Chen, S.-M.; He, L.-Y. Welfare loss of China's air pollution: How to make personal vehicle transportation policy. *China Econ. Rev.* **2014**, *13*, 106–118. [CrossRef]
10. Mukhopadhyay, A.; Pandit, V. Control of industrial air pollution through sustainable development. *Environ. Dev. Sustain.* **2014**, *16*, 35–48. [CrossRef]
11. Zhang, X.; Zhang, X.; Chen, X. Happiness in the air: How does a dirty sky affect mental health and subjective well-being? *J. Environ. Econ. Manag.* **2017**, *85*, 81–94. [CrossRef]
12. Du, G.; Shin, K.J.; Managi, S. Variability in impact of air pollution on subjective well-being. *Atmos. Environ.* **2018**, *183*, 175–208. [CrossRef]
13. Naboni, E.; Woźniak-Szpakiewicz, E.; Doray, C. Future cities: Design opportunities for the improvement of air quality. *Czas. Tech. Archit.* **2014**, *111*, 235–250. Available online: https://repozytorium.biblos.pk.edu.pl/resources/30538 (accessed on 16 December 2021).
14. Bereitschaft, B.; Debbage, K. Urban form, air pollution, and CO_2 emissions in large U.S. metropolitan areas. *Prof. Geogr.* **2013**, *65*, 612–635. [CrossRef]
15. McCarty, J.; Kaza, N. Urban form and air quality in the United States. *Landsc. Urban. Plan.* **2015**, *139*, 168–179. [CrossRef]
16. Cárdenas Rodríguez, M.; Dupont-Courtade, L.; Oueslati, W. Air pollution and urban structure linkages: Evidence from european cities. *Renew. Sustain. Energy Rev.* **2016**, *53*, 1–9. [CrossRef]
17. Stone, B. Urban sprawl and air quality in large US cities. *J. Environ. Manag.* **2008**, *86*, 688–698. [CrossRef]
18. Clark, L.P.; Millet, D.B.; Marshall, J.D. Air quality and urban form in U.S. urban areas: Evidence from regulatory monitors. *Environ. Sci. Technol.* **2011**, *45*, 7028–7035. [CrossRef]
19. Wang, Z.; Fang, C. Spatial-temporal characteristics and determinants of $PM_{2.5}$ in the Bohai Rim urban agglomeration. *Chemosphere* **2016**, *148*, 148–162. [CrossRef]
20. Li, F.; Zhou, T. Effects of urban form on air quality in China: An analysis based on the spatial autoregressive model. *Cities* **2019**, *89*, 130–140. [CrossRef]
21. She, Q.; Peng, X.; Xu, Q.; Long, L.; Wei, N.; Liu, M.; Jia, W.; Zhou, T.; Han, J.; Xiang, W. Air quality and its response to satellite-derived urban form in the Yangtze River delta, China. *Ecol. Indic.* **2017**, *75*, 297–306. [CrossRef]
22. Zhang, C.; Zhang, S. Study on urban form and air quality in metropolitan area: Identify relations and research framework. *Urban Dev. Stud.* **2014**, *21*, 47–53. [CrossRef]
23. Feng, H.; Zou, B.; Tang, Y. Scale- and region-dependence in landscape-$PM_{2.5}$ correlation: Implications for urban planning. *Remote Sens.* **2017**, *9*, 918. [CrossRef]
24. Shi, K.; Li, Y.; Chen, Y.; Li, L.; Huang, C. How does the urban form-$PM_{2.5}$ concentration relationship change seasonally in chinese cities? A comparative analysis between national and urban agglomeration scales. *J. Clean. Prod.* **2019**, *239*, 118088. [CrossRef]
25. Song, Y.; Huang, B.; He, Q.; Chen, B.; Wei, J.; Mahmood, R. Dynamic assessment of $PM_{2.5}$ exposure and health risk using remote sensing and geo-spatial big data. *Environ. Pollut.* **2019**, *253*, 288–296. [CrossRef]
26. Xu, X.; Xiao, T.; Ma, S. The features analysis on divisions of season in southwest China. *Plateau Mt. Meteorol. Res.* **2010**, *30*, 35–40.
27. Larkin, A.; van Donkelaar, A.; Geddes, J.A.; Martin, R.V.; Hystad, P. Relationships between changes in urban characteristics and air quality in east Asia from 2000 to 2010. *Environ. Sci. Technol.* **2016**, *50*, 9142–9149. [CrossRef]
28. Van Donkelaar, A.; Martin, R.V.; Brauer, M.; Boys, B.L. Use of satellite observations for long-term exposure assessment of global concentrations of fine particulate matter. *Environ. Health Perspect.* **2015**, *123*, 135–143. [CrossRef]
29. Ministry of Ecology and Environment the People's Republic of China. *Specifications and Test Procedures for Ambient Air Quality Continuous Automated Monitoring System for PM10 and PM2.5*; Ministry of Ecology and Environment the People's Republic of China: Beijing, China, 2013. Available online: https://www.mee.gov.cn/ywgz/fgbz/bz/bzwb/jcffbz/202201/t20220129_968586.shtml (accessed on 16 December 2021).
30. National Bureau of Statistics of China. *China Statistical Yearbook*; National Bureau of Statistics of China: Beijing, China. Available online: http://www.stats.gov.cn/tjsj/ndsj/ (accessed on 16 December 2021).
31. Chen, Y.; Li, X.; Zheng, Y.; Guan, Y.; Liu, X. Estimating the relationship between urban forms and energy consumption: A case study in the Pearl River delta, 2005–2008. *Landsc. Urban Plan.* **2011**, *102*, 33–42. [CrossRef]
32. Wu, J.; Xie, W.; Li, W.; Li, J. Effects of urban landscape pattern on $PM_{2.5}$ pollution—A beijing case study. *PLoS ONE* **2015**, *10*, e0142449. [CrossRef]
33. Jaeger, J.; Nazarnia, N. Social and ecological impacts of the exponential increase of urban sprawl in Montréal. *Can. Commun. Dis. Rep.* **2016**, *42*, 207–208. [CrossRef]

34. Li, J.; Huang, X. Impact of land-cover layout on particulate matter 2.5 in urban areas of China. *Int. J. Digit. Earth* **2020**, *13*, 474–486. [CrossRef]
35. Tao, Y.; Zhang, Z.; Ou, W.; Guo, J.; Pueppke, S.G. How does urban form influence $PM_{2.5}$ concentrations: Insights from 350 different-sized cities in the rapidly urbanizing Yangtze River delta region of China, 1998–2015. *Cities* **2020**, *98*, 102581. [CrossRef]
36. Dong, L.; Liang, H. Spatial analysis on China's regional air pollutants and CO_2 emissions: Emission pattern and regional disparity. *Atmos. Environ.* **2014**, *92*, 280–291. [CrossRef]
37. Zhao, L.; Sun, C.; Liu, F. Interprovincial two-stage water resource utilization efficiency under environmental constraint and spatial spillover effects in China. *J. Clean. Prod.* **2017**, *164*, 715–725. [CrossRef]
38. Lee, C. Impacts of urban form on air quality in metropolitan areas in the United States. *Comput. Environ. Urban Syst.* **2019**, *77*, 101362. [CrossRef]
39. Fan, C.; Tian, L.; Zhou, L.; Hou, D.; Song, Y.; Qiao, X.; Li, J. Examining the impacts of urban form on air pollutant emissions: Evidence from China. *J. Environ. Manag.* **2018**, *212*, 405–414. [CrossRef]
40. Fang, C.; Li, G.; Wang, S. Changing and differentiated urban landscape in China: Spatiotemporal patterns and driving forces. *Environ. Sci. Technol.* **2016**, *50*, 2217–2227. [CrossRef]
41. Danek, T.; Zaręba, M. The use of public data from low-cost sensors for the geospatial analysis of air pollution from solid fuel heating during the COVID-19 pandemic spring period in Krakow, Poland. *Sensors* **2021**, *21*, 5208. [CrossRef]
42. Zhang, Y.-L.; Cao, F. Fine particulate matter ($PM_{2.5}$) in China at a city level. *Sci. Rep.* **2015**, *5*, 14884. [CrossRef]
43. Wang, S.; Zhou, C.; Wang, Z.; Feng, K.; Hubacek, K. The characteristics and drivers of fine particulate matter ($PM_{2.5}$) distribution in China. *J. Clean. Prod.* **2017**, *142*, 1800–1809. [CrossRef]
44. Burton, R.M.; Suh, H.H.; Koutrakis, P. Spatial variation in particulate concentrations within metropolitan Philadelphia. *Environ. Sci. Technol.* **1996**, *30*, 400–407. [CrossRef]
45. Han, L.; Zhou, W.; Li, W.; Li, L. Impact of urbanization level on urban air quality: A case of fine particles ($PM_{2.5}$) in Chinese cities. *Environ. Pollut.* **2014**, *194*, 163–170. [CrossRef]
46. Liu, Y.; Wu, J.; Yu, D.; Ma, Q. The relationship between urban form and air pollution depends on seasonality and city size. *Environ. Sci. Pollut. Res.* **2018**, *25*, 15554–15567. [CrossRef] [PubMed]
47. Bechle, M.J.; Millet, D.B.; Marshall, J.D. Effects of income and urban form on urban NO_2: Global evidence from satellites. *Environ. Sci. Technol.* **2011**, *45*, 4914–4919. [CrossRef] [PubMed]
48. Yin, H.; Xu, T. The mismatch between population urbanization and land urbanization in China. *Urban Plan. Forum* **2013**, *2*, 10–15. [CrossRef]
49. Loo, B.P.Y.; Chow, A.S.Y. Spatial restructuring to facilitate shorter commuting: An example of the relocation of Hong Kong international airport. *Urban Stud.* **2011**, *48*, 1681–1694. [CrossRef]
50. Namdeo, A.; Goodman, P.; Mitchell, G.; Hargreaves, A.; Echenique, M. Land-use, transport and vehicle technology futures: An air pollution assessment of policy combinations for the Cambridge sub-region of the UK. *Cities* **2019**, *89*, 296–307. [CrossRef]

Article

What Are the Sectors Contributing to the Exceedance of European Air Quality Standards over the Iberian Peninsula? A Source Contribution Analysis

Pedro Jiménez-Guerrero [1,2]

1 Department of Physics, Regional Campus of International Excellence Campus Mare Nostrum, University of Murcia, 30100 Murcia, Spain; pedro.jimenezguerrero@um.es; Tel.: +34-868-88-8175
2 Biomedical Research Institute of Murcia (IMIB-Arrixaca), 30120 Murcia, Spain

Abstract: The Iberian Peninsula, located in southwestern Europe, is exposed to frequent exceedances of different threshold and limit values of air pollution, mainly related to particulate matter, ozone, and nitrous oxide. Source apportionment modeling represents a useful modeling tool for evaluating the contribution of different emission sources or sectors and for designing useful mitigation strategies. In this sense, this work assesses the impact of various emission sectors on air pollution levels over the Iberian Peninsula using a source contribution analysis (zero-out method). The methodology includes the use of the regional WRF + CHIMERE modeling system (coupled to EMEP emissions). In order to represent the sensitivity of the chemistry and transport of gas-phase pollutants and aerosols, several emission sectors have been zeroed-out to quantify the influence of different sources in the area, such as on-road traffic or other mobile sources, combustion in energy generation, industrial emissions or agriculture, among others. The sensitivity analysis indicates that large reductions of precursor emissions (coming mainly from energy generation, road traffic, and maritime-harbor emissions) are needed for improving air quality and attaining the thresholds set in the European Directive 2008/50/EC over the Iberian Peninsula.

Keywords: air pollution; sensitivity; aerosols; zero-out; Iberian Peninsula

Citation: Jiménez-Guerrero, P. What Are the Sectors Contributing to the Exceedance of European Air Quality Standards over the Iberian Peninsula? A Source Contribution Analysis. *Sustainability* 2022, 14, 2759. https://doi.org/10.3390/su14052759

Academic Editors: Weixin Yang, Guanghui Yuan and Yunpeng Yang

Received: 24 January 2022
Accepted: 24 February 2022
Published: 26 February 2022

Publisher's Note: MDPI stays neutral with regard to jurisdictional claims in published maps and institutional affiliations.

Copyright: © 2022 by the author. Licensee MDPI, Basel, Switzerland. This article is an open access article distributed under the terms and conditions of the Creative Commons Attribution (CC BY) license (https://creativecommons.org/licenses/by/4.0/).

1. Introduction

Atmospheric pollution has become one of the most important health and environmental problems worldwide, affecting industrialized and developing countries around the world. Its importance and implications for sustainability have been recognized by the United Nations in their Sustainable Development Goals (SDGs) [1]. Health-relevant indicators of household and ambient pollution exposure and disease burden are included in the formal system of SDG indicators. Targets of particular relevance to ambient and household air pollution include SDG target 3.9.1, which calls for a substantial reduction in the number of deaths and illnesses from air pollution [2,3], or SDG target 11.6.2, which aims to reduce the environmental impact of cities by improving air quality [4,5].

The exposure of humans to air pollution (both photochemical and particulate matter) may be the source of many health problems ([6–12], among many others). The use of chemistry transport models (CTMs) can be a useful tool for assessing these air quality-related health problems. Recently, the premature deaths and the costs of the health impacts of air pollution in Europe were calculated by using ground-level concentrations from different CTMs, indicating that the total number of premature deaths (acute and chronic) ranges from 500,000 to 800,000; their associated costs are around EUR 300 billion [11,13,14].

The Iberian Peninsula (IP), especially, presents serious problems that are related mainly to tropospheric ozone (O_3) [15], sulphur dioxide (SO_2), nitrogen dioxide (NO_2), and particles of different diameters: particulate matter with a diameter of less than 10 (PM_{10}) and more than 2.5 μm ($PM_{2.5}$) [16]. In this sense, a number of studies have covered the entire IP using modeling techniques [17–22]. The results of these previous works indicate that achieving the objectives proposed by the EU directives are more difficult in the IP when compared northern countries, partly due to their particular emission distribution [23,24], and partly due to different meteorological situations, namely: (1) a lower precipitation rate (and, hence, a higher resuspension rate due to soil dryness); (2) the increased formation of secondary aerosols associated with the higher temperatures; (3) an enhanced frequency of African dust outbreaks; and (4) the higher occurrence of the recirculation of air masses that prevent air renovation [20,25].

Moreover, air pollution problems will become even more severe under future climates [26–31]. Therefore, reliable estimations of air pollution for present-day conditions and an enhanced understanding of the chemico-physical processes occurring in the atmosphere become essential, not only for informing and alerting the population, but also to understand the causes of those episodes and to implement effective abatement policies.

For that purpose, CTMs are essential for defining, evaluating, and implementing emission abatement plans through the use of sensitivity analysis strategies [32,33]. These strategies have, as a first step, the accurate identification of pollution sources and their individual contributions to the concentrations of atmospheric pollutants. To this end, a wide range of modeling methodologies has been proposed and applied for the apportionment of atmospheric pollutants [34–37]. Particularly, source apportionment relies on the determination of the contribution of different sources to pollutant concentrations by establishing the mass continuity relationships between emissions and concentrations at receptor locations. Sensitivity analyses measure how pollutant concentrations at receptors respond to perturbations at sources. Most of the sensitivity questions are left to modelers since the experimental approach is difficult and expensive.

The traditional approach to sensitivity consists in performing "twin simulations", with one parameter perturbed [34]. In the case of the most straightforward method to assess sensitivity (brute-force method, BFM), the perturbed parameter is emissions. In the BFM, a model simulation is conducted and repeated with modified emissions, comparing the outputs of the simulations [38,39]. This method is limited because the computational cost depends, in a linear way, on the number of perturbations to examine and the strong influence of the numerical errors when the changes in the concentrations are small. Related to the BFM, the zero-out method [40,41] sets a specific emissioin sector to zero and measures the change produced in the output concentrations. In this sense, it can be considered as an extreme case of the BFM.

Since the management of air pollutant emissions is one of the predominant factors for abating urban air quality, this work assesses the source contribution of different emitting sectors to the air pollution levels in the IP, taking a particular look at the number of exceedances of air quality limits and thresholds related to health issues. For that, the WRF (meteorology) + CHIMERE (chemistry transport) modeling system has been used for a summer and a winter period over the IP in order to assess air quality-related problems in the area.

2. Materials and Methods

2.1. Modeling System

The modeling system applied consists in the Weather Research and Forecasting (WRF, meteorology) + CHIMERE (chemistry transport model) + EMEP (emissions) methods. The simulations cover the entire IP (excluding a blending area of five grid points), have a resolution of 9 km, and have been run and evaluated on an hourly basis during a period covering a summer and a winter scenario (months of June–July–August 2011, JJA, and December 2011–January–February 2012, DJF). Precisely, the simulation period ranges from 24 May 2011 to 1 September 2011, and from 23 November 2011 to 1 March 2012, with the first week being the spin-up period. The election of the 9-km resolution was conditioned by a compromise between the use of high resolutions and the computational time needed for the ensemble of simulations to be conducted in this analysis.

The regional modeling system consists of the Advanced Research Weather Research and Forecasting (WRF-ARW) Model v3.9.1 [42,43], which provides the meteorology to the CTM. WRF is a fully compressible, Eulerian, non-hydrostatic model that solves the equations that govern the atmospheric motions. A total of 33 vertical layers on sigma coordinates cover the region from the ground level up to 10 hPa. The boundary conditions used for driving the WRF simulations are obtained from the ERA-Interim reanalysis [44] every six hours. WRF fields have been coupled off-line on an hourly basis to CHIMERE CTM [45]. With respect to the CHIMERE configuration, the MELCHIOR2 gas-phase mechanism has been used [46].

Regarding the inclusion of particles within the CTM, CHIMERE includes aerosol and heterogeneous chemistry. Different chemical aerosol components have been included in the model configuration, namely, ammonium, nitrate, sulphate, and organic and elemental carbon with three subcomponents: (1) primary aerosol, (2) secondary anthropogenic, and (3) secondary biogenic subcomponents. Marine aerosols (sea salt) have also been included in the simulation. The aerosol microphysical description is based on a sectional aerosol approach that includes 6 bins using a geometrical progression and ranging from 10 nm to 40 μm. Table 1 summarizes the physico-chemical options for the regional modeling system.

Table 1. Parameterizations of the meteorological and chemistry transport model used in the simulations for the IP.

WRF (Meteorological Model) [42,43]	CHIMERE (Chemistry Transport Model) [45]
Microphysics: WSM6 [47]	Chemical Mechanism: MELCHIOR2 [46]
PBL: Yonsei University [48]	Aerosol chemistry: Inorganic (thermodynamic equilibrium with ISORROPIA module) [49]
Radiation: CAM [50]	Organic aerosol chemistry: [51]
Soil: Noah LSM [52]	Natural aerosols: dust, re-suspended, and inert sea-salt [45]
Cumulus: Kain–Fritsch [53]	Emissions: anthropogenic emissions EMEP [54] + biogenic emissions MEGAN (Model of Emissions of Gases and Aerosols from Nature) [55]
Boundary conditions: ERA-Interim [44]	Boundary conditions: LMDz-INCA+GOCART [56]

Here, the climatological boundary conditions for the CTM are based on the LMDz-INCA global chemistry/climate model [57]. Other considerations to bear in mind, with respect to the boundary conditions, are that (1) the changes in stratospheric ozone are very limited and, hence, are neglected in the simulations, and (2) it has been assumed that long-range transport over the IP is limited and overwhelmed by local processes [58]. This assumption is hampered by the persistent outbreaks of Saharan dust over the IP, which may exert an important influence on the regional PM_{10} levels over Spain and Portugal [59,60]. However, this contribution focuses on a sensitivity analysis of anthropogenic emissions, and hence, the impact of desert sources and their influence on the air quality of the IP is beyond the scope of this work.

Anthropogenic emissions are obtained from the EMEP database [54] and cover the entire period of simulations. Natural emissions have been estimated with the MEGAN model [55] and include species such as monoterpene, isoprene, and other biogenic volatile organic compounds (BVOCs). The meteorological inputs needed for the estimation of emissions are obtained from the WRF simulations previously described.

2.2. Sensitivity Analysis: The Zero-Out Method

The sensitivity analysis methods perturb inputs to the modeling system (e.g., modify the emissions of sulphur oxides) and quantify the response of the model output (e.g., change in sulphate concentration). As commented on before, there are several approaches for a sensitivity analysis based on the BFM in order to study the contribution from different sources; a zero-out method has been applied in this study because of its simplicity. Here, the methodology includes a base model, run with all emission sources (BC), and ten emission scenarios in which emissions from anthropogenic sources (classified according to the SNAP) are excluded, analogously to previous works [37,61,62].

The zero-out method has been extensively used for source attribution because it seems intuitive and obvious that the removal of an emission source should quantify the corresponding impact of that emission source [40,41,63]. Despite that this methodology is valid and widely used for sensitivity analysis (as in our case), it should be carefully considered for areas with a strong secondary production, because the sum of zero-out impacts over all sources may not be exactly equal to the total concentration when considering non-linear systems as those represented by atmospheric processes [34]. In this sense, Clappier et al. [36] warn that, when the non-linearity of the relationship between concentrations and emissions is noticeable, source apportionment methods may not be appropriate to assess the impact of mitigation or abatement strategies. When non-linearity is limited or negligible, source apportionment methods may be acceptable, bearing in mind the complexity of the models involved in the representation of air pollution.

Since our objective is to conduct a source apportionment analysis for the IP, the zero-out method has been applied to all the SNAP activities, including anthropogenic sources (Table 2). The sensitivity to air pollution levels of these sources is covered and identified in the simulations (harbors and ships, industries, road transport, central heating, agriculture, etc.).

Table 2. Tags for the different simulations included in this contribution. The scenarios are run while zeroing-out the emissions specified by the SNAP sector.

SNAP	Emissions Zeroed-Out
SNAP1	Combustion in energy and transformation industries
SNAP2	Non-industrial combustion plants, including private wood combustion
SNAP3	Combustion in manufacturing industry
SNAP4	Production processes
SNAP5	Extraction and distribution of fossil fuels and geothermal energy
SNAP6	Solvents and other product use
SNAP7	Road transport
SNAP8	Other mobile sources and machinery (excl. international ship traffic)
SNAP9	Waste treatment and disposal
SNAP10	Agriculture
Base Case (BC)	No emissions zeroed-out

3. Results

3.1. Evaluation of the Modeling Results

Despite that the goal of this contribution is not to provide a comprehensive evaluation of the air quality concentrations simulated by WRF + CHIMERE, the results from the monitoring network EMEP have been used to characterize the skill of the model for reproducing the concentrations of air pollutants (EMEP data available online at: http://www.emep.int (accessed on 8 May 2012); see [64] for further details). The ten stations with simultaneous data of tropospheric O_3, NO_2, and PM_{10} in the IP (SO_2 and $PM_{2.5}$ have been excluded because of the scarcity of data for the target period) have been used for the model evaluation. Their location is shown in Figure 1.

Figure 1. EMEP stations included for the model validation.

The available EMEP measurements were filtered before comparing the model results with EMEP data in order to remove uncertain data (for instance, those data before a calibration of equipment or after an interruption was eliminated). In addition, after the EMEP data is filtered, the criteria of temporal coverage >85% were selected for measurement sites. Since EMEP stations are located far from large emission sources (more than 10 km), the data are assumed to fit the resolution of the model used for regional background concentrations ([64] and references therein).

A number of common metrics were used to examine the model skills, differencing between gas-phase and particulate matter. For gases, two scores have been selected: mean normalized gross error (MNGE)—which indicates the performance of the simulations to represent the magnitude of the observation—and the mean normalized bias error (MNBE)—another common parameter that reveals the departure between observations and modeling data. These provide a useful quantification of the overall under- or overestimations of the model.

As for the particulate matter evaluation, a number of authors (e.g., [16,65–67], among many others) suggested using the mean fractional bias (MFB) and the mean fractional error (MFE) instead of MNBE or MNGE (Table 3). Boylan and Russell [65] propose that a model performance goal is met when both the MFE and MBE are less than or equal to 50% and ±30%, respectively, and a model performance criterion is met when the MFE ≤75% and MFB is less than or equal to ±60%.

Table 3. Statistical figures used in the evaluation of the WRF + CHIMERE modeling system. N: number of observations available. C_{mod}: model concentration. C_{obs}: observation concentration.

Value	Formula	Range
Model Mean (MM)	$\frac{1}{N}\sum C_{mod}$	$0, +\infty$
Observations Mean (OM)	$\frac{1}{N}\sum C_{obs}$	$0, +\infty$
Mean Normalized Bias Error (MNBE)	$\frac{1}{N}\sum \frac{(C_{mod}-C_{obs})}{C_{obs}}$	$-\infty, +\infty$
Mean Normalized Gross Error (MNGE)	$\frac{1}{N}\sum \frac{|C_{mod}-C_{obs}|}{C_{obs}}$	$0, +\infty$
Mean Fractional Bias (MFB)	$\frac{1}{N}\sum \left(\frac{(C_{mod}-C_{obs})}{\left(\frac{C_{mod}+C_{obs}}{2}\right)} \right)$	$-200, +200$
Mean Fractional Error (MFE)	$\frac{1}{N}\sum \left(\frac{|C_{mod}-C_{obs}|}{\left(\frac{C_{mod}+C_{obs}}{2}\right)} \right)$	$0, +200$

Therefore, MNBE and MNGE have been used for gaseous pollutants, while for particulate matter, the MFB and MFE have been utilized. A general pattern of the air pollution levels provided by WRF + CHIMERE simulations can be found in Figure 2. Maximum O_3 concentrations are modeled for summertime in the easternmost part of the IP, with ground levels that exceed 120 µg m^{-3} as the daily mean in Catalonia (northeastern IP). For NO_2, monthly means can be as high as 50 µg m^{-3} in the largest cities of the peninsula (e.g., Madrid, Lisbon, Porto) and in an industrial area such as Algeciras Bay (southernmost part of the IP), where industrial emissions are increased by port and maritime activity. The Algeciras port (the second most important port of Spain), located at the head of the bay, has a strategic importance in terms of the maritime traffic of fuel and general supplies [68]. Hence, the presence of this port makes the area of the Algeciras Bay a high risk environment for pollution derived from its commercial activities. For SO_2, besides Algeciras, levels are over 20 µg m^{-3} downwind of several power plants (As Pontes, in northern Spain; and Andorra (Teruel), in the eastern IP) that burn coal for the generation of electricity [69,70]. Last, particulate matter does not exhibit a clear spatial pattern in the IP. The spatial patterns depend both on the industrialization of the regions, especially regarding inorganic particulate matter, and the Saharan dust outbreaks [20]. In this sense, $PM_{2.5}$ and PM_{10} seasonal patterns showed maximum concentrations during summertime, as is also indicated by the scientific literature.

Regarding model validation, overall, negative fractional biases are calculated for PM_{10} and NO_2, while positive deviations for O_3 are obtained when comparing the base-case simulation to EMEP stations (Table 4).

Table 4. Model evaluation against EMEP stations. (Top) Summer (JJA) and (bottom) winter (DJF).

Summer Code	JJA 2011 Station Name Performance Criteria	PM_{10} MFB (%) $\leq \pm 60\%$ [1]	MFE (%) $\leq +75\%$ [1]	NO_2 MNBE (%)	MNGE (%) $\leq +50\%$ [2]	O_3 MNBE (%)	MNGE (%) $\leq +50\%$ [2]
ES07	Víznar	−38.8	68.8	−28.9	41.9	23.1	24.9
ES08	Niembro	−9.0	42.6	−19.0	41.6	22.1	22.1
ES09	Campisábalos	−54.0	54.1	−35.8	49.0	5.0	25.7
ES10	Cabo de Creus	−41.9	43.7	−15.2	33.6	1.5	26.7
ES11	Barcarrota	−58.9	68.9	−45.5	46.1	22.3	26.4
ES12	Zarra	−52.5	53.0	−46.8	49.3	20.3	24.2
ES13	Peñausende	−55.1	57.0	−28.5	59.4	11.2	12.5
ES14	Els Torms	−48.6	49.2	−34.7	44.9	20.3	21.0
ES15	Risco Llano	−52.5	62.5	−47.3	47.3	24.9	25.6
ES16	O Saviñao	6.8	41.8	18.3	40.9	32.3	33.9
Winter Code	DJF 2011 Station Name Performance Criteria	PM_{10} MFB (%) $\leq \pm 60\%$ [1]	MFE (%) $\leq +75\%$ [1]	NO_2 MNBE (%)	MNGE (%) $\leq +50\%$ [2]	O_3 MNBE (%)	MNGE (%) $\leq +50\%$ [2]
ES07	Víznar	−55.7	55.8	−41.8	48.3	23.0	25.9
ES08	Niembro	−17.1	21.7	−17.1	21.7	8.3	8.3
ES09	Campisábalos	−28.7	57.2	−36.1	48.0	2.0	15.0
ES10	Cabo de Creus	−34.8	35.2	−17.8	27.7	2.2	25.5
ES11	Barcarrota	−21.6	34.9	−13.0	31.9	23.6	24.9
ES12	Zarra	−20.7	34.2	−20.7	34.2	29.8	29.8
ES13	Peñausende	−8.3	35.4	0.3	38.5	19.8	20.2
ES14	Els Torms	−34.3	45.8	−31.3	44.2	30.2	32.4
ES15	Risco Llano	−37.7	58.0	−39.6	48.9	26.0	26.6
ES16	O Saviñao	−11.3	28.2	−7.8	26.3	17.1	17.3

MFB: Mean Fractional Bias; MFE: Mean Fractional Error; MNGE: Mean Normalized Gross Error; MNBE: Mean Normalized Bias Error. [1] Boylan and Russell [65]; [2] EU Directive 2008/50/EC Uncertainty.

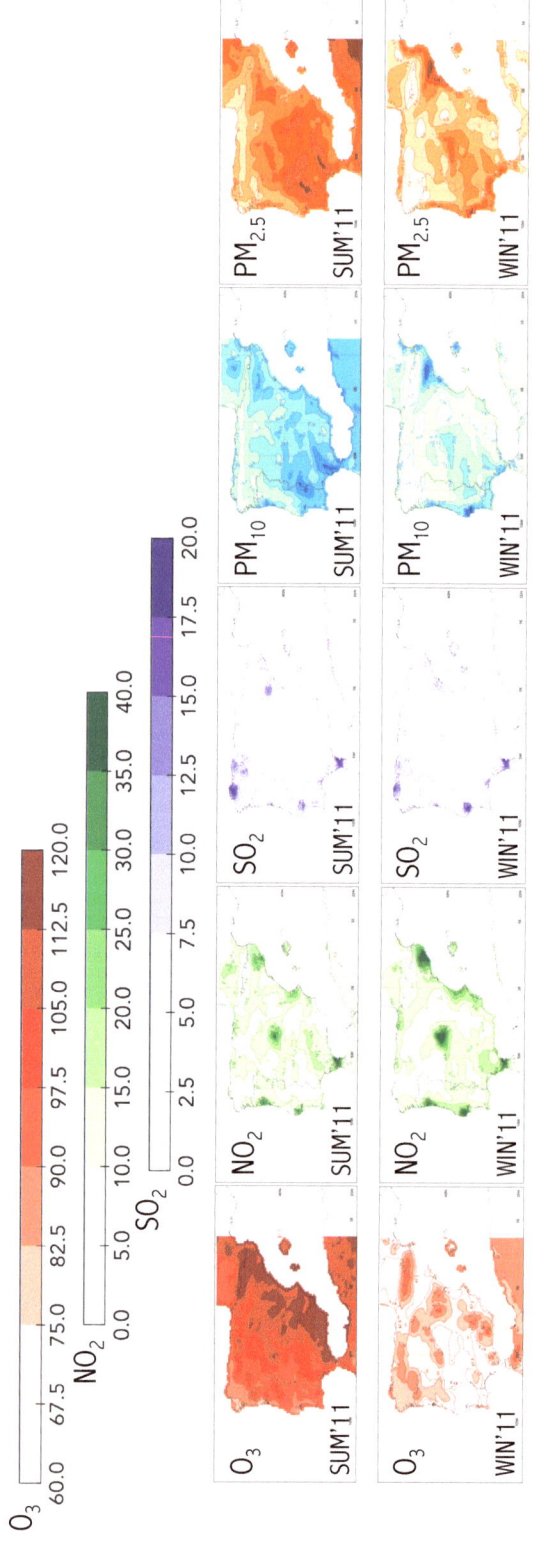

Figure 2. Summer (top) and winter (bottom) 2011 average concentration of tropospheric ozone (red), nitrogen dioxide (green), sulphur dioxide (purple), PM$_{10}$ (blue), and PM$_{2.5}$ (orange). All units in µg m^{-3}.

With respect to gaseous pollutants, the WRF + CHIMERE model presents a MNGE under 50% for NO_2, which is the value set by the EU Directive 2008/50/EC uncertainty criteria. However, this pollutant is underestimated in both seasons and in all stations (except for in summer in ES16-O Saviñao and winter in ES13-Peñausende), possibly due to uncertainties in emission inventories [71] and the relatively coarse horizontal resolution used, which represents only partially the spatial gradient of the emissions [72]. Negative biases vary between −8% in wintertime in ES16-O Saviñao (northwestern Spain) and −47% in ES12-Zarra (at the Levantine Spanish coast). Tropospheric O_3 is generally overestimated (bias under +20% in summer and under +30% during wintertime). This is related to the NO_2 underestimation, limiting the titration of tropospheric O_3 by NO_2. Moreover, the CHIMERE lateral boundary conditions for O_3 are overestimated [57,72], especially during wintertime, and therefore, the positive biases during the cold season (ranging from 2% at ES09-Campisábalos to 30% at ES14-Els Torms, northeastern Spain) are attributable to the overestimation of the background concentrations at the boundaries of the domain.

For particulate matter (PM_{10}), the magnitude of the MFB and MFE are similar in both seasons, meeting the performance criteria established by Boylan and Russell [65] for all stations and during all seasons. There is a pervasive tendency to underestimate PM_{10} levels (negative MFB in all stations and both seasons, except for station ES16-O Saviñao, northwestern Spain, in summer). This summer MFB ranges from −9% in ES08-Niembro station (northern Spain) to −59% in ES11-Barcarrota (southwestern Spain). In wintertime, the maximum MFB is −56% in ES07-Víznar (southern Spain), while the minimum MFB is estimated in ES13-Peñausende (western Spain, near the Portuguese border) as −8%. More interesting is the fact that high MFEs are found in ES07-Víznar station for both seasons (68% in summer and 56% in winter). The MFB is strongly negative and almost coincident with the MFE (e.g., −56% for the MFB error in wintertime and 56% for the MFE during this season). This could be caused by the high contribution of Saharan dust at this location [25,73], which is pervasively underestimated by CTMs in southern Mediterranean stations, especially regarding the peak levels [74–76].

3.2. Source Contribution

Figures 3 and 4 represent the results of the source contribution experiment for summertime and wintertime, respectively. The information shown in those Figures is quantified in Table 5, which indicates the relative reductions in the areas with the worst air quality in the entire IP (that is, reductions in those locations of the target domain where the daily mean and the daily mean of max. 1-hr ground-level air quality concentrations are the highest). The results are shown with respect to the base-case scenario (BC), and focus only on anthropogenic sectors (that is, excluding, for instance, the contribution of background concentrations or external transport, which cannot be controlled in abatement strategies). Overall, Table 5 indicates that the maximum reductions in air pollution levels are achieved when zeroing-out three SNAP sectors, as expected from the scientific literature: combustion in energy and transformation industries (SNAP1), road transport (SNAP7), and other mobile sources (SNAP8). The most important added value of this contribution, nonetheless, is the quantification of the respective contributions of these aforementioned sectors. For the sake of brevity, our analysis below focuses only on the assessment of the contribution from these sectors (despite that agriculture, SNAP10, may play also an important role for SO_2 and particulate matter).

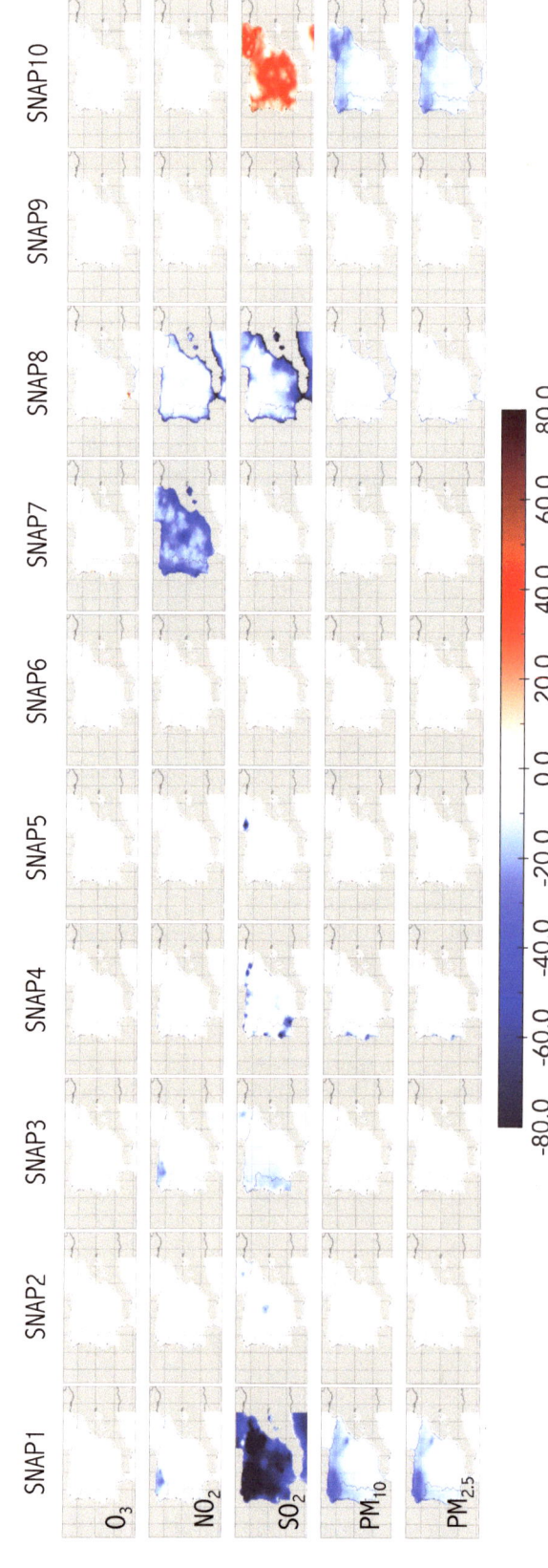

Figure 3. Relative contribution (%) of each anthropogenic SNAP sector to the daily mean levels of pollutants over the IP during summertime (JJA) 2011.

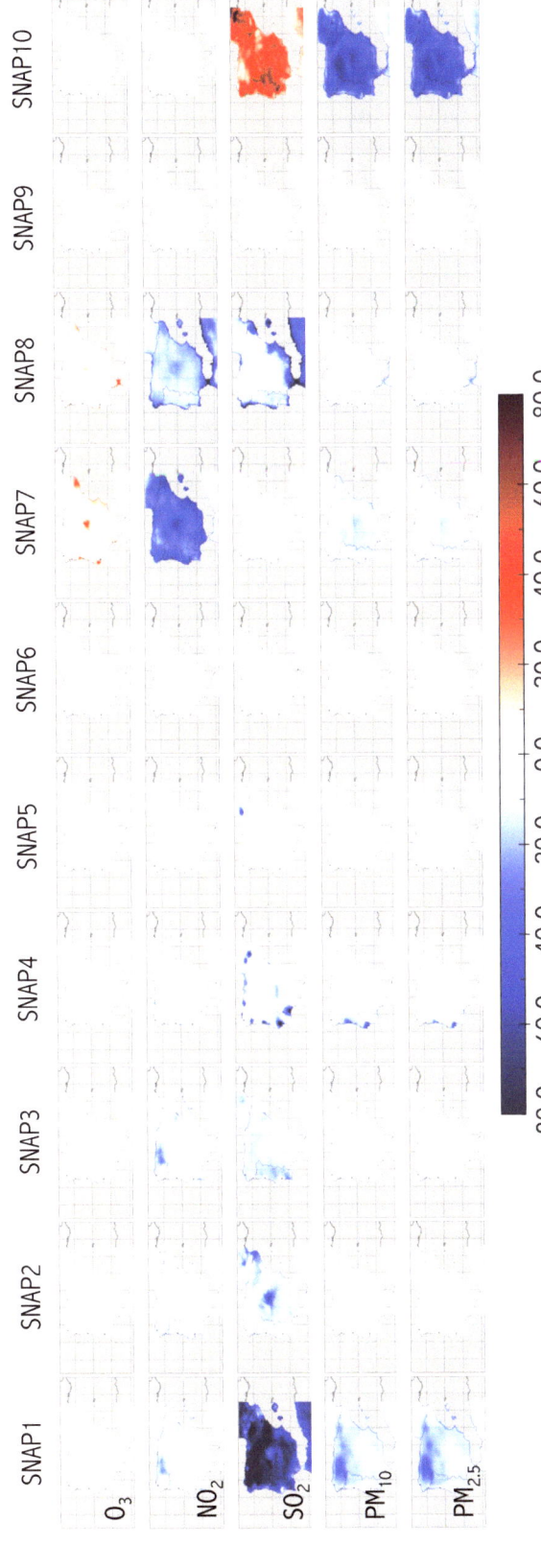

Figure 4. Id. Figure 3 but for wintertime (DJF).

Table 5. Variation in the mean and maximum levels of atmospheric pollutants over the entire IP when zeroing-out the different SNAP sectors (base case minus zeroed-out SNAP sector simulation; hence, a positive value indicates an improvement in air quality).

Summer	Summer (JJA)				
	Concentration	Base Case	w/o SNAP	Reduction	
Pollutant	Mean ($\mu g\ m^{-3}$)	Max ($\mu g\ m^{-3}$)	Zero-out sector	Mean	Max
Tropospheric ozone, O_3	132.5	164.6	SNAP7	2.3%	5.7%
			SNAP8	5.0%	1.9%
Nitrogen dioxide, NO_2	66.6	124.2	SNAP8	47.4%	37.1%
Sulphur dioxide, SO_2	33.0	70.7	SNAP1	2.0%	2.4%
			SNAP8	40.9%	40.3%
Particulate matter $\phi < 10\ \mu m$, PM_{10}	38.7	62.2	SNAP1	6.2%	4.3%
			SNAP8	7.0%	2.6%
			SNAP10	5.7%	2.6%
Particulate matter $\phi < 2.5\ \mu m$, $PM_{2.5}$	19.7	29.3	SNAP1	0.0%	4.8%
			SNAP8	0.0%	2.4%
			SNAP10	5.1%	4.8%
Winter	Winter (DJF)				
	Concentration	Base Case	w/o SNAP	Reduction	
Pollutant	Mean ($\mu g\ m^{-3}$)	Max ($\mu g\ m^{-3}$)	Zero-out sector	Mean	Max
Tropospheric ozone, O_3	95.8	103.7	SNAP7	−1.2%	−2.3%
Nitrogen dioxide, NO_2	60.0	95.4	SNAP7	32.9%	17.7%
			SNAP8	12.3%	9.7%
Sulphur dioxide, SO_2	33.0	70.7	SNAP1	4.5%	3.6%
			SNAP8	2.5%	22.2%
Particulate matter $\phi < 10\ \mu m$, PM_{10}	54.3	93.5	SNAP4	6.6%	17.5%
			SNAP7	3.9%	3.3%
			SNAP10	14.0%	13.8%
Particulate matter $\phi < 2.5\ \mu m$, $PM_{2.5}$	21.0	34.4	SNAP1	4.7%	3.7%
			SNAP7	5.3%	4.0%
			SNAP10	16.1%	14.3%

For tropospheric O_3, on-road traffic (SNAP7) is the most important contributor in summertime. The highest daily mean levels of tropospheric O_3 during summer (133 $\mu g\ m^{-3}$) reduce by 2%, while 1-h maximum concentrations (165 $\mu g\ m^{-3}$) decrease by 6%. In addition, zeroing-out other mobile sources (SNAP8) reduces the highest daily mean and 1-h maximum O_3 summertime levels by 5% and 2%, respectively. On the contrary, zeroing-out on-road traffic (SNAP7) during winter slightly contributes to an increase in tropospheric O_3 concentrations (1% and 2% in wintertime, mean and maximum concentration, 96 and 104 $\mu g\ m^{-3}$, respectively), but this increase does not involve the exceedance of the objective value, as will be shown later in Section 3.4.

The response of tropospheric O_3 to changes in their precursors (nitrogen oxides, NOx, and volatile organic compounds (VOCs)) has been widely covered in the scientific literature, and particularly over the IP [77,78]. Overall, under certain conditions, O_3 concentrations are reduced when NOx emissions decrease. This chemical regime is denoted as NOx-sensitive conditions. Conversely, under other conditions, tropospheric O_3 reduces its levels when VOC emissions (particularly, non-methane volatile organic compounds, NMVOCs) are reduced, and might even increase its concentration when NOx emissions are mitigated. This regime is known as VOC-sensitive conditions. These O_3 sensitivity regimes can help with explaining the variations in the levels of this pollutant over the Iberian Peninsula. Namely, the increase in winter O_3 mean levels in the Algeciras Bay when zeroing-out the

SNAP8 emissions and the shipping route of the Strait of Gibraltar is a direct consequence of the high NO_2 concentrations over this target area, associated with the important NOx emissions of the SNAP8 sector. When removing shipping emissions, mostly NOx emissions are removed, and hence, the increase of tropospheric O_3 reveals the strong VOC-limited chemical regime for O_3 formation in that area. At low $NMVOC/NO_x$ ratios, the results are sensitive to the concentrations of volatile compounds [77,79,80], and hence, an accurate amount of NMVOC ship emissions is essential for studying and understanding their possible impact on the O_3 levels, especially in such polluted areas as the Mediterranean Sea.

The most important pollutant coming from on-road traffic (SNAP7) is NO_2, and this sector is the dominant source in the largest populated areas of the IP. For NO_2, reductions in the highest daily mean levels in the target domain are around 10 µg m^{-3} in wintertime (up to 30 µg m^{-3} as daily mean levels in summertime), especially in the Barcelona and Madrid Greater Areas, and the axis of highways covering the Levantine and Western areas of the IP (Barcelona–Murcia and Porto–Lisbon, in that order), representing almost 50% of the NO_2 levels for this pollutant in summertime (Figure 3) and over 60% in wintertime at those sites and roads (Figure 4).

Other mobile sources (SNAP8) also largely contribute to NO_2 and SO_2 over the peninsula (playing also a role regrding the PM_{10} levels). In this sense, SNAP8 is responsible for 47% and 37% of the daily mean (67 µg m^{-3}) and maximum (124 µg m^{-3}) levels of NO_2 in the target domain in summer (12% and 10% in winter; the concentrations are 60 and 95 µg m^{-3} for mean and maxima, in that order). For wintertime, on-road traffic contributes to highest mean and maximum NO_2 concentrations by 33% and 18%, respectively. Last, as shown in Figure 3, combustion in energy and transformation industries (SNAP1) can add up to 4 µg m^{-3} in the area close to power plants, representing up to 10% of NO_2 levels in those areas. However, Table 5 indicates that the contribution of this SNAP to maximum values is not significant when considering the entire IP.

For SO_2, combustion in energy and transformation industries (SNAP1) represents an important source of the contribution to the levels of this pollutant. The simulations shown in Figure 3 for summertime and Figure 4 for wintertime feature strong reductions in SO_2 ground-level concentrations over land when zeroing-out SNAP1 (mean reduction, 2.5 µg m^{-3}, reaching 7 µg m^{-3} in large emitting areas associated with coal combustion). These results are in agreement with Valverde et al. [70], who indicate that the contribution to SO_2 from power plants in the IP ranges from 2 to 25 µg m^{-3}.

This energy sector contribution can be as much as 60% over the IP, except in the Mediterranean coastal areas, where the reduction is around 30-40%. In summertime, the contribution of energy facilities can add up to 2% to the mean and maximum levels (39 and 141 µg m^{-3}, in that order) of SO_2 simulated by the model. It is, however, SNAP8 (other mobile sources) which contributes most to summer SO_2 highest mean and maximum levels (41% and 40%, respectively). The winter contribution is much lower, with SNAP8 representing only 3% and 22% of the highest winter SO_2 mean and maxima (33 and 71 µg m^{-3}, in that order). Analogous contributions of SNAP1 can be found for winter in the target domain (5 and 4%). The contribution of harbor emissions to sulphur dioxide levels may reach 50% in the Iberian Levantine coast, both for summertime and wintertime (Figures 3 and 4), reaching up to 2 µg m^{-3} in the western Mediterranean areas, and around 5 µg m^{-3} in the Algeciras harbor and Gibraltar (southern IP) during summertime, highlighting the importance of this sector.

With respect to PM_{10}, Table 5 indicates that, albeit for summertime the sector with the largest contribution to highest daily mean and maximum levels (39 and 62 µg m^{-3}) is combustion in energy and transformation industries (SNAP1) (6.2% and 4.3%), production processes (SNAP4) is the source that contributes most during wintertime to the PM_{10} highest mean and maxima (54 and 93 µg m^{-3}), representing 7% and 18% of those levels. The second largest contributor to PM_{10} is SNAP8 (other mobile sources) in summer (7% and 3% to highest mean and maxima) and SNAP7 (road traffic) in winter (4% and 3%). It is noticeable that removing agriculture emissions (SNAP10) contributes to a decrease in

PM levels and a simultaneous increase in SO_2 concentrations both for summer (Figure 3) and winter (Figure 4), since zeroing-out the most important contributor to NH_3 emission hampers the formation of ammonium sulphate, and hence, more SO_2 is available in the gas-phase [20,27,81]. Analogous results can be found for $PM_{2.5}$, but with an enhanced contribution of agriculture (SNAP10) to the $PM_{2.5}$ daily mean and maxima, which can reach 16% and 14%, respectively.

3.3. Source Contribution at Critical Selected Sites

Figure 5 shows the Air Quality Index (AQI) in the IP (estimated from EPA Air Quality Index [82]) in order to assess the most critical areas in the target domain regarding air pollution. In this index, the concentrations that correspond to an AQI value of 100 are those established as the standards of the European Union, compiled in Directive 2008/50/EC. The election of the AQI in this contribution is not critical, since only the areas with the poorest air quality are searched to calculate the source contribution at those particular locations.

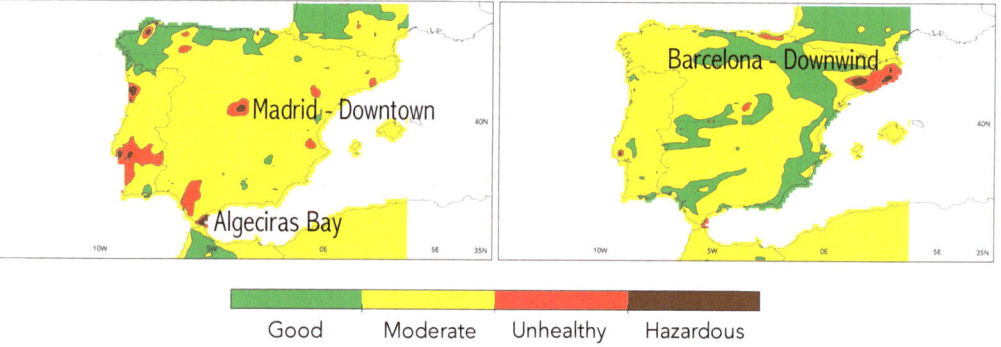

Figure 5. Total air quality indexes (AQI_{total}) for summer (JJA) (**left**) and winter (DJF) (**right**), indicating the most polluted areas of the IP (AQI = hazardous).

The AQI has been estimated individually for all pollutants with regulatory values included in this contribution (O_3, NO_2, SO_2, PM_{10}, and $PM_{2.5}$) and the AQI_{total} (shown in Figure 5) has been estimated as the highest value among all individual indexes. During the summer and winter periods, air quality was hazardous in the two largest Spanish cities (Madrid and Barcelona) and the industrial-harbor area of Algeciras Bay, located in southern Spain (Figure 5). Therefore, this section is devoted to the analysis of the source apportionment at these locations in order to shed some light on the causes of the strategy to abate those pollutants. For that, the point with the worst air quality in a domain of 100 km², centred over Madrid, Barcelona, and Algeciras, respectively, has been selected for further analysis.

For gas-phase pollutants, Figure 6 (left) indicates that most of summertime tropospheric O_3 comes from the "Other" sector at all the three sites. This "Other" contribution is not estimated by zeroing-out any emission sector, but estimated as the difference between the BC and the addition of all anthropogenic sources. Therefore, it includes the contribution of different processes (e.g., long-range transport, background levels, stratosphere–troposphere exchange, etc.).

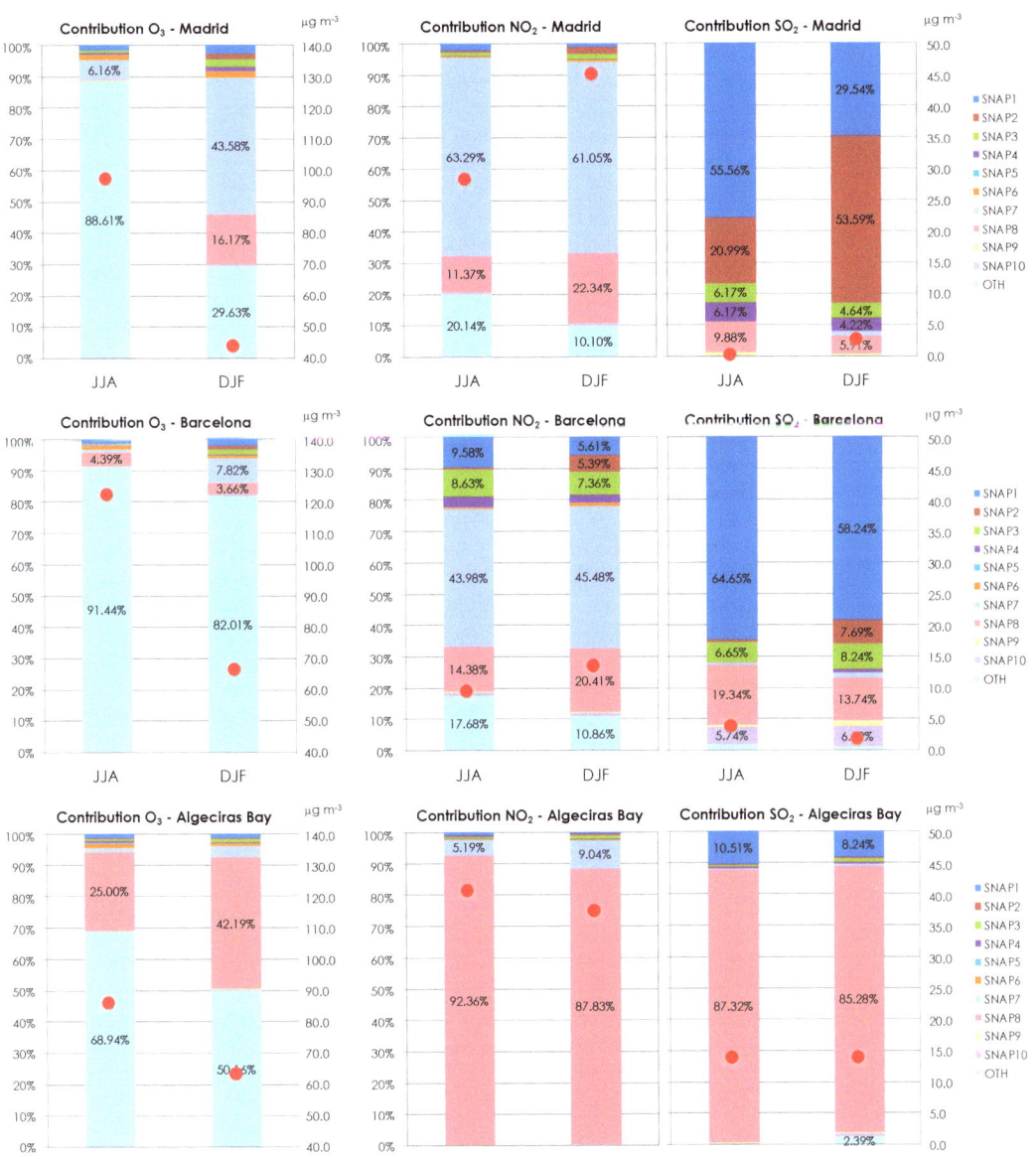

Figure 6. (Left axis) Relative contribution (%) of each anthropogenic SNAP sector to the daily mean levels of O_3 (left), NO_2 (center), and SO_2 (right) over the most polluted areas of the IP (Madrid, top; Barcelona, center; Algeciras Bay, bottom). (Right axis) Red dot stands for the mean concentrations of O_3 (left), NO_2 (center), and SO_2 (right) in µg m^{-3}.

During summer (winter), this contribution can be as large as 88% (30%) in Madrid, 91% (82%) in Barcelona. and 69% (50%) in Algeciras Bay. These numbers are in agreement with previous works. For instance, the background values contribute with more than 50% to the O_3 concentration measured in the westernmost region of the IP [83]. Moreover, the importance of intercontinental ozone transport in the ground levels of ozone over Europe has been highlighted [84], and can be as high as 10–16 ppb (20–32 µg m^{-3}). In Barcelona and the Algeciras Bay, the anthropogenic sector contributing most to tropospheric O_3 levels is SNAP8 (other mobile sources), especially related to shipping emissions in the area. SNAP8 adds up 4% (25%) and 4% (42%) of summer and wintertime O_3, respectively, in Barcelona (Algeciras). These results are in agreement with those of the literature [85,86]. These works find out that shipping emissions increase ground levels of summer tropospheric O_3 by 5 to 10% in the Mediterranean sea. This may be caused by the large NO_2 emissions of ships, which can enhance the production of ozone [87]. Last, SNAP7 (road traffic) has a limited contribution to summertime O_3 levels in Madrid and Barcelona, around 8%, which is in a strong agreement with previous works [88].

With respect to NO_2 (Figure 6, center), on-road traffic (SNAP7) is the sector with the highest contribution to the surface levels of NO_2 in Madrid and Barcelona (over 60% in Madrid and over 44% in Barcelona for both seasons), followed by SNAP8 (other mobile sources). While for Barcelona, it is the shipping and maritime activity that contributes most to SNAP8 (being responsible for 14% and 20% of summer and winter NO_2 levels in the city), in Madrid, the contribution of SNAP8 (11% in summer and 22% in winter) comes mainly from the activity of the Madrid airport. In Algeciras, around 90% of NO_2 levels can be attributed to the shipping sector, both in summertime and wintertime. The contribution of SNAP8 is very similar in Algeciras Bay for SO_2 levels (the source apportionment indicates that over 85% of SO_2 mean levels in Algeciras come from SNAP8) (Figure 6, right). However, in the city of Madrid, most of the summer (winter) SO_2 has an origin in combustion during energy-generation activities (SNAP1): 56% (30%) of monthly means for summertime (wintertime), followed by non-industrial combustion plants, including private wood combustion—SNAP2—(21%/54% of summer/winter levels). In Barcelona, SNAP1 is also responsible for around 60% of SO_2 levels, with a limited contribution of shipping emissions (19% for summertime and 14% during winter) and agriculture—SNAP10—(around 6% for both seasons). It should be highlighted that the levels of SO_2 in the urban areas of Madrid and Barcelona are very low, with mean monthly concentrations under 5 $\mu g\ m^{-3}$.

Figure 7 indicates the results regarding the contribution of each SNAP sector to the daily mean levels of $PM_{2.5}$ (left) and PM_{10} (right). The most important contributor to $PM_{2.5}$ and PM_{10} concentrations in Madrid, Barcelona, and Algeciras is the sector "Other", highlighting the importance of external sources to the domain during summertime (e.g., Saharan dust transport). In this sense, the outside contribution represents 72% (73%), 59% (63%), and 52% (57%) of summertime $PM_{2.5}$ (PM_{10}) levels in Madrid, Barcelona, and Algeciras, respectively. However, this contribution is much lower for wintertime, when the external contribution accounts for only 16% (7%), 31% (29%), and 35% (29%) of $PM_{2.5}$ (PM_{10}) levels at the aforementioned sites. The fact that the PM_{10} contribution is larger than $PM_{2.5}$ for summertime, but lower for wintertime, points to an important role of dust outbreaks over the IP during the summer months, as aforementioned [25,73].

Agriculture (SNAP10) effects on particulate matter levels are much larger in wintertime than during summertime. SNAP10 has a larger contribution to summer particles in Barcelona (18% for $PM_{2.5}$ and 16% for PM_{10}) than in the case of Madrid (6% for $PM_{2.5}$ and PM_{10}) or Algeciras (14% and 10% for $PM_{2.5}$ and PM_{10}, respectively). These contributions increase notably for wintertime, with agriculture being the most important contributor to wintertime $PM_{2.5}$ and PM_{10} levels in Madrid (49% and 52%, respectively) and Barcelona (39% and 40%).

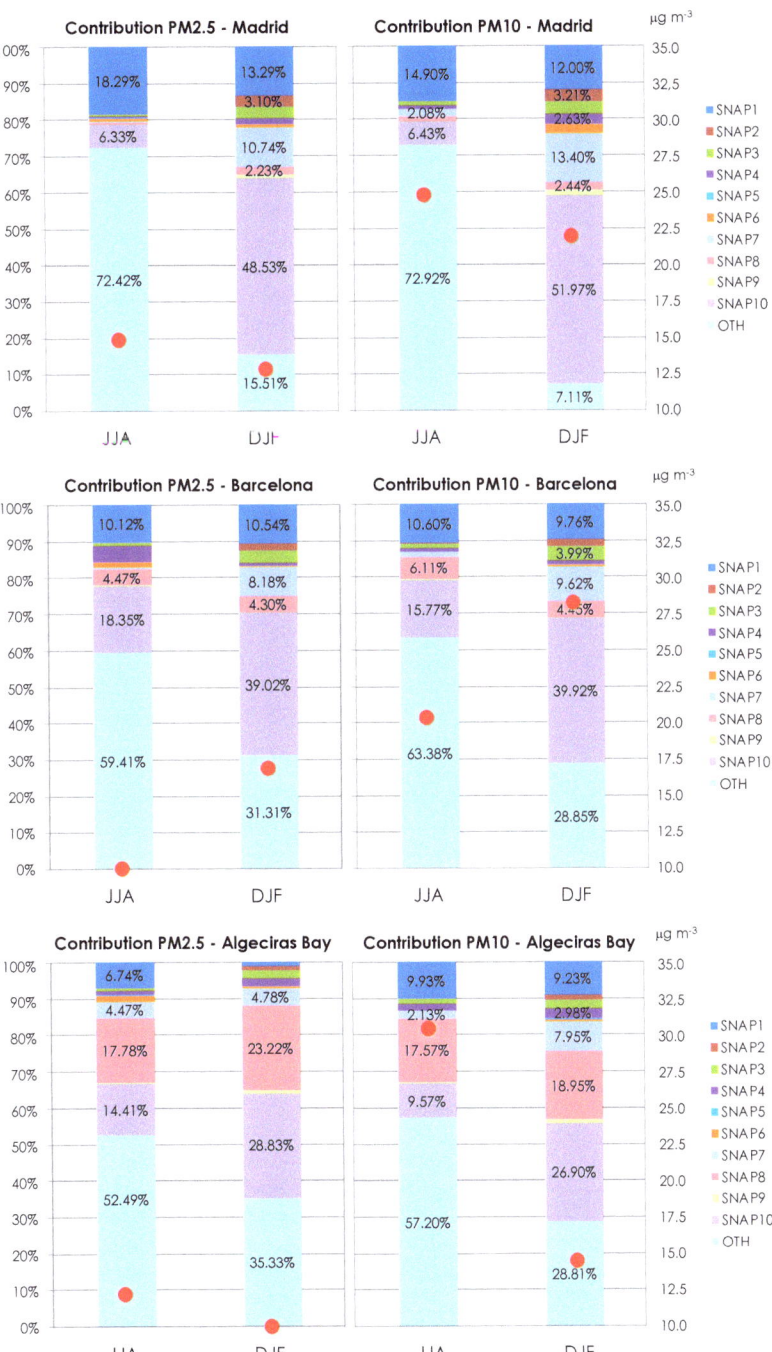

Figure 7. (Left axis) Relative contribution (%) of each anthropogenic SNAP sector to the daily mean levels of PM$_{2.5}$ (left) and PM$_{10}$ (right) over the most polluted areas of the IP (Madrid, top; Barcelona, center; Algeciras Bay, bottom). (Right axis) Red dot stands for the mean monthly concentrations of PM$_{2.5}$ (left) and PM$_{10}$ (right) in µg m^{-3}.

Combustion in energy and transformation industries (SNAP1) also notably contributes to particle levels in the city of Madrid ($PM_{2.5}$: 18% for summer and 13% for winter; PM_{10}: 15% and 12% in summer and winter, in that order), Barcelona ($PM_{2.5}$: 10% for summer and 11% for winter; PM_{10}: 11% and 10% in summer and winter, respectively), and Algeciras ($PM_{2.5}$: 7% and 1% for summer/winter; PM_{10}: 10% and 9% in summer and winter, in that order). On-road traffic (SNAP7) is only noticeable for wintertime $PM_{2.5}$(PM_{10}) concentrations, being 11% (13%), 8% (10%), and 5% (8%) in Madrid, Barcelona, and Algeciras, while the contributions of SNAP8 (other mobile sources) are very high in Algeciras, being the second largest contributor for particulate matter both in summer (18% for $PM_{2.5}$ and PM_{10}) and winter (23% and 19% for $PM_{2.5}$ and PM_{10}, respectively), due to the presence of important harbor/industrial activity in the area [89,90]. Over a coastal area such as Barcelona, the estimated contribution of harbor emissions to the urban background reached 9–12% for PM_{10} and 11–15% for $PM_{2.5}$ [91]. Our results are in agreement with those numbers (despite being slightly lower), since the estimations of the contribution of SNAP8 to $PM_{2.5}$(PM_{10}) background levels in Barcelona is around 4–6%. This contribution is linked both to primary emissions from fuel oil combustion but also to the formation of secondary aerosols from gas-phase precursors.

3.4. Response of Air Quality Exceedances to Zeroed-Out Emissions

It is important to characterize the contribution of each emitting sector to air pollution not only from the point of view of the percent contribution to mean air quality levels, but also to attribute the role of those sources in the exceedances of limit values for the protection of human health. In this sense, Table 6 summarizes the contribution over the entire IP of each SNAP sector (only for those sectors with significant variations with respect to the BC) to the number of exceedances of different target values selected: objective value for O_3, 120 $\mu g\ m^{-3}$, 8 h; limit value for NO_2, 200 $\mu g\ m^{-3}$, 1 h, not to be exceeded (n.t.b.e.) more than 3 times a calendar year; limit value for SO_2, 125 $\mu g\ m^{-3}$, 1 day, n.t.b.e. more than 3 times a calendar year; limit value for PM_{10}, 50 $\mu g\ m^{-3}$, 1 day, n.t.b.e. more than 35 times a calendar year. Additionally, the limit value for $PM_{2.5}$, 25 $\mu g\ m^{-3}$, 1 calendar year, was explored, but as we have only simulated summer and winter periods, this latter limit value cannot be assessed.

With respect to the exceedance of the target, limit, and threshold values set in the Directive 2008/50/EC, Table 6 indicates a clear improvement in the O_3 objective value (120 $\mu g\ m^{-3}$, max. 8 h) when zeroing-out the on-road traffic emissions (SNAP7) for summertime (days with exceedances reduce from 23 to 16 in summer; no exceedances are simulated for winter in the base case); however, this management strategy is hard to take into practice because of the socio-economical implications of road traffic reduction. Moreover, other mobile sources (SNAP8) contribute to 5 days with exceedances of the object value for O_3 (23 days in BC vs. 18 in noSNAP8).

Additionally, other mobile sources (SNAP8) is the sector causing most of the exceedances of the limit values related to NO_2 (200 $\mu g\ m^{-3}$, 1 h) and SO_2 (125 $\mu g\ m^{-3}$, daily mean) over the IP (playing also a role on PM_{10} exceedances). In this sense, SNAP8 causes the two exceedances of the limit value of modeled NO_2 and is responsible for six out of the eight exceedances of the daily limit value for SO_2 (125 $\mu g\ m^{-3}$) over the domain for summertime (no values over the limit value for NO_2 or SO_2 are modeled during wintertime). SO_2 concentrations over the limit value are found over the Algeciras Bay, and are caused mainly from the contribution of the high sulphur emissions coming from ship fuels. It is noteworthy that the contribution of shipping emissions to the exceedances of the limit value for PM_{10} is not as large as for SO_2 (in agreement with [92]), since there are components of particulate matter from shipping not directly affected by the sulphur content in the fuels. In this sense, just 2 of the 18 summertime exceedances of the daily mean 50-$\mu g\ m^{-3}$ limit value for PM_{10} are caused by SNAP8 (no exceedances of the PM_{10} limit value are caused by other mobile sources in wintertime). For particles, combustion in energy generation (SNAP1) is responsible of 5 out of the 18 (27) exceedances of the PM_{10}

limit value for summertime (wintertime), while agriculture (SNAP10) contributes to 2 (6) exceedances of the daily mean 50-μg m^{-3} limit value for summertime (wintertime).

Table 6. Variation in the number of exceedances over the entire IP when zeroing-out the different SNAP sectors (base case minus zeroed-out SNAP sector simulation).

Summer	Summer 2011			
	Concentration	w/o SNAP		
Pollutant	Limit value	Zero-out sector	N exc. BC	N. exc. noSNAP
O_3	Objective value for O_3, 120 μg m^{-3}, 8 h	SNAP7	23	16
		SNAP8		18
NO_2	Limit value for NO_2, 200 μg m^{-3}, 1 h	SNAP8	2	0
SO_2	Limit value for SO_2, 125 μg m^{-3}, 1 day	SNAP1	8	5
		SNAP8		2
PM_{10}	Limit value for PM_{10}, 50 μg m^{-3}, 1 day	SNAP1	18	13
		SNAP8		16
		SNAP10		16
Winter	December 2011			
	Concentration	w/o SNAP		
Pollutant	Limit value	Zero-out sector	N exc. BC	N. exc. noSNAP
O_3	Objective value for O_3, 120 μg m^{-3}, 8 h	SNAP7	0	0
NO_2	Limit value for NO_2, 200 μg m^{-3}, 1 h	SNAP8	0	0
SO_2	Limit value for SO_2, 125 μg m^{-3}, 1 day	SNAP1	0	0
		SNAP8		0
PM_{10}	Limit value for PM_{10}, 50 μg m^{-3}, 1 day	SNAP4	27	22
		SNAP7		26
		SNAP10		21

4. Discussion and Conclusions

Efficient air quality management requires an accurate identification of pollution sources and of their individual contributions to the ambient pollutant concentrations. To this end, the zero-out methodology has been proposed and applied for the apportionment of atmospheric pollutants in the IP. This method is based on the application of WRF + CHIMERE chemistry transport model coupled to EMEP emissions.

Regarding tropospheric O_3, on-road traffic is the only anthropogenic sector with a noticeable contribution to maximum O_3 levels during summertime (6%) and is responsible for 7 summer days with exceedances in the objective value of 120 μg m^{-3} (max. 8-hr mean) established by the 2008/50/EC directive. These results are in agreement with those of the scientific literature [37,62]. These authors found out that the on-road transport sector (SNAP7) was the largest overall anthropogenic source sector contributing to July 2011 O_3 concentrations in Europe, with non-road transport (SNAP8) contributions ranking second, as in our case (2% contribution to summertime maximum O_3 levels and five exceedances of the objective value). An analogous analysis can be completed for SNAP8 (other mobile sources) with respect to NO_2, with this sector prevailing in the contribution to mean ground-level concentrations during summertime and contributing to the two exceedances of the limit value for the protection of human health for NO_2 (200 μg m^{-3}, 1 h) modeled over the IP. The importance of this sector in the IP is larger closer to the major shipping routes and main harbors, with relative contributions varying from 10 to 50% depending on the pollutant (the lowest contribution for particulate matter, the largest for SO_2 and NO_2).

Last, the other anthropogenic sector with a noticeable impact is agriculture. Removing agriculture emissions (SNAP10) contributes to a decrease in PM levels and a simultaneous increase of SO_2 concentrations. The reduction of the most important source contributing to ammonia emissions controls the formation of ammonium sulphate. Therefore, reducing the levels of ammonia permits the SO_2 to remain in the gas phase. Agriculture contributes to the limit value for the protection of human health regarding PM_{10} (50 µg m^{-3}, daily mean) with 2 exceedances out of 18, while this number increases to 6 out of 27 wintertime exceedances.

With respect to the temporal pattern, in general, the source contribution does not exhibit a strong seasonality, except for particulate mater under the "Other" sector, which includes the external contribution to particle levels. Despite this seasonal behaviour for particulate matter, both gas-phase pollutants and particles exhibit a strong spatial uniformity, since background concentrations in the modeling system are provided by coarse resolution chemistry/climate models that do not allow for a sharp gradient in the background concentrations.

The external contribution of particles to the "Other" sector is mainly composed of mineral matter from Saharan dust. The fact that the boundary contributions to PM_{10} are larger than for $PM_{2.5}$ for summertime, but lower for wintertime, points to an important role of dust outbreaks over the IP during the summer months, which contributes mainly with large particles. These results are in line with those of Karachamdani et al. [37] for 16 European cities, who indicate that the boundary condition contributions for the Mediterranean cities are larger than for other European cities, ranging from about 40–50% during summertime to 10–15% in wintertime, because those Mediterranean cities were largely influenced by the long-range transport of dust emissions from northern Africa in the summer months.

Locally, the IP undergoes diverse problems related to air quality both during summer and winter. Focusing on the most polluted areas of the target domain (the cities of Madrid, Barcelona, and Algeciras Bay), the impact of road transport (SNAP7) emissions is high for NO_2 ground levels over largely populated areas (Madrid or Barcelona areas), but the concentration of this pollutant is dominated by other mobile sources (such as maritime or airport emissions included in SNAP8). Over coastal areas of the target domain, a poor air quality caused by large NO_2 concentrations can be attributed to shipping routes. In this sense, Merico et al. [87] also highlight the influence of harbor and shipping emissions on air quality of the nearby coastal areas of the Mediterranean.

For SO_2, energy generation (SNAP1) controls the mean levels of this pollutant over most of the areas considered. Valverde et al. [70] indicate that the contribution of power plants to the surface concentration of SO_2 occurs mainly close to the source (<20 km) related to a fumigation process when the emission injection takes place within the planetary boundary layer, but those plumes can reach long distances (>250 km) from the sources.

In the Algeciras Bay, maritime emissions largely contribute to the levels of SO_2. The implementation of low-sulphur fuels in shipping may contribute to substantially reducing the number of exceedances of the limit values for the protection of human health and to reduce several pathologies such as cardiovascular and cancer deaths, childhood asthma, or premature mortality and morbidity [93]. Summertime PM_{10} and $PM_{2.5}$ levels are dominated by the external contribution of Saharan dust, while for wintertime, agriculture can have a dominant position in Madrid and Barcelona. The important contribution of agriculture to PM levels was highlighted by Lelieveld et al. [94], who stated that this sector is the largest contributor to $PM_{2.5}$ levels in Europe.

Hence, this evaluated contribution has allowed us to identify which sectors contribute most to air pollution problems in the IP. However, it should be borne in mind that the uncertainties associated with several factors (principally, the boundary conditions in the CTMs and emission inventories) can condition the accuracy of the obtained results [37,95]. For instance, Jiménez et al. [17] analyze the impact of initial and boundary conditions over the Levantine coast of the IP, indicating that, despite the influence of initial condition reduces with the spin-up time (a 48-h spin-up time is sufficient to reduce the impact factor of initial conditions to 10% or less), the importance of having accurate boundary conditions becomes essential, since its influence on the results increases with the time of the simulation, reaching up to 5 µg m^{-3} for certain pollutants. With respect to the emission inventories, Baldasano et al. [96] point to industrial facilities as the main sources of uncertainties in emission inventories over the target area.

Nonetheless, this work can provide a very useful contribution to a better understanding of the sensitivity of air pollutants in a complex area such as the IP, and can provide valuable information for the design of mitigation strategies or plans that lead to an improvement in European air quality and the attainment of the SDG over the target area.

Funding: The authors acknowledge the ECCE project (PID2020-115693RB-I00) of the Ministerio de Ciencia e Innovación/Agencia Estatal de Investigación (MCIN/AEI/10.13039/501100011033/) and the European Regional Development Fund (ERDF/FEDER Una manera de hacer Europa). Additionally, the authors thanks the reviewers for their valuable contributions and fruitful discussions.

Institutional Review Board Statement: Not applicable.

Informed Consent Statement: Not applicable.

Data Availability Statement: Data are available upon resonable request from the corresponding author (pedro.jimenezguerrero@um.es)

Conflicts of Interest: The author declares no conflict of interest.

Abbreviations

The following abbreviations are used in this manuscript:

AQI	Air Quality Index
BC	Base Case
BFM	Brute Force Method
CTM	Chemistry Transport Model
EMEP	European Monitoring and Evaluation Programme
IP	Iberian Peninsula
MFB	Mean Fractional Bias
MFE	Mean Fractional Error
MNBE	Mean Normalized Bias Error
MNGE	Mean Normalized Gross Error
NMVOC	Non-Methane Volatile Organic Compounds
SNAP	Selected Nomenclature for Air Pollution
VOC	Volatile Organic Compounds
WRF	Weather Research and Forecasting

References

1. United Nations. *Transforming Our World: The 2030 Agenda for Sustainable Development (A/RES/70/1)*; UN General Assembly: New York, NY, USA, 2015.
2. Tsai, W.T.; Lin, Y.Q. Trend Analysis of Air Quality Index (AQI) and Greenhouse Gas (GHG) Emissions in Taiwan and Their Regulatory Countermeasures. *Environments* **2021**, *8*, 29. [CrossRef]
3. Kahraman, A.C.; Sivri, N. Comparison of metropolitan cities for mortality rates attributed to ambient air pollution using the AirQ model. *Environ. Sci. Pollut. Res.* **2022**, *29*, 1. doi: 10.1007/s11356-021-18341-1. [CrossRef] [PubMed]
4. Osman, T.; Kenawy, E.; Abdrabo, K.I.; Shaw, D.; Alshamndy, A.; Elsharif, M.; Salem, M.; Alwetaishi, M.; Aly, R.M.; Elboshy, B. Voluntary Local Review Framework to Monitor and Evaluate the Progress towards Achieving Sustainable Development Goals at a City Level: Buraidah City, KSA and SDG11 as A Case Study. *Sustainability* **2021**, *13*, 9555. doi: 10.3390/su13179555. [CrossRef]
5. Pateman, R.; Tuhkanen, H.; Cinderby, S. Citizen Science and the Sustainable Development Goals in Low and Middle Income Country Cities. *Sustainability* **2021**, *13*, 9534. doi: 10.3390/su13179534. [CrossRef]
6. Cuvelier, C.; Thunis, P.; Vautard, R.; Amann, M.; Bessagnet, B.; Bedogni, M.; Berkowicz, R.; Brandt, J.; Brocheton, F.; Builtjes, P.; et al. CityDelta: A model intercomparison study to explore the impact of emission reductions in European cities in 2010. *Atmos. Environ.* **2007**, *41*, 189–207. [CrossRef]
7. Ballester, F.; Medina, S.; Boldo, E.; Goodman, P.; Neuberger, M.; Iñiguez, C.; Künzli, N. Reducing ambient levels of fine particulates could substantially improve health: A mortality impact assessment for 26 European cities. *J. Epidemiol. Community Health* **2008**, *62*, 98–105. doi: 10.1136/jech.2007.059857. [CrossRef]
8. Pope, C.A.; Ezzati, M.; Dockery, D.W. Fine-Particulate Air Pollution and Life Expectancy in the United States. *N. Engl. J. Med.* **2009**, *360*, 376–386. [CrossRef]

9. Beelen, R.; Raaschou-Nielsen, O.; Stafoggia, M.; Andersen, Z.J.; Weinmayr, G.; Hoffmann, B.; Wolf, K.; Samoli, E.; Fischer, P.; Nieuwenhuijsen, M.; et al. Effects of long-term exposure to air pollution on natural-cause mortality: An analysis of 22 European cohorts within the multicentre ESCAPE project. *Lancet* **2014**, *383*, 785–795. doi: 10.1016/S0140-6736(13)62158-3. [CrossRef]
10. Héroux, M.E.; Anderson, H.R.; Atkinson, R.; Brunekreef, B.; Cohen, A.; Forastiere, F.; Hurley, F.; Katsouyanni, K.; Krewski, D.; Krzyzanowski, M.; et al. Quantifying the health impacts of ambient air pollutants: recommendations of a WHO/Europe project. *Int. J. Public Health* **2015**, *60*, 619–627. doi: 10.1007/s00038-015-0690-y. [CrossRef]
11. Tarín-Carrasco, P.; Im, U.; Geels, C.; Palacios-Peña, L.; Jiménez-Guerrero, P. Contribution of fine particulate matter to present and future premature mortality over Europe: A non-linear response. *Environ. Int.* **2021**, *153*, 106517. [CrossRef]
12. Guzmán, P.; Tarín-Carrasco, P.; Morales-Suárez-Varela, M.; Jiménez-Guerrero, P. Effects of air pollution on dementia over Europe for present and future climate change scenarios. *Environ. Res.* **2022**, *204*, 112012. doi: 10.1016/j.envres.2021.112012. [CrossRef]
13. Im, U.; Brandt, J.; Geels, C.; Hansen, K.M.; Christensen, J.H.; Andersen, M.S.; Solazzo, E.; Kioutsioukis, I.; Alyuz, U.; Balzarini, A.; et al. Assessment and economic valuation of air pollution impacts on human health over Europe and the United States as calculated by a multi-model ensemble in the framework of AQMEII3. *Atmos. Chem. Phys.* **2018**, *18*, 5967–5989. doi: 10.5194/acp-18-5967-2018. [CrossRef] [PubMed]
14. Tarín-Carrasco, P.; Morales-Suárez-Varela, M.; Im, U.; Brandt, J.; Palacios-Peña, L.; Jiménez-Guerrero, P. Isolating the climate change impacts on air-pollution-related-pathologies over central and southern Europe—A modelling approach on cases and costs. *Atmos. Chem. Phys.* **2019**, *19*, 9385–9398. doi: 10.5194/acp-19-9385-2019. [CrossRef]
15. Monteiro, A.; Carvalho, A.; Ribeiro, I.; Scotto, M.; Barbosa, S.; Alonso, A.; Baldasano, J.; Pay, M.; Miranda, A.; Borrego, C. Trends in ozone concentrations in the Iberian Peninsula by quantile regression and clustering. *Atmos. Environ.* **2012**, *56*, 184–193. [CrossRef]
16. Baldasano, J.; Pay, M.; Jorba, O.; Gassó, S.; Jiménez-Guerrero, P. An annual assessment of air quality with the CALIOPE modeling system over Spain. *Sci. Total Environ.* **2011**, *409*, 2163–2178. doi: 10.1016/j.scitotenv.2011.01.041. [CrossRef]
17. Jiménez, P.; Parra, R.; Baldasano, J.M. Influence of initial and boundary conditions for ozone modeling in very complex terrains: A case study in the northeastern Iberian Peninsula. *Environ. Model. Softw.* **2007**, *22*, 1294–1306. [CrossRef]
18. Vivanco, M.G.; Palomino, I.; Vautard, R.; Bessagnet, B.; Martín, F.; Menut, L.; Jiménez, S. Multi-year assessment of photochemical air quality simulation over Spain. *Environ. Model. Softw.* **2009**, *24*, 63–73. doi: 10.1016/j.envsoft.2008.05.004. [CrossRef]
19. Borge, R.; López, J.; Lumbreras, J.; Narros, A.; Rodríguez, E. Influence of boundary conditions on CMAQ simulations over the Iberian Peninsula. *Atmos. Environ.* **2010**, *44*, 2681–2695. doi: 10.1016/j.atmosenv.2010.04.044. [CrossRef]
20. Pay, M.T.; Jiménez-Guerrero, P.; Baldasano, J.M. Assessing sensitivity regimes of secondary inorganic aerosol formation in Europe with the CALIOPE-EU modeling system. *Atmos. Environ.* **2012**, *51*, 146–164. doi: 10.1016/j.atmosenv.2012.01.027. [CrossRef]
21. Vedrenne, M.; Borge, R.; Lumbreras, J.; Conlan, B.; Rodríguez, M.E.; de Andrés, J.M.; de la Paz, D.; Pérez, J.; Narros, A. An integrated assessment of two decades of air pollution policy making in Spain: Impacts, costs and improvements. *Sci. Total Environ.* **2015**, *527–528*, 351–361. doi: 10.1016/j.scitotenv.2015.05.014. [CrossRef]
22. Palacios-Peña, L.; Baró, R.; Guerrero-Rascado, J.J.L.; Alados-Arboledas, L.; Brunner, D.; Jiménez-Guerrero, P. Evaluating the representation of aerosol optical properties using an online coupled model over the Iberian Peninsula. *Atmos. Chem. Phys.* **2017**, *17*, 277–296. doi: 10.5194/acp-17-277-2017 [CrossRef]
23. Guevara, M.; Martínez, F.; Arévalo, G.; Gassó, S.; Baldasano, J.M. An improved system for modelling Spanish emissions: HERMESv2.0. *Atmos. Environ.* **2013**, *81*, 209–221. doi: 10.1016/j.atmosenv.2013.08.053. [CrossRef]
24. Nunes, R.A.O.; Alvim-Ferraz, M.C.M.; Martins, F.G.; Calderay-Cayetano, F.; Durán-Grados, V.; Moreno-Gutiérrez, J.; Jalkanen, J.P.; Hannuniemi, H.; Sousa, S.I.V. Shipping emissions in the Iberian Peninsula and the impacts on air quality. *Atmos. Chem. Phys.* **2020**, *20*, 9473–9489. [CrossRef]
25. Querol, X.; Pey, J.; Pandolfi, M.; Alastuey, A.; Cusack, M.; Pérez, N.; Moreno, T.; Viana, M.; Mihalopoulos, N.; Kallos, G.; et al. African dust contributions to mean ambient PM10 mass-levels across the Mediterranean Basin. *Atmos. Environ.* **2009**, *43*, 4266–4277. [CrossRef]
26. Carvalho, A.; Monteiro, A.; Solman, S.; Miranda, A.; Borrego, C. Climate-driven changes in air quality over Europe by the end of the 21st century, with special reference to Portugal. *Environ. Sci. Policy* **2010**, *13*, 445–458. [CrossRef]
27. Jiménez-Guerrero, P.; Montávez, J.P.; Gómez-Navarro, J.J.; Jerez, S.; Lorente-Plazas, R. Impacts of climate change on ground level gas-phase pollutants and aerosols in the Iberian Peninsula for the late XXI century. *Atmos. Environ.* **2012**, *55*, 483–495. [CrossRef]
28. Jiménez-Guerrero, P.; Gómez-Navarro, J.J.J.J.; Baró, R.; Lorente, R.; Ratola, N.; Montávez, J.P.J. Is there a common pattern of future gas-phase air pollution in Europe under diverse climate change scenarios? *Clim. Chang.* **2013**, *121*, 661–671. [CrossRef]
29. Jiménez-Guerrero, P.; Jerez, S.; Montávez, J.P.; Trigo, R.M. Uncertainties in future ozone and PM10 projections over Europe from a regional climate multiphysics ensemble. *Geophys. Res. Lett.* **2013**, *40*, 5764–5769. [CrossRef]
30. Monteiro, A.; Sá, E.; Fernandes, A.; Gama, C.; Sorte, S.; Borrego, C.; Lopes, M.; Russo, M.A. How healthy will be the air quality in 2050? *Air Qual. Atmos. Health* **2018**, *11*, 353–362. [CrossRef]
31. Vautard, R.; Colette, A.; van Meijgaard, E.; Meleux, F.; van Oldenborgh, G.J.; Otto, F.; Tobin, I.; Yiou, P. Attribution of Wintertime Anticyclonic Stagnation Contributing to Air Pollution in Western Europe. *Bull. Am. Meteorol. Soc.* **2018**, *99*, S70–S75. [CrossRef]
32. Yarwood, G.; Emery, C.; Jung, J.; Nopmongcol, U.; Sakulyanontvittaya, T. A method to represent ozone response to large changes in precursor emissions using high-order sensitivity analysis in photochemical models. *Geosci. Model Dev.* **2013**, *6*, 1601–1608. [CrossRef]

33. Pisoni, E.; Albrecht, D.; Mara, T.; Rosati, R.; Tarantola, S.; Thunis, P. Application of uncertainty and sensitivity analysis to the air quality SHERPA modelling tool. *Atmos. Environ.* **2018**, *183*, 84–93. [CrossRef]
34. Koo, B.; Wilson, G.M.; Morris, R.E.; Dunker, A.M.; Yarwood, G. Comparison of Source Apportionment and Sensitivity Analysis in a Particulate Matter Air Quality Model. *Environ. Sci. Technol.* **2009**, *43*, 6669–6675. [CrossRef] [PubMed]
35. Kranenburg, R.; Segers, A.J.; Hendriks, C.; Schaap, M. Source apportionment using LOTOS-EUROS: Module description and evaluation. *Geosci. Model Dev.* **2013**, *6*, 721–733. [CrossRef]
36. Clappier, A.; Belis, C.A.; Pernigotti, D.; Thunis, P. Source apportionment and sensitivity analysis: Two methodologies with two different purposes. *Geosci. Model Dev.* **2017**, *10*, 4245–4256. [CrossRef]
37. Karamchandani, P.; Long, Y.; Pirovano, G.; Balzarini, A.; Yarwood, G. Source-sector contributions to European ozone and fine PM in 2010 using AQMEII modeling data. *Atmos. Chem. Phys.* **2017**, *17*, 5643–5664. [CrossRef]
38. Wang, L.; Wei, Z.; Wei, W.; Fu, J.S.; Meng, C.; Ma, S. Source apportionment of PM2.5 in top polluted cities in Hebei, China using the CMAQ model. *Atmos. Environ.* **2015**, *122*, 723–736. [CrossRef]
39. Huang, Y.; Deng, T.; Li, Z.; Wang, N.; Yin, C.; Wang, S.; Fan, S. Numerical simulations for the sources apportionment and control strategies of PM2.5 over Pearl River Delta, China, part I: Inventory and PM2.5 sources apportionment. *Sci. Total Environ.* **2018**, *634*, 1631–1644. [CrossRef]
40. Baker, K.; Woody, M.; Tonnesen, G.; Hutzell, W.; Pye, H.; Beaver, M.; Pouliot, G.; Pierce, T. Contribution of regional-scale fire events to ozone and PM2.5 air quality estimated by photochemical modeling approaches. *Atmos. Environ.* **2016**, *140*, 539–554. [CrossRef]
41. Han, X.; Zhang, M.; Zhu, L.; Skorokhod, A. Assessment of the impact of emissions reductions on air quality over North China Plain. *Atmos. Pollut. Res.* **2016**, *7*, 249–259. [CrossRef]
42. Klemp, J.B.; Skamarock, W.C.; Dudhia, J. Conservative Split-Explicit Time Integration Methods for the Compressible Nonhydrostatic Equations. *Mon. Weather. Rev.* **2007**, *135*, 2897–2913. [CrossRef]
43. Skamarock, W.C.; Klemp, J.B.; Dudhia, J.; Gill, D.O.; Barker, D.M.; Wang, W.; Powers, J.G. *A Description of the Advanced Research WRF Version 3*; note-475+ STR; NCAR Technical Report: Boulder, CO, USA, 2008.
44. Dee, D.P.; Uppala, S.M.; Simmons, A.J.; Berrisford, P.; Poli, P.; Kobayashi, S.; Andrae, U.; Balmaseda, M.A.; Balsamo, G.; Bauer, P.; et al. The ERA-Interim reanalysis: Configuration and performance of the data assimilation system. *Q. J. R. Meteorol. Soc.* **2011**, *137*, 553–597. [CrossRef]
45. Menut, L.; Bessagnet, B.; Khvorostyanov, D.; Beekmann, M.; Blond, N.; Colette, A.; Coll, I.; Curci, G.; Foret, G.; Hodzic, A.; et al. CHIMERE 2013: A model for regional atmospheric composition modelling. *Geosci. Model Dev.* **2013**, *6*, 981–1028. [CrossRef]
46. Derognat, C.; Beekmann, M.; Baeumle, M.; Martin, D.; Schmidt, H. Effect of biogenic volatile organic compound emissions on tropospheric chemistry during the Atmospheric Pollution Over the Paris Area (ESQUIF) campaign in the Ile-de-France region. *J. Geophys. Res.* **2003**, *108*, 8560. [CrossRef]
47. Hong, S.; Lim, J.O.J. The WRF Single-Moment 6-Class Microphysics Scheme (WSM6). *Asia-Pac. J. Atmos. Sci.* **2006**, *42*, 129–151.
48. Hong, S.Y.; Noh, Y.; Dudhia, J. A New Vertical Diffusion Package with an Explicit Treatment of Entrainment Processes. *Mon. Weather. Rev.* **2006**, *134*, 2318–2341. [CrossRef]
49. Nenes, A.; Pandis, S.N.; Pilinis, C. ISORROPIA: A New Thermodynamic Equilibrium Model for Multiphase Multicomponent Inorganic Aerosols. *Aquat. Geochem.* **1998**, *4*, 123–152. [CrossRef]
50. Collins, W.D.; Rasch, P.J.; Boville, B.A.; Hack, J.J.; McCaa, J.R.; Williamson, D.L.; Kiehl, J.T.; Briegleb, B.; Bitz, C.; Lin, S.J.; et al. *Description of the NCAR Community Atmosphere Model (CAM 3.0)*; Note NCAR/TN-464+ STR; NCAR Technical Report; NCAR: Boulder, CO, USA, 2004; Volume 226, 1326–1334.
51. Bessagnet, B.; Menut, L.; Curci, G.; Hodzic, A.; Guillaume, B.; Liousse, C.; Moukhtar, S.; Pun, B.; Seigneur, C.; Schulz, M. Regional modeling of carbonaceous aerosols over Europe—Focus on secondary organic aerosols. *J. Atmos. Chem.* **2008**, *61*, 175–202. [CrossRef]
52. Chen, F.; Dudhia, J. Coupling an Advanced Land Surface-Hydrology Model with the Penn State-NCAR MM5 Modeling System. Part I: Model Implementation and Sensitivity. *Mon. Weather. Rev.* **2001**, *129*, 569–585. [CrossRef]
53. Kain, J.S. The Kain-Fritsch convective parameterization: An update. *J. Appl. Meteorol.* **2004**, *43*, 170–181. doi: 10.1175/1520-0450(2004)043<0170:TKCPAU>2.0.CO;2. [CrossRef]
54. Vestreng, V.; Ntziachristos, L.; Semb, A.; Reis, S.; Isaksen, I.S.; Tarrasón, L. Evolution of NOx emissions in Europe with focus on road transport control measures. *Atmos. Chem. Phys.* **2009**, *9*, 1503–1520. [CrossRef]
55. Guenther, A.; Karl, T.; Harley, P.; Wiedinmyer, C.; Palmer, P.I.; Geron, C. Estimates of global terrestrial isoprene emissions using MEGAN (Model of Emissions of Gases and Aerosols from Nature). *Atmos. Chem. Phys.* **2006**, *6*, 3181–3210. [CrossRef]
56. Folberth, G.A.; Hauglustaine, D.A.; Lathière, J.; Brocheton, F. Interactive chemistry in the Laboratoire de Météorologie Dynamique general circulation model: Model description and impact analysis of biogenic hydrocarbons on tropospheric chemistry. *Atmos. Chem. Phys.* **2006**, *6*, 2273–2319. [CrossRef]
57. Szopa, S.; Foret, G.; Menut, L.; Cozic, A. Impact of large scale circulation on European summer surface ozone and consequences for modelling forecast. *Atmos. Environ.* **2009**, *43*, 1189–1195. [CrossRef]
58. Jiménez-Guerrero, P.; Ratola, N. Influence of the North Atlantic oscillation on the atmospheric levels of benzo[a]pyrene over Europe. *Clim. Dyn.* **2021**, *57*, 1173–1186. [CrossRef]

59. Borge, R.; Lumbreras, J.; Vardoulakis, S.; Kassomenos, P.; Rodríguez, E. Analysis of long-range transport influences on urban PM10 using two-stage atmospheric trajectory clusters. *Atmos. Environ.* **2007**, *41*, 4434–4450. doi: [CrossRef]
60. Salvador, P.; Artíñano, B.; Molero, F.; Viana, M.; Pey, J.; Alastuey, A.; Querol, X. African dust contribution to ambient aerosol levels across central Spain: Characterization of long-range transport episodes of desert dust. *Atmos. Res.* **2013**, *127*, 117–129. [CrossRef]
61. Brandt, J.; Silver, J.D.; Christensen, J.H.; Andersen, M.S.; Bønløkke, J.H.; Sigsgaard, T.; Geels, C.; Gross, A.; Hansen, A.B.; Hansen, K.M.; et al. Contribution from the ten major emission sectors in Europe and Denmark to the health-cost externalities of air pollution using the EVA model system – an integrated modelling approach. *Atmos. Chem. Phys.* **2013**, *13*, 7725–7746. [CrossRef]
62. Tagaris, E.; Sotiropoulou, R.E.P.; Gounaris, N.; Andronopoulos, S.; Vlachogiannis, D. Effect of the Standard Nomenclature for Air Pollution (SNAP) Categories on Air Quality over Europe. *Atmosphere* **2015**, *6*, 1119–1128. [CrossRef]
63. Arunachalam, S.; Valencia, A.; Akita, Y.; Serre, M.L.; Omary, M.; Garcia, V.; Isakov, V. A Method for Estimating Urban Background Concentrations in Support of Hybrid Air Pollution Modeling for Environmental Health Studies. *Int. J. Environ. Res. Public Health* **2014**, *11*, 10518–10536. [CrossRef]
64. Tørseth, K.; Aas, W.; Breivik, K.; Fjæraa, A.M.; Fiebig, M.; Hjellbrekke, A.G.; Lund Myhre, C.; Solberg, S.; Yttri, K.E. Introduction to the European Monitoring and Evaluation Programme (EMEP) and observed atmospheric composition change during 1972–2009. *Atmos. Chem. Phys.* **2012**, *12*, 5447–5481. [CrossRef]
65. Boylan, J.W.; Russell, A.G. PM and light extinction model performance metrics, goals, and criteria for three-dimensional air quality models. *Atmos. Environ.* **2006**, *40*, 4946–4959. Special issue on Model Evaluation: Evaluation of Urban and Regional Eulerian Air Quality Models, [CrossRef]
66. Zhang, H.; Chen, G.; Hu, J.; Chen, S.H.; Wiedinmyer, C.; Kleeman, M.; Ying, Q. Evaluation of a seven-year air quality simulation using the Weather Research and Forecasting (WRF)/Community Multiscale Air Quality (CMAQ) models in the eastern United States. *Sci. Total Environ.* **2014**, *473–474*, 275–285. [CrossRef] [PubMed]
67. Gao, M.; Guttikunda, S.K.; Carmichael, G.R.; Wang, Y.; Liu, Z.; Stanier, C.O.; Saide, P.E.; Yu, M. Health impacts and economic losses assessment of the 2013 severe haze event in Beijing area. *Sci. Total Environ.* **2015**, *511*, 553–561. [CrossRef]
68. Sammartino, S.; García Lafuente, J.; Sánchez Garrido, J.; De los Santos, F.; Álvarez Fanjul, E.; Naranjo, C.; Bruno, M.; Calero, C. A numerical model analysis of the tidal flows in the Bay of Algeciras, Strait of Gibraltar. *Cont. Shelf Res.* **2014**, *72*, 34–46. [CrossRef]
69. Dios, M.; Souto, J.; Casares, J. Experimental development of CO_2, SO_2 and NOx emission factors for mixed lignite and subbituminous coal-fired power plant. *Energy* **2013**, *53*, 40–51. [CrossRef]
70. Valverde, V.; Pay, M.T.; Baldasano, J.M. A model-based analysis of SO_2 and NO_2 dynamics from coal-fired power plants under representative synoptic circulation types over the Iberian Peninsula. *Sci. Total Environ.* **2016**, *541*, 701–713. [CrossRef]
71. Kushta, J.; Georgiou, G.K.; Proestos, Y.; Christoudias, T.; Thunis, P.; Savvides, C.; Papadopoulos, C.; Lelieveld, J. Evaluation of EU air quality standards through modeling and the FAIRMODE benchmarking methodology. *Air Qual. Atmos. Health* **2019**, *12*, 73–86. [CrossRef]
72. Terrenoire, E.; Bessagnet, B.; Rouïl, L.; Tognet, F.; Pirovano, G.; Létinois, L.; Beauchamp, M.; Colette, A.; Thunis, P.; Amann, M.; et al. High-resolution air quality simulation over Europe with the chemistry transport model CHIMERE. *Geosci. Model Dev.* **2015**, *8*, 21–42. [CrossRef]
73. Mateos, D.; Cachorro, V.; Toledano, C.; Burgos, M.; Bennouna, Y.; Torres, B.; Fuertes, D.; González, R.; Guirado, C.; Calle, A.; et al. Columnar and surface aerosol load over the Iberian Peninsula establishing annual cycles, trends, and relationships in five geographical sectors. *Sci. Total Environ.* **2015**, *518–519*, 378–392. [CrossRef]
74. Jiménez-Guerrero, P.; Pérez, C.; Jorba, O.; Baldasano, J.M. Contribution of Saharan dust in an integrated air quality system and its on-line assessment. *Geophys. Res. Lett.* **2008**, *35*, L03814. [CrossRef]
75. Stein, A.F.; Wang, Y.; de la Rosa, J.D.; de la Campa, A.M.S.; Castell, N.; Draxler, R.R. Modeling PM10 Originating from Dust Intrusions in the Southern Iberian Peninsula Using HYSPLIT. *Weather. Forecast.* **2011**, *26*, 236 – 242. [CrossRef]
76. de la Paz, D.; Vedrenne, M.; Borge, R.; Lumbreras, J.; Manuel de Andrés, J.; Pérez, J.; Rodríguez, E.; Karanasiou, A.; Moreno, T.; Boldo, E.; et al. Modelling Saharan dust transport into the Mediterranean basin with CMAQ. *Atmos. Environ.* **2013**, *70*, 337–350. [CrossRef]
77. Jiménez, P.; Baldasano, J.M. Ozone response to precursor controls in very complex terrains: Use of photochemical indicators to assess O3-NOx-VOC sensitivity in the northeastern Iberian Peninsula. *J. Geophys. Res. Atmos.* **2004**, *109*, 1–20. [CrossRef]
78. Castell, N.; Stein, A.F.; Mantilla, E.; Salvador, R.; Millán, M. Evaluation of the use of photochemical indicators to assess ozone—NOx—VOC sensitivity in the Southwestern Iberian Peninsula. *J. Atmos. Chem.* **2009**, *63*, 73–91. [CrossRef]
79. Marmer, E.; Dentener, F.; Aardenne, J.v.; Cavalli, F.; Vignati, E.; Velchev, K.; Hjorth, J.; Boersma, F.; Vinken, G.; Mihalopoulos, N.; et al. What can we learn about ship emission inventories from measurements of air pollutants over the Mediterranean Sea? *Atmos. Chem. Phys.* **2009**, *9*, 6815–6831. [CrossRef]
80. Carrillo-Torres, E.R.; Hernández-Paniagua, I.Y.; Mendoza, A. Use of Combined Observational- and Model-Derived Photochemical Indicators to Assess the O3-NOx-VOC System Sensitivity in Urban Areas. *Atmosphere* **2017**, *8*, 22. doi: [CrossRef]
81. Renner, E.; Wolke, R. Modelling the formation and atmospheric transport of secondary inorganic aerosols with special attention to regions with high ammonia emissions. *Atmos. Environ.* **2010**, *44*, 1904–1912. [CrossRef]
82. Kyrkilis, G.; Chaloulakou, A.; Kassomenos, P.A. Development of an aggregate Air Quality Index for an urban Mediterranean agglomeration: Relation to potential health effects. *Environ. Int.* **2007**, *33*, 670–676. [CrossRef]

83. Borrego, C.; Monteiro, A.; Martins, H.; Ferreira, J.; Fernandes, A.P.; Rafael, S.; Miranda, A.I.; Guevara, M.; Baldasano, J.M. Air quality plan for ozone: An urgent need for North Portugal. *Air Qual. Atmos. Health* **2016**, *9*, 447–460. [CrossRef]
84. Derwent, R.G.; Utembe, S.R.; Jenkin, M.E.; Shallcross, D.E. Tropospheric ozone production regions and the intercontinental origins of surface ozone over Europe. *Atmos. Environ.* **2015**, *112*, 216–224. [CrossRef]
85. Gencarelli, C.N.; De Simone, F.; Hedgecock, I.M.; Sprovieri, F.; Pirrone, N. Development and application of a regional-scale atmospheric mercury model based on WRF/Chem: A Mediterranean area investigation. *Environ. Sci. Pollut. Res.* **2014**, *21*, 4095–4109. [CrossRef] [PubMed]
86. Aksoyoglu, S.; Baltensperger, U.; Prévôt, A.S.H. Contribution of ship emissions to the concentration and deposition of air pollutants in Europe. *Atmos. Chem. Phys.* **2016**, *16*, 1895–1906. [CrossRef]
87. Merico, E.; Donateo, A.; Gambaro, A.; Cesari, D.; Gregoris, E.; Barbaro, E.; Dinoi, A.; Giovanelli, G.; Masieri, S.; Contini, D. Influence of in-port ships emissions to gaseous atmospheric pollutants and to particulate matter of different sizes in a Mediterranean harbour in Italy. *Atmos. Environ.* **2016**, *139*, 1–10. [CrossRef]
88. Valverde, V.; Pay, M.T.; Baldasano, J.M. Ozone attributed to Madrid and Barcelona on-road transport emissions: Characterization of plume dynamics over the Iberian Peninsula. *Sci. Total Environ.* **2016**, *543*, 670–682. doi: [CrossRef] [PubMed]
89. Cesari, D.; Amato, F.; Pandolfi, M.; Alastuey, A.; Querol, X.; Contini, D. An inter-comparison of PM10 source apportionment using PCA and PMF receptor models in three European sites. *Environ. Sci. Pollut. Res.* **2016**, *23*, 15133–15148. [CrossRef]
90. Palomares-Salas, J.C.; González-de-la Rosa, J.J.; Agüera-Pérez, A.; Sierra-Fernández, J.M.; Florencias-Oliveros, O. Forecasting PM10 in the Bay of Algeciras Based on Regression Models. *Sustainability* **2019**, *11*, 968. [CrossRef]
91. Pérez, N.; Pey, J.; Reche, C.; Cortés, J.; Alastuey, A.; Querol, X. Impact of harbour emissions on ambient PM10 and PM2.5 in Barcelona (Spain): Evidences of secondary aerosol formation within the urban area. *Sci. Total Environ.* **2016**, *571*, 237–250. [CrossRef]
92. Jonson, J.E.; Jalkanen, J.P.; Johansson, L.; Gauss, M.; Denier van der Gon, H.A.C. Model calculations of the effects of present and future emissions of air pollutants from shipping in the Baltic Sea and the North Sea. *Atmos. Chem. Phys.* **2015**, *15*, 783–798. [CrossRef]
93. Sofiev, M.; Winebrake, J.J.; Johansson, L.; Carr, E.W.; Prank, M.; Soares, J.; Vira, J.; Kouznetsov, R.; Jalkanen, J.P.; Corbett, J.J. Cleaner fuels for ships provide public health benefits with climate tradeoffs. *Nat. Commun.* **2018**, *9*, 406. [CrossRef]
94. Lelieveld, J.; Evans, J.S.; Fnais, M.; Giannadaki, D.; Pozzer, A. The contribution of outdoor air pollution sources to premature mortality on a global scale. *Nature* **2015**, *525*, 367–371. [CrossRef] [PubMed]
95. Solazzo, E.; Bianconi, R.; Hogrefe, C.; Curci, G.; Tuccella, P.; Alyuz, U.; Balzarini, A.; Baro, R.; Bellasio, R.; Bieser, J.; et al. Evaluation and error apportionment of an ensemble of atmospheric chemistry transport modeling systems: Multivariable temporal and spatial breakdown. *Atmos. Chem. Phys.* **2017**, *17*, 3001–3054. [CrossRef] [PubMed]
96. Baldasano, J.M.; Güereca, L.P.; López, E.; Gassó, S.; Jimenez-Guerrero, P. Development of a high-resolution (1 km × 1 km, 1h) emission model for Spain: The High-Elective Resolution Modelling Emission System (HERMES). *Atmos. Environ.* **2008**, *42*, 7215–7233. [CrossRef]

Article

Research on the Impact of Environmental Regulation on Green Technology Innovation from the Perspective of Regional Differences: A Quasi-Natural Experiment Based on China's New Environmental Protection Law

Qin Liu [†], Ying Zhu [†], Weixin Yang [*,†] and Xueyu Wang [†]

Business School, University of Shanghai for Science and Technology, Shanghai 200093, China; liuq828@hotmail.com (Q.L.); zhuying9827@163.com (Y.Z.); 673403483@163.com (X.W.)
* Correspondence: iamywx@outlook.com; Tel.: +86-21-5596-0082
† These authors contributed equally to this work.

Abstract: Environmental regulations have a certain impact on regional green technology innovation affected by regional differences. Using the panel data of 30 provincial-level administrative regions in China (excluding Tibet, Hong Kong, Macao, and Taiwan) from 2011 to 2019, we consider China's new environmental protection law (NEPL) as a quasi-natural experiment to evaluate the impact of environmental regulation on green technology innovation in a difference-in-differences (DID) framework and further analyze the influences of regional differences. The results indicate that environmental regulations can promote regional green technology innovation, and that regional differences have a significant impact on this issue. Furthermore, environmental regulations in regions with high and low levels of economic development and education, and regions with medium and low levels of energy consumption have a significant impact on green technology innovation. The government should reasonably formulate environmental regulation policies on the basis of regional differences, encourage cross-regional exchanges and cooperation, and more efficiently stimulate regional green technology innovation to achieve sustainable development.

Keywords: environmental regulations; green technology innovation; regional differences; difference-in-differences

1. Introduction

The extensive economic development model in the past decades has made environmental pollution one of the factors restricting the sustainable development of China's economy. The deterioration of resource and environmental conditions has increased the uncertainty of global development and the unprecedented challenges of the global governance system. Many countries have promulgated environmental laws and regulations to promote environmental protection [1]. Since the green development model has been the basic strategy to promote the harmonious coexistence of man and nature in the report of the 19th National Congress of the Communist Party of China (CPC), ecological civilization construction has become an important strategy for China's development. The construction of ecological civilization is not only related to the sustainable development of China's social economy, but also to global ecological security and the healthy development of human beings. The concept of an ecological civilization is increasingly rooted in the hearts of the people, and while pollution control efforts continue to increase, improving environmental quality is urgent [2]. The Chinese government, aware of the seriousness of environmental problems and the importance of green technology innovation, has formulated a series of environmental regulation policies to urge enterprises to reduce their pollution emissions, and to encourage enterprises to carry out green technology innovation through capital investment in recent years.

The Chinese government released the Environmental Protection Law of the People's Republic of China (Old Environmental Protection Law) in 1989. The revised Environmental Protection Law of the People's Republic of China (New Environmental Protection Law, NEPL) was implemented in 2015. Compared to the Old Environmental Protection Law, the NEPL implemented more severe penalties and significant supervision, emphasized information disclosure and encouraged public participation. Therefore, the NEPL has been described as the strictest environmental protection law in China's history. The promulgation of the NEPL marked a new stage of Chinese environmental legislation. The policy impact of environmental regulations has been of wide concern to many scholars.

In April 2019, the National Development and Reform Commission of the People's Republic of China and the Ministry of Science and Technology of China jointly issued guidance on the construction of a market-oriented green technology innovation system, which further refined the road map and timetable of green technology innovation system construction [3]. Since then, green technology innovation has become the key task of the current national ecological civilization construction. In a brief report on the green patent classification system construction and a green patent statistical analysis of the China National Intellectual Property Administration, the connotation and standard of the green patent were preliminarily clarified. Under the guidance of policies and funds, green industries were represented by energy conservation and environmental protection. Moreover, clean production, clean energy and a circular economy have increasingly become the focus of investment.

On the one hand, environmental regulation, as an important part of public regulation is an effective way to correct market failure. Researching the impact of environmental regulations on the coordinated development of the environment and economy is conducive to the design of the most suitable environmental management system for the Chinese government [4]. On the other hand, green technology innovation is an important means to guide enterprises to improve production technology and to achieve energy conservation and emissions reduction. Therefore, understanding the relationship between the environmental regulations and green technology innovation is helpful to clarify the relationship between environmental protection and economic development, so as to seek the way of a balanced development of the environment and the economy.

China is vast, with some regional differences in social and economic development and natural resources. Whether environmental regulation policies will have different effects in different types of regions is an important research topic. Therefore, based on the study of the impact of environmental regulation policies on green technology innovation, this study further analyzes the differences of environmental regulation policy effects from the perspective of regional differences, with a quasi-natural experiment based on China's NEPL. In this study, the NEPL is treated as an exogenous policy shock to identify its policy impact on green technology innovation via a difference-in-differences (DID) framework. The impact mechanisms were also investigated to clarify how the NEPL affects green technology innovation and how regional differences affect the policy impact. It is hoped this will provide suggestions for the government to formulate environmental regulation policies according to the actual situation of a region, and for enterprises to carry out green technology innovation according to their own individual situations.

2. Literature Review
2.1. Research on Environmental Regulation

Environmental regulation as an effective means to restrain corporate behavior and protect the environment, and the study of its connotations and effects have been the focus of academic attention. Scholars' researches on environmental regulation have mainly focused on the following three aspects. First, many scholars have studied the definition and evolution of environmental regulation. There are a number of classifications of environmental regulation according to definitions from different angles. Command-and-control regulation (CCR) and market-based incentive regulation (MIR) are two commonly mentioned environmental regulations [5–7]. Some researchers have also proposed environmental regulations such as informal environmental regulation, implicit environmental regulation and public participation environmental regulation that are not implemented by the government [8–10]. Second, the research on the measurement of the level of environmental regulation, index selection and measurement method has been one of the most important themes. Considering the quality of relevant variable data, many empirical studies have certain limitations, and there is no unified measurement standard. At present, domestic and foreign scholars mainly measure environmental regulation from the following two perspectives. One is from the perspective of the specific implementation of environmental regulation, the selection of pollution control costs, the proportion of pollution control investment in the total cost or output value of enterprises, or the amount of policy supervision [11–13]. The other is to consider pollutant treatment efficiency to construct a comprehensive environmental index from the perspective of an environmental governance effect under environmental regulation [14–16]. In addition, environmental regulation efficiency may vary across regions due to being influenced by external environmental factors such as the economic base, industrial structure and education levels [17]. The third aspect reveals that an increasing number of scholars have studied the impact and driving mechanism of environmental regulation and the relationships between environmental regulations, green development and enterprises' innovation have been the main research aspect. Moreover, these relationships may vary in different regions and periods because of other factors and effects [18,19].

2.2. Research on Green Technology Innovation

The methods for measuring green technology innovation have varied across this research area due to the limitation of data availability. For one thing, some scholars have constructed the green technology innovation index and calculated the green innovation efficiency (GIE) to measure green innovation development [20,21], and for another, green patents have been increasingly used to measure green technology innovation in recent years [22]. According to the relevant definition of the State Intellectual Property Office of China, green technology refers to technologies that are conducive to saving resources, improving energy efficiency, preventing and controlling pollution, and achieving sustainable development [23]. It mainly includes alternative energy, environmental materials, energy conservation and emissions reduction, pollution control and governance, and recycling technology. A green patent refers to the invention, utility model and design patent with the theme of green technology. Patent documents also provide information such as patent inventors, claims, patent families and citations, which is conducive to identifying the type of innovation subject and the quality of the innovation.

2.3. The Impact of Environmental Regulation on Green Technology Innovation

Many scholars have researched the impact of environmental regulation tools and intensity on green technology innovation from the perspective of environmental regulation. At present, there is still much controversy about the research in this area, mainly manifested as a question of whether environmental regulations will positively promote or negatively inhibit the development of green technology innovation in the long run. There are mainly three different views. First is that environmental regulation promotes green technology innovation. The Porter hypothesis claimed that environmental protection policies actually increase the net output of enterprises, and finally improve enterprises' competitive advantages [24]. Lanjouw and Mody (1996) [25] expanded the study to the United States, Japan, and Germany, and verified the positive effect of environmental regulation on green technology innovation. In addition, Domazlicky (2004) [26], Yang et al. (2012) [27], Mazzanti (2009) [28] and other studies have also verified that environmental regulation has certain technical effects. Xing et al. (2019) [29] found that environmental commitment and sustainability exploitation innovation are fundamental for realizing the positive effects of environmental regulation on firm performance and provided deeper insight into the effect of ambidextrous sustainability innovation in the 'strong' Porter hypothesis. Yuan and Xiang (2018) [30] found that in the long run, environmental regulation inhibits patent output and does not support the 'weak' Porter hypothesis, while improving energy efficiency hinders labor productivity and does not support the 'strong' version of the Porter Hypothesis. Second, environmental regulations may hinder green technological innovation. For example, Chintrakam and Weber (2008) [31] selected the relevant data on the American manufacturing industry to study and concluded that the government's environmental regulations caused enterprises to lack sufficient funds for the invention of environmental protection technology patents. The third view holds that there is no simple linear relationship between environmental regulations and green technological innovation. On the one hand, some scholars have found that there is a U-shaped relationship between environmental regulations and green technological innovation [32,33], while on the other hand, some scholars have found that there is an inverted U-shaped relationship between environmental regulation and green technology innovation [10,34].

In recent years, scholars have analyzed this problem from different angles. First, from the perspective of different environmental regulation means and tools, different means and tools of environmental regulation have different effects on green technology innovation. For example, the flexible environmental policies have a significant positive impact on technological innovation, and the implementation of environmental regulation has actively alleviated the relationship between flexible environmental policies and technological innovation [35]. The non-linear impact of formal and informal environmental regulations on green growth, and formal and informal environmental regulations have showed different effects at different stages [36]. Government direct funding and tax incentives may promote green technology innovation [37]. The productivity effect driven by market-based incentive regulation is much stronger than that of command-and-control by investigating how different regulatory instruments and the relative stringency impact green productivity based on China's reality [38]. Second, from the perspective of international technology transfer, the research on the mechanism of foreign direct investment (FDI), the environmental regulation effect, and green technology innovation is one of the most commonly considered aspects. For example, environmental regulations may have a positive effect on enterprise ecological technology innovation through FDI [39]. The influence of environmental regulation and FDI exerted on green innovation efficiency may be different for different manufacturing industries [40]. In addition, the impact of trade structure upgrading on green technology innovation is closely related to environmental regulation [41]. Third, from the perspective of heterogeneity, industry heterogeneity has been the most popular angle. Scholars have usually studied the influence of industry heterogeneity from the aspects of pollution-intensive industries, cleaning industries, technology-intensive industries and labor-intensive industries. For example, Cai et al. (2020) [42] have found that direct environmental regulation has a strong and significant incentive effect on green technology innovation in pollution-intensive industries, and direct environmental regulation can effectively encourage green technology innovation in technology-intensive industries compared with labor-intensive industries. In addition, heterogeneity of the enterprises' ownership may influence the relationship between the environmental regulations and green technology innovation [43].

2.4. Innovation of This Study

In conclusion, the relationship between environmental regulations and green technological innovation is a complex problem and researching it from different perspectives is conducive to a more profound understanding of this problem. This study focuses on the impact of environmental regulation policies represented by the promulgation and implementation of China's new environmental protection law (NEPL) on green technological innovation and innovates and supplements the research content on the basis of existing research. Due to the vast territory of China, there are some differences in the scale of economic development, industrial development models and natural resources between different regions, so whether environmental regulation policies will have different effects in different types of regions is an important research topic. In view of this, the innovation of this study is that this study researches the relationship between the environmental regulation policy and green technology innovation from the perspective of regional differences by using the generalized difference method, and further analyzes the influence of regional differences in it, hoping to provide a reference for the formulation of government policies.

Further, the research steps of this study are as follows: First, the generalized difference-in-differences method is used to group the empirical research referring to Cai et al. (2016) [44]. According to the policy effect of the new environmental protection law, 30 provinces in China are divided into 14 treated groups and 16 control groups. Second, the environmental regulation policy selected in this study is representative. According to the impact of the NEPL on the number of green patent applications, this study determines its policy effects and dynamic effects test. Third, this study innovatively sets the grouping and dummy variables according to regional differences, and divides the provinces into three levels, namely, at the level of economic development, the level of education and the level of energy consumption, so as to further identify the impact of environmental regulation policies in the different regions on green technological innovation at the different levels. Fourth, robustness tests are used to further illustrate the reliability of this study. This study changes the explained variables, with the amount of green patent as an indicator for regression testing, and then changes the explanatory variables, testing the effect of the environmental regulation policy with fixed regional differences, with multiplication terms conducted by dummy variables according to the level of the regional differences and environmental regulation policy effect. Fifth, this study summarizes conclusions, analyzes reasons, provides corresponding recommendations for the decision-making of government and enterprises, and finally illustrates the limitations and future research prospects.

3. Mechanism Analysis and Research Hypotheses

This study analyzes the mechanism of environmental regulation affecting enterprise green technology innovation from three aspects: government, enterprise, and regional differences.

3.1. The Mechanism of the New Environmental Protection Law on Green Innovation Activities

Environmental regulation is a policy tool for governments to use mandatory means to reduce environmental pollution. The NEPL is an ordered environmental regulation policy. Compared with other incentive environmental regulation means, its scope of action is more extensive, and it has more stringent mandatory guidelines. Faced with the severe constraints caused by the mandatory environmental regulation policy, enterprises will make decisions according to their own situation, showing heterogeneous self-selection behavior, namely, transfer, upgrading or transformation [45,46]. For small businesses, due to their financial and technical constraints, the cost of environmental regulation cannot be internalized in a short period of time, and relocation or being shut down become the main responses for dealing with the environmental regulation policy. For medium-sized enterprises with certain financial and technical support, green technology innovation is carried out with the goal of energy conservation and emission reduction, and the production line can be transformed, so as to move towards upgrading their business. For large enterprises with strong comprehensive strength, they can carry out all-round resource reconfiguration in the technical space, geographical space and industrial space according to their own characteristics, so as to cope with the environmental regulation through the three ways of transfer, upgrading and transformation [22]. When the compensation effect of the environmental regulation on enterprise innovation exceeds the offset effect of the cost internalization caused by the environmental regulation, then enterprise innovation obtains sustainable conditions, that is, the environmental regulation plays a positive role in promoting enterprise innovation. As with the opinions of the Porter hypothesis, environmental regulation can improve enterprises' innovation abilities and enhance their competitiveness [24].

With the transfer of time, when mandatory constraints become routine, the society will generally recognize the green development path. At this time, green innovation becomes the main development mode of enterprises. When the enterprise is guided by the concept of green innovation throughout the whole process of output, this process requires a certain period of accumulation and precipitation, and this upgrading and transformation cannot be completed in a short timeframe. Therefore, there is a certain lag in the positive effect from environmental regulation on green technological innovation.

Hypothesis 1 (H1). *The New Environmental Protection Law (NEPL) of China has a positive impact on green technology innovation, but the policy effect is lagging behind.*

3.2. The Mechanism of the Regional Differences Influencing Environmental Regulation and Enterprises' Green Innovation

With regard to environmental protection, in the process of carrying out the national policy, provinces often formulate local laws and regulations according to local conditions from the aspects of economic development, industrial development, and the technical level of their respective provinces. China has a vast territory, and the difference in economic level, education level and energy consumption between the regions will lead to different intensities and types of environmental regulation policies, leading to different effects on green technology innovation.

3.2.1. Differences in Economic Development Levels

This study analyzes the influence of the difference in economic development level on the green technology innovation of the government and from enterprises.

From the perspective of the government, the government's policy objectives and focus will transform according to the trend of economic development. When regional economic development reaches a certain level, the government often transfers the working focus to industrial transformation and upgrading. Moreover, the relevant literature reveals that environmental regulation can promote the adjustment and optimization of industrial structures in the region. Therefore, the government will actively innovate green technology to accelerate green and clean industrial development. From the perspective of enterprises, the better the regional economic development, the more active the innovation of enterprises is.

3.2.2. Differences in Education Levels

The difference in education level between regions also affects the policy effect of environmental regulation. This study explains the mechanism of environmental regulation in two ways.

First, human capital plays an important role in technological innovation, especially in the R&D ability of employees. The higher education level in the region can often cultivate more high-quality innovative talent, thus providing the necessary human capital support for the green technology innovation of enterprises. Second, as a new engine of economic development, the industry–university–institute cooperation model can promote the R&D innovation activities of enterprises by integrating the tripartite resources of industry, university, and research institutions [47]. In regions with high levels of education, research institutions and universities are more intensive, and the combination of production, education, and research is more active, which can provide the necessary talents and technical support for green technology research and development and reduce the cost of information.

3.2.3. Differences in Energy Consumption Levels

In addition to the difference in economic development levels and education levels, the difference in energy consumption between regions will also affect the green technology innovation of enterprises. Specifically, a regional industrial structure with more energy consumption is generally characterized by a large proportion of the first and second industries, and a small proportion of the tertiary industry, while environmental pollution in the region is often more serious. The government will formulate stringent environmental regulation policies to reduce pollutant emissions to transform the mode of economic development.

If the proportion of tertiary industry in the region with less energy consumption is larger, then the pollutants emitted by enterprises in the region are lower, therefore, the environmental protection policy is more relaxed, and the pressure of green technology research and development is reduced.

Hypothesis 2 (H2). *The impact of environmental regulation policies on green technological innovation is different under different economic levels, education levels and energy consumption levels.*

4. Materials and Methods

4.1. Model Building Econometric Strategy

4.1.1. Benchmark Difference-in-Differences

This study uses the difference-in-differences (DID) model to analyze the impact of environmental regulation on green technology innovation. Therefore, this study constructs the following regression model and selects the following control variables based on the theoretical analysis:

$$Y_{i,t} = \alpha_0 + \alpha_1 \cdot treat_{i,t} \times post_{i,t} + \delta \cdot x_{i,t} + \mu_{i,t} + \gamma_{i,t} + \varepsilon_{i,t} \quad (1)$$

In the upper formula, i denotes a province, t denotes the year. The explained variable $Y_{i,t}$ measures the growth of green technology innovation activities, represented by green invention patent applications and $post_{i,t}$ denotes the time-determination variable. This variable is 0 before the policy shock year and is taken as 1 after the policy shock, while $treat_{i,t}$ denotes the virtual variable of whether each province strengthens the environmental regulation in response to policy shocks. The indicator variable $treat_{i,t} \times post_{i,t}$ denotes a cross variable determined by the value of the annual $post_{i,t}$ and $treat_{i,t}$. The coefficient of the cross variable reflects the effect of environmental regulation policy, that is, after a policy shock, whether the enhancement of government environmental regulation will effectively promote green technology innovation. If the implementation of the province's policy was set as the following year, $treat_{i,t} \times post_{i,t}$ is 1, otherwise $treat_{i,t} \times post_{i,t}$ is 0. $x_{i,t}$ represents other control variables that also affect green technology innovation (GTI). Additionally, $\mu_{i,t}$ is the city fixed effect, $\gamma_{i,t}$ is the time fixed effect, and $\varepsilon_{i,t}$ is a random disturbance.

4.1.2. Parallel Trend Assumption and Time Trend Analysis

The parallel trend assumption is the basic premise of DID analysis. Therefore, this study conducted a dynamic effect analysis to test whether the benchmark regression met the parallel trend assumption, as well as to identify the time effect of the environmental regulation policy. An event study approach was employed to study the dynamic effect of the environmental regulation policy on green technology innovation. In order to observe how the promulgation of the new environmental law affects the behavior of green technology innovation over time, reference is made to Chen (2017) [48] and Tao Feng et al. (2021) [22]. The model is described as follows:

$$Y_{i,t} = \beta_0 + \beta_t \cdot \sum\nolimits_{t=2011 \; t \neq 2014}^{2019} treat_{i,t} \times post_{i,t} + \delta \cdot x_{i,t} + \mu_{i,t} + \varepsilon_{i,t} \quad (2)$$

where $post_{i,t}$ is the time dummy variable, $treat_{i,t} \times post_{i,t}$ is the interaction term of the grouping variable $treat_{i,t}$ and the time dummy variable $post_{i,t}$, and β_t represents the policy effect of the new environmental law on the quantity and quality of green patents in this year. Here, the first year (2014) of the formal implementation of the new environmental law was taken as the reference group, and the corresponding interaction term was not introduced. This model can also be used for the key parallel trend test in a DID estimation. If the estimated coefficient β_t of $treat_{i,t} \times post_{i,t}$ is not significant before 2015, it means that the parallel trend condition is satisfied.

4.1.3. Regional Differences Analysis

In addition, this study intended to analyze the environmental regulation policy effect on green technology innovation with regional differences. The model is further described as follows:

$$Y_{i,t} = \lambda + \gamma \cdot treat_{i,t} \times post_{i,t} \times H, M, Lgdp_{i,t} + \delta \cdot x_{i,t} + \mu_{i,t} + \varepsilon_{i,t} \qquad (3)$$

$$Y_{i,t} = \lambda + \gamma \cdot treat_{i,t} \times post_{i,t} \times H, M, Ledu_{i,t} + \delta \cdot x_{i,t} + \mu_{i,t} + \varepsilon_{i,t} \qquad (4)$$

$$Y_{i,t} = \lambda + \gamma \cdot treat_{i,t} \times post_{i,t} \times H, M, Lenergy_{i,t} + \delta \cdot x_{i,t} + \mu_{i,t} + \varepsilon_{i,t} \qquad (5)$$

where $Hgdp_{i,t}$, $Mgdp_{i,t}$ and $Lgdp_{i,t}$ represent high, medium and low levels of economic development, respectively, and $Hedu_{i,t}$, $Medu_{i,t}$, and $Ledu_{i,t}$ represent high, medium and low education levels, respectively. This study also used local unit GDP energy consumption to measure the level of local energy consumption, grouped by high, medium and low, generating three virtual variables: $Henergy_{i,t}$, $Menergy_{i,t}$, and $Lenergy_{i,t}$, represent high, medium and low levels of energy consumption, respectively.

4.1.4. Determination on the Time-Point of Policy Shocks

By consulting the policies and regulations promulgated by the relevant departments in China in the last 10 years, this study selected the revised Environmental Protection Law (New Environmental Protection Law, NEPL) adopted by vote on 24 April 2014 and formally implemented on 1 January 2015, as the time point of the policy impact. The reasons for choosing the NEPL as a policy shock were as follows: first, the NEPL differs from local laws and regulations, and its influence is national, dominant and authoritative. The NEPL also has far-reaching implications because it is accompanied by a large number of legal documents and technical standards updates. Second, the NEPL is the most significant since the implementation of China's Environmental Protection Law, defining the mission and responsibility of government departments for environmental supervision, and making environmental regulation operational and enforceable [49]. Finally, the NEPL has exerted considerable pressure on enterprises, including limiting the emissions standards of some pollutants, and updating the environmental protection indicators of some products in some industries. On the one hand, the implementation of these measures has increased the cost of sewage from enterprises. On the other hand, it also encourages enterprises to develop green technology innovation, reduce pollutant emissions in their production processes, and improve the green level of products. Therefore, in terms of the severity of the policy, the promulgation of the NEPL has an obvious environmental regulation effect, which in turn has a certain impact on enterprises' green technology innovation. Considering the accuracy of the study, this study uses the panel data of 30 provincial-level administrative regions in China (excluding Tibet, Hong Kong, Macao, and Taiwan) from 2011 to 2019 to verify whether the promulgation of the NEPL has a significant policy impact effect.

Figure 1 shows the change of total sulfur dioxide emissions in China from 2011 to 2019. Overall, it can be seen that the promulgation of the new Environmental Protection Law has a more obvious inhibitory effect on SO_2 emissions and industrial SO_2 emissions, and that the emissions of various provinces in China have been reduced to varying degrees. Therefore, it was reasonable to choose the new Environmental Protection Law as the time point of the policy shocks.

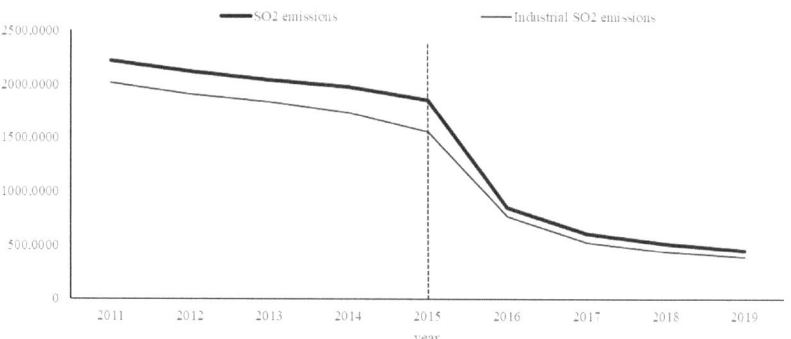

Figure 1. SO$_2$ emissions and industrial SO$_2$ emissions in China from 2010 to 2019.

4.1.5. Selection of Treatment Group and Control Group

For empirical analysis, the samples needed to be divided into a treated group and a control group. Since the implementation of the NEPL is of national significance, it was impossible to distinguish between the provinces that do not implement policies and provinces that do implement policies. Therefore, this study adopted the generalized difference-in-differences method to divide the treated group and the control group according to the effect after implementation. Among them, the treatment group was the province with significantly enhanced environmental regulation after the implementation of the NEPL, and the control group was the region with insignificant enhancement of environmental regulation after the implementation of the NEPL. In this study, the degree of decline in total SO$_2$ emissions was the basis for classification. Therefore, the average reduction ratio of total SO$_2$ emissions in each province from 2015 to 2019 and 2011 to 2014 was calculated. Taking 61.24% as the dividing standard, the provinces with higher emissions than 61.24% were set as the treated group, and the provinces with lower emissions than 61.24% were set as the control group. Table 1 presents the grouping results, with 14 provinces in the treated group and 16 provinces in the control group.

Table 1. Results of treated group and control group.

Groups	Provinces
Treated group	Beijing (78.23%), Tianjin (75.07%), Shanxi (61.77%), Shanghai (75.40%), Zhejiang (69.12%), Shandong (61.31%), Henan (70.23%), Hubei (61.78%), Guangdong (64.51%), Guangxi (64.96%), Chongqing (67.15%), Sichuan (61.45%), Shaanxi (65.07%), Gansu (62.40%)
Control group	Hebei (58.32%), Inner Mongolia (55.25%), Liaoning (56.05%), Jilin (56.69%), Heilongjiang (54.36%), Jiangsu (50.72%), Anhui (50.62%), Fujian (49.48%), Jiangxi (36.54%), Hunan (50.64%), Qinghai (30.26%), Hainan (57.31%), Guizhou (58.11%), Yunnan (46.57%), Ningxia (51.62%), Xinjiang (49.62%)

4.2. Variable Setting

This part introduces the selection of explained variables, explanatory variables and control variables. The main variables and the associated definitions are in Table 2.

Table 2. Main variables and the associated definitions.

Type of Variable	Definition	Variable	Description
Explained variable	Level of enterprises' green technology innovation	$app\_invention_{i,t}$	The logarithm of the number of patent applications for green technology inventions for each year in each province is used as a measurement index.
Explanatory variable of policy effect	Time effect	$post_{i,t}$	The variable value is set to 0 before 2015 and taken as 1 after 2015 (including 2015).
	Policy effect	$treat_{i,t}$	The variable value is set to 0 in the control group and taken as 1 after 2015 in the treated group.
	Cross term	$treat_{i,t} \times post_{i,t}$	The variable value is determined by the value of $treat_{i,t}$ and $post_{i,t}$.
Explanatory variable of regional differences	Economic development levels	$gdp_{i,t}$	GDP per capita with 2000 as base is used as a group of indicators, divided into high, medium and low groups: $Hgdp_{i,t}, Mgdp_{i,t}, Lgdp_{i,t}$.
	Education levels	$edu_{i,t}$	The number of years of education is used as a grouping indicator and di-vided into three groups: $Hedu_{i,t}, Medu_{i,t}, Ledu_{i,t}$.
	Energy consumption levels	$energy_{i,t}$	Energy consumption per unit GDP as a grouping indicator, divided into high, medium and low groups: $Henergy_{i,t}, Menergy_{i,t}, Lenergy_{i,t}$.
	Cross term	$treat_{i,t} \times post_{i,t} \times H, M, Lgdp_{i,t}$ $treat_{i,t} \times post_{i,t} \times H, M, Ledu_{i,t}$ $treat_{i,t} \times post_{i,t} \times H, M, Lenergy_{i,t}$	The variable value is determined by the value of $treat_{i,t}$, $post_{i,t}$ and $H, M, Lgdp_{i,t}$ ($edu_{i,t}$, $energy_{i,t}$).
Control variable	Government subsidies	$subsidy_{i,t}$	Proportion of environmental protection subsidy to total fiscal expenditure in each province is used as a measurement index.
	Collection of sewage charges	$tax_{i,t}$	Proportion of collection of sewage to total fiscal expenditure per year in each province is used as a measurement index.
	R&D Investment of the enterprise	$rd_{i,t}$	The logarithm of the internal expenditure of R&D funds per year in each province is used as a measurement index.
	Foreign trade dependence	$trade_{i,t}$	The ratio of the total export to GDP in each province per year is used as a measurement index.
	Ownership structure	$sharehold_{i,t}$	The ratio of state-owned and collective-owned enterprises to total number of enterprises in each province per year.

4.2.1. Explained Variable

The explained variable is the level of green technological innovation. Constructing the green technological innovation index and green patents are currently the two most popular methods to measure the level of green technological innovation. The number of patent applications reflects the active degree of innovation, which is mainly divided into invention patents and utility model patents. According to the Chinese patent survey report, compared with utility model patents, invention patents have more R&D investment, longer R&D cycles, more stringent audits and thus more difficulties to apply. Therefore, invention patents can better reflect the innovation ability [50]. This study mainly used the number of green invention patents to measure the number of green technological innovation activities, and then reflected the level of green technological innovation. This study used the logarithm of the number of green invention patent applications in each province from 2011 to 2019 as the explained variable.

4.2.2. Explanatory Variable

The core explanatory variable is the environmental regulation policy effect. On the connotation and measurement of environmental regulation, it is mainly analyzed from the perspective of command-and-control regulation (CCR) and market-based incentive regulation (MIR). Different authors have measured the intensity of environmental regulations in a variety of ways, including the method of comprehensive index and the method of single index. The DID model is one of the most common methods for measuring the policy effect. Some scholars have taken the implementation of environmental regulation policy as a policy impact and used the DID method to verify the effect of command-and-control environmental regulation policy [7]. In recent years, some scholars have begun to measure and analyze from the perspective of implicit environmental regulation, and to measure the intensity of environmental regulation from the perspective of public awareness of environmental protection and participation. This study's authors believe that it is important to improve public awareness of environmental protection and participation, but in the current situation, such implicit environmental regulation cannot achieve the expected significant effect, therefore the measurement of such environmental regulation was not included in the main research scope of this study. This study took the implementation of the NEPL as a policy shock point, set up a treatment group and a control group, constructed a time dummy variable and a policy dummy variable, and took the product of the two variables as the core explanatory variable to measure the policy effect of mandatory environmental regulation.

Moreover, this study took the explanatory variables reflecting regional differences. First, this study used local per capita GDP data to measure the level of local economic development reflecting differences in economic development, and grouped them by high, medium and low, generating three virtual variables: $Hgdp_{i,t}$, $Mgdp_{i,t}$, and $Lgdp_{i,t}$. Second, this study measured the level of local education with the per capita years of education, and grouped them by high, medium and low, generating three virtual variables: $Hedu_{i,t}$, $Medu_{i,t}$, and $Ledu_{i,t}$. Third, this study used local unit GDP energy consumption to measure the level of local energy consumption, and grouped them by high, medium and low, generating three virtual variables: $Henergy_{i,t}$, $Menergy_{i,t}$, and $Lenergy_{i,t}$.

4.2.3. Control Variable

This study set the control variables from both government and enterprise aspects. Here follows the five control variables.

Government subsidy reward and government pollution tax punishment. Environmental regulation is a policy tool for governments to use mandatory means to reduce environmental pollution. Generally speaking, the ultimate goal of environmental regulation is not to stimulate green technological innovation in enterprises. Previous studies have found that environmental regulation policies have positive technical effects and negative distortion effects. On the one hand, the policy compulsion of environmental regulation urges enterprises to perform green technology innovation to reduce environmental pollution. On the other hand, the high environmental tax in environmental regulation may crowd out the R&D investment in green technology innovation, and enterprises can obtain green technology through patent purchases rather than through their own R&D; in summary, the effect of environmental regulation policy is different from that of the technical and distortion effects.

Green technology R&D investment. R&D investment has a direct positive impact on enterprise innovation performance [51]. Green R&D investment is the premise and foundation of green technology innovation, and related research points out that the relationship between them is not a simple linear relationship. In particular, only reasonable and sustained R&D investment can make the innovation activities of enterprises achieve obvious results. If R&D investment is insufficient and innovation activities lack the necessary financial support, the technological R&D progress of enterprises will be affected. If R&D investment is excessive, the unreasonable allocation of resources within the enterprise will reduce the economic benefits of the enterprise and ultimately have a negative impact on the enterprises' green technology innovation.

Foreign trade dependence of enterprises. There are two reasons why the level of foreign trade affects enterprises' green technology innovation. First, the level of green technology and environmental standards for products are different, so countries import and export products in accordance with their own standards. Second, based on traditional trade protection and trade sanctions, an increasing number of countries have begun to adopt 'green trade barriers' to protect their markets [52]. Specifically, green trade barriers refer to laws, regulations, standards, and other means to limit the import of products with higher pollution. To achieve trade protection, this standard is generally not lower than the national environmental standards. From the perspective of theoretical research, many scholars have analyzed and demonstrated from various industries that domestic enterprises should increase the level of green environmental protection of their own products in order to deal with the constraints of 'environmental trade barriers' in various countries.

Ownership structure of enterprises. China's corporate ownership forms are diverse, and the impact of environmental regulation on green technology innovation may differ under different ownership structures. First, state-owned enterprises and collectively owned enterprises have the attributes of policy tools [53], and they need to play the role of policy tools in environmental protection. Therefore, under the background of environmental regulation, the green technology innovation effect of this type of enterprise is more obvious. Second, to alleviate the pressure of capital investment in enterprises' innovative activities, the government often subsidizes these enterprises; however, it is difficult to obtain government subsidies for enterprises with different forms of ownership. Related documents point out that compared with state-owned and collectively owned enterprises, other types of ownership enterprises have more difficulties in applying for government subsidies. Third, green technology innovation has the characteristics of large investment and a lagging return on income so that it cannot bring economic benefits to enterprises in the short term. Therefore, compared with state-owned and collectively owned enterprises, other ownership types of enterprises lack the initiative to perform green technology innovation.

4.3. Data Sources

This study used the panel data of 30 provincial-level administrative regions in China (excluding Tibet, Hong Kong, Macao, and Taiwan) from 2011 to 2019. For the purpose of ensuring the accuracy and rigor of this study, the data used in this study were from the following official sources: the data on GDP, government's fiscal expenditure, fiscal revenue, and population education level derived from the National Bureau of Statistics of the People's Republic of China. The data on SO_2 emissions, environmental subsidies and sewage charges were derived from the *China Environmental Statistics Yearbook*. The energy consumption was derived from the *China Energy Statistics Yearbook*; and the relevant data of enterprises were derived from the *China Industrial Economic Statistics Yearbook*. The green patent data was taken from the China Research Data Service Platform (CNRDS) and filtered by province. Table 3 provides some descriptive statistical results for the variables.

Table 3. The statistical description of the main variables.

Variables	Observation	Mean	Std. Dev.	Min	Max
$app\_invention_{i,t}$	270	7.4402	1.4531	2.6391	10.7811
$treat_{i,t}$	270	0.4667	0.4998	0.0000	1.0000
$post_{i,t}$	270	0.5556	0.49069	0.0000	1.0000
$treat_{i,t} \times post_{i,t}$	270	0.2593	0.4390	0.0000	1.0000
$treat_{i,t} \times post_{i,t} \times Hgdp_{i,t}$	270	0.1111	0.3149	0.0000	1.0000
$treat_{i,t} \times post_{i,t} \times Mgdp_{i,t}$	270	0.0741	0.2624	0.0000	1.0000
$treat_{i,t} \times post_{i,t} \times Lgdp_{i,t}$	270	0.0741	0.2624	0.0000	1.0000
$treat_{i,t} \times post_{i,t} \times Hedu_{i,t}$	270	0.0926	0.2904	0.0000	1.0000
$treat_{i,t} \times post_{i,t} \times Medu_{i,t}$	270	0.0926	0.2904	0.0000	1.0000
$treat_{i,t} \times post_{i,t} \times Ledu_{i,t}$	270	0.0741	0.2624	0.0000	1.0000
$treat_{i,t} \times post_{i,t} \times Henergy_{i,t}$	270	0.0556	0.2295	0.0000	1.0000
$treat_{i,t} \times post_{i,t} \times Menergy_{i,t}$	270	0.0926	0.2904	0.0000	1.0000
$treat_{i,t} \times post_{i,t} \times Lenergy_{i,t}$	270	0.1111	0.3149	0.0000	1.0000
$subsidy_{i,t}$	270	0.0299	0.00098	0.0118	0.0681
$tax_{i,t}$	270	0.0029	0.0021	0.0001	0.0151
$rd_{i,t}$	270	14.7163	1.3118	11.5494	17.2490
$trade_{i,t}$	270	0.1396	0.1381	0.0069	0.6602
$sharehold_{i,t}$	270	0.1058	0.0749	0.0171	0.2953

Note: Table 3 is a statistical description of the standard numerical values (no logarithm) of the main variables in this study. This study used exponential smoothing to interpolate sewage charges data due to data missing for individual years in individual provinces.

5. Empirical Results and Discussions

5.1. Benchmark Regression Results

This study first conducted a regression analysis of the full sample data without considering the regional differences to prove that the green technology innovation in provinces with an enhanced environmental regulation also improved after the policy shock. As mentioned above, this study used the DID method for empirical analysis. The benchmark regression results are shown in Table 4.

Table 4. Benchmark regression results.

	$app\_invention_{i,t}$					
	(1)	(2)	(3)	(4)	(5)	(6)
$treat_{i,t} \times post_{i,t}$	1.0189 ***	0.9586 ***	0.8010 ***	0.2432 ***	0.2236 ***	0.2153 ***
	(0.1013)	(0.1034)	(0.0957)	(0.0632)	(0.0645)	(0.0627)
$subsidy_{i,t}$		14.0406 **	15.7923 ***	0.6152	−0.2664	0.9730
		(5.8608)	(5.2874)	(3.2439)	(3.2959)	(3.2147)
$tax_{i,t}$			−181.1505 ***	−62.8807 ***	−63.9463 ***	−49.4819 ***
			(24.2092)	(15.5571)	(15.5417)	(15.5286)
$rd_{i,t}$				1.2434 ***	1.5540 ***	1.6553 ***
				(0.0663)	(0.0795)	(0.0814)
$trade_{i,t}$					−0.9124	−0.3797
					(0.6421)	(−0.6379)
$sharehold_{i,t}$						4.9062 ***
						(1.2492)
Constant	7.1760 ***	14.0406 **	7.2930 ***	−15.7796 ***	−15.1643 ***	−17.3253 ***
	(0.0433)	(5.8608)	(0.1715)	(1.1205)	(1.1990)	(1.2873)
Year effects	Yes	Yes	Yes	Yes	Yes	Yes
City effects	Yes	Yes	Yes	Yes	Yes	Yes
Observations	270	270	270	270	270	270
R−squared	0.2973	0.3138	0.4449	0.8026	0.8043	0.8164

Note: Standard errors in parentheses; ** and *** represent 5% and 1% significance levels, respectively.

According to the regression results of the model (6), the coefficient of interaction term $treat_{i,t} \times post_{i,t}$ was significantly positive at the 1% level, indicating that the environmental regulation policy had a positive effect on green technology innovation. Furthermore, the

$treat_{i,t} \times post_{i,t}$ increased the $app\_invention_{i,t}$ statistically significantly, by approximately 21.53% with all the control variables. Meanwhile, the promotion effect of government subsidies on green technology innovation was not obvious, but the negative effect of a government pollution tax on green technology innovation was obvious. As expected, R&D investment promoted green technology innovation, and state-owned or collective-owned enterprises had stronger green technology innovation capabilities. The coefficient of the foreign trade dependence was negative but not obvious, indicating that foreign trade dependence is not the main reason affecting green technology innovation.

5.2. Parallel Trend Assumption and Time Trend Analysis

In addition to the premise of randomness, the DID method also needed to verify the parallel trend assumption, that is, it needed to verify that if the treated group was not affected by policy shocks, then the change trend should be the same as the control group. Therefore, if this assumption does not hold, it cannot be explained that the impact on the treated group was caused by policy shocks. At present, for the assumption of a parallel trend, the treated group and the plotting method can be generally used to observe the change trend of the two groups of data. In addition, the year before the policy shock could also be selected as the time dummy variable to observe whether the corresponding interaction terms were significant. If they are not significant, this indicates that the data basically meets the parallel trend assumption. The latter detection method is also more common in practical applications. In this study, we generated the interaction terms between the year virtual variables and the processing group virtual variables, and then regressed these interaction terms as explanatory variables. The coefficients of the interaction terms reflected the difference between the treated and control groups in a specific year. We needed to confirm whether the coefficients of the interaction terms were significant or not, and if not significant, then the data basically met the parallel trend assumption.

Figure 2 shows the dynamic effect of the promulgation and implementation of the NEPL on the green technology innovation activities. The regression results of the parallel trend assumption and time trend are in Appendix A. It can be seen that the estimated coefficients of the years before the implementation of the NEPL were negative and basically insignificant, which indicated that the parallel trend assumption of the DID estimation was satisfied. After the promulgation and implementation of the NEPL, the number of green patent applications had increased significantly from 2015 to 2019. Specifically, the coefficients of the interaction terms in 2015 to 2019 were significantly positive, while the coefficients of the interaction terms from 2013 to 2015 were basically not significant. The above results show that the promulgation and implementation of the NEPL had increased the number of green innovation patents, and that the policy effect was lagging behind. Therefore, Hypothesis 1 was verified.

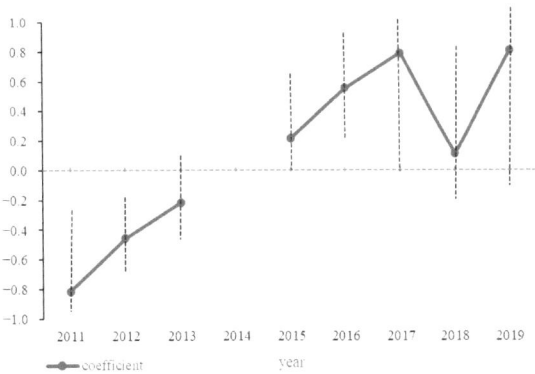

Figure 2. Dynamic effect analysis.

5.3. Regression Results Based on Regional Differences

In addition to the two direct entities of government and enterprise, the differences between the different categories are also worth discussing. This study holds that from the three angles of economic development difference, education difference, and energy consumption difference, the regional interaction terms are constructed to perform regression analysis. The regression results are presented in Table 5. The coefficients of control variables are not showed in Table 5 for simplicity.

Table 5. Regression results based on regional differences.

	$app\_invention_{i,t}$		
Difference in economy levels	(1)	(2)	(3)
$treat_{i,t} \times post_{i,t} \times Hgdp_{i,t}$	0.2278 ** (0.1037)		
$treat_{i,t} \times post_{i,t} \times Mgdp_{i,t}$		0.1020 (0.1082)	
$treat_{i,t} \times post_{i,t} \times Lgdp_{i,t}$		data	0.2533 ** (0.1042)
Observations	270	270	270
R-squared	0.8110	0.8079	0.8119
Difference in education levels	(4)	(5)	(6)
$treat_{i,t} \times post_{i,t} \times Hedu_{i,t}$	0.1885 * (0.1100)		
$treat_{i,t} \times post_{i,t} \times Medu_{i,t}$		0.0627 (0.0962)	
$treat_{i,t} \times post_{i,t} \times Ledu_{i,t}$			0.3361 *** (0.1034)
Observations	270	270	270
R-squared	0.8095	0.8075	0.8155
Difference in energy consumption	(7)	(8)	(9)
$treat_{i,t} \times post_{i,t} \times Henergy_{i,t}$	0.0290 (0.1216)		
$treat_{i,t} \times post_{i,t} \times Menergy_{i,t}$		0.2371 ** (0.0950)	
$treat_{i,t} \times post_{i,t} \times Lenergy_{i,t}$			0.2639 ** (0.1020)
Observations	270	270	270
R-squared	0.8072	0.8121	0.8125

Note: Standard errors in parentheses; *, ** and *** represent 10%, 5%, and 1% significance levels, respectively.

According to the regression results of the environmental regulation policy effect with differences in economic development, the coefficients of $treat_{i,t} \times post_{i,t} \times Hgdp_{i,t}$, $treat_{i,t} \times post_{i,t} \times Mgdp_{i,t}$, and $treat_{i,t} \times post_{i,t} \times Lgdp_{i,t}$ were positive. To be exact, the coefficients of $treat_{i,t} \times post_{i,t} \times Hgdp_{i,t}$ and $treat_{i,t} \times post_{i,t} \times Lgdp_{i,t}$ were significantly positive at the 5% level. Furthermore, $treat_{i,t} \times post_{i,t} \times Hgdp_{i,t}$ and $treat_{i,t} \times post_{i,t} \times Lgdp_{i,t}$ separately increased $app\_invention_{i,t}$ statistically by approximately 22.78% and 25.33% with all the control variables, indicating that the environmental regulation policy was more likely to promote green technology innovation in regions with high or low economic development levels. Conversely, the coefficient of $treat_{i,t} \times post_{i,t} \times Mgdp_{i,t}$ was not obvious, indicating that the environmental regulation policy had little positive effect on green technology innovation in areas with a medium economic development level.

As far as the difference in education level is concerned, the coefficients of the $treat_{i,t} \times post_{i,t} \times Hedu_{i,t}$, $treat_{i,t} \times post_{i,t} \times Medu_{i,t}$, and $treat_{i,t} \times post_{i,t} \times Ledu_{i,t}$ were positive. To be exact, the coefficients of $treat_{i,t} \times post_{i,t} \times Hedu_{i,t}$ and $treat_{i,t} \times post_{i,t} \times Ledu_{i,t}$ were significantly positive at the 10% level and 1% level, respectively. Furthermore, $treat_{i,t} \times post_{i,t} \times Hedu_{i,t}$ and $treat_{i,t} \times post_{i,t} \times Ledu_{i,t}$ increased $app\_invention_{i,t}$ statistically by nearly 18.85% and 33.61% with all the control variables, indicating that the environmental regulation policy was more likely to promote green technology innovation in regions with high or low education levels. Additionally, the coefficient of $treat_{i,t} \times post_{i,t} \times Medu_{i,t}$ was not obvious, indicating that the environmental regulation policy had little positive effect on green technology innovation in areas with a medium education level.

In terms of the difference in energy consumption, the coefficients of $treat_{i,t} \times post_{i,t} \times Henergy_{i,t}$, $treat_{i,t} \times post_{i,t} \times Menergy_{i,t}$, and $treat_{i,t} \times post_{i,t} \times Lenergy_{i,t}$ were positive. To be exact, the coefficients of $treat_{i,t} \times post_{i,t} \times Menergy_{i,t}$ and $treat_{i,t} \times post_{i,t} \times Lenergy_{i,t}$ were significantly positive at the 5% level. Furthermore, $treat_{i,t} \times post_{i,t} \times Menergy_{i,t}$ and $treat_{i,t} \times post_{i,t} \times Lenergy_{i,t}$ increased $app\_invention_{i,t}$ statistically by approximately 23.71% and 26.39% with all the control variables, indicating that the environmental regulation policy was more likely to promote green technology innovation in regions with a lower energy consumption levels. The coefficient of $treat_{i,t} \times post_{i,t} \times Henergy_{i,t}$ was not obvious, indicating that environmental regulation policy had little positive effect on green technology innovation in areas with a high energy consumption level.

Therefore, Hypothesis 2 was verified.

5.4. Robust Check

5.4.1. Replace Explained Variable into Green Invention Patent Acquisition

Although the number of patent applications can reflect the degree of activity and ability of green technology innovation to some extent, not all the patents applied for can be approved. Therefore, this study further selected the quantitative index of patent acquisition as the explained variable and took its logarithm for the regression test to verify the reliability of the research conclusion. The regression results are shown in Table 6.

Table 6. Results of the robustness checks 1.

	\multicolumn{6}{c}{$acq\_invention_{i,t}$}					
	(1)	(2)	(3)	(4)	(5)	(6)
$treat_{i,t} \times post_{i,t}$	0.7782 ***	0.7380 ***	0.6264 ***	0.2086 ***	0.1945 ***	0.841 ***
	(0.0768)	(0.0787)	(0.0738)	(0.0513)	(0.0525)	(8.060)
$subsidy_{i,t}$		9.3644 ***	10.6058 ***	−0.7620	−1.3932	−0.4343
		(4.4564)	(4.0823)	(2.6348)	(2.6796)	(2.6217)
$tax_{i,t}$			−128.3800 ***	−39.7943 ***	−40.5571 ***	−29.3662 ***
			(18.6917)	(12.6359)	(12.6355)	(12.6641)
$rd_{i,t}$				1.1871 ***	1.1650 ***	1.2434 ***
				(0.0623)	(0.0646)	(0.0663)
$trade_{i,t}$					−0.6531	−0.2410
					(0.5221)	(0.5202)
$sharehold_{i,t}$						3.7959 ***
						(1.0187)
Constant	5.8751 ***	5.6059 ***	5.9748 ***	−11.3069 ***	−10.8664 ***	12.5383 ***
	(0.0328)	(0.1322)	(0.1324)	(0.9101)	(0.9748)	(1.0498)
Year effects	Yes	Yes	Yes	Yes	Yes	Yes
City effects	Yes	Yes	Yes	Yes	Yes	Yes
Observations	270	270	270	270	270	270
R-squared	0.3033	0.3131	0.4271	0.7745	0.7760	0.7886

Note: Standard errors in parentheses; *** represents 1% significance level.

In models (1) to (6), the coefficients of the interaction terms $treat_{i,t} \times post_{i,t}$ were significantly positive at the 1% level, indicating that the promulgation of the NEPL had a certain positive promotion effect on green technology innovation, which was basically consistent with the benchmark regression results, indicating that the research conclusions are robust.

5.4.2. Replace Explanatory Variables into Interaction Terms with Regional Differences

According to the regional economic development level, education levels and energy consumption levels, the different provinces were divided into high level and low level. Among them, the provinces with a high economic development level corresponded to the dummy variable 1, and the provinces with a low economic development level corresponded to the dummy variable 0; the provinces with a high education level corresponded to the dummy variable 1, and the provinces with a low education level corresponded to the dummy variable 0; provinces with a high energy consumption level corresponded to the dummy variable 0; and provinces with a low energy consumption level corresponded to the dummy variable 1. The regression results are shown in Table 7.

Table 7. Results of the robustness checks 2.

	$app\_invention_{i,t}$			$acq\_invention_{i,t}$		
	(1)	(2)	(3)	(4)	(5)	(6)
$treat_{i,t} \times post_{i,t} \times gdp_{i,t}$	0.2398 ** (0.0928)			0.1665 ** (0.0762)		
$treat_{i,t} \times post_{i,t} \times edu_{i,t}$		0.1491 * (0.0869)			0.1294 * (0.0711)	
$treat_{i,t} \times post_{i,t} \times energy_{i,t}$			0.2373 *** (0.0686)			0.2069 *** (0.0559)
$subsidy_{i,t}$	0.8468 (3.2614)	0.7513 (3.3253)	0.9939 (3.2129)	−0.3660 (2.6799)	−0.6213 (2.7193)	−0.4148 (2.6204)
$tax_{i,t}$	−55.9687 *** (15.8030)	−50.3484 *** (15.8145)	−55.3665 *** (15.5676)	−34.1441 *** (12.9853)	−30.1273 *** (12.9328)	−34.5012 *** (12.6965)
$rd_{i,t}$	1.7175 *** (0.0784)	1.7372 *** (0.0784)	1.6520 *** (0.0816)	1.3034 *** (0.0644)	1.3150 *** (0.0641)	1.2406 *** (0.0665)
$trade_{i,t}$	−0.0215 (0.7031)	−0.4556 (0.6711)	−0.2694 (0.6446)	−0.0717 (0.5777)	−0.3095 (0.5488)	−0.1454 (0.5257)
$sharehold_{i,t}$	5.1061 *** (1.2618)	5.1012 *** (1.2719)	5.1992 *** (1.2487)	3.9607 *** (1.0368)	3.9660 *** (1.0401)	4.0517 *** (1.0184)
Constant	−18.2645 *** (1.2512)	−18.4954 *** (1.2554)	−17.3028 *** (1.2880)	−13.4239 *** (1.0281)	−13.5614 *** (1.0267)	12.5206 *** (1.0505)
Year effects	Yes	Yes	Yes	Yes	Yes	Yes
City effects	Yes	Yes	Yes	Yes	Yes	Yes
Observations	270	270	270	270	270	270
R-squared	0.8125	0.8095	0.8165	0.7808	0.7794	0.7887

Note: Standard errors in parentheses; *, ** and *** represent 10%, 5%, and 1% significance levels, respectively.

In models (1) to (3), the number of green invention patent applications was used as the explained variable indicator, and the coefficients of interaction terms were significantly positive. In models (4) to (6), the number of green invention patents acquisitions was used as the explained variable indicator, and the interaction coefficients were also significantly positive. The coefficients of the product terms of environmental regulation policy effects were significantly positive, indicating that on the basis of controlling regional differences, the promulgation of the NEPL can have a certain positive promotion effect on green technology innovation, which is basically consistent with the benchmark regression results. It also further explains the impact of regional differences on the impact of the environmental regulation on green technology innovation effect and verifies the empirical results' robustness.

6. Conclusions

6.1. Research Findings

While examining the policy effect of the environmental regulation on green technology innovation, this study further introduced the influence of regional differences on the basis of previous literature research, clarified the mechanism from three levels of economy, education and energy consumption, and deepened the research on the impact of environmental regulation on green technology innovation. From the results of the mechanism analysis and empirical analysis, the key findings of this study are as follows:

First, the New Environmental Protection Law (NEPL) of China has a positive impact on green technology innovation. According to the benchmark regression results and robustness checks, the coefficients interaction terms are significantly positive. That is, the weak Porter hypothesis is established, which shows that the implementation of environmental regulation policy in China has a certain effect at this stage; however, it should be emphasized that the empirical results of this study cannot judge the competitiveness of enterprises so that this study cannot verify whether a 'strong' Porter hypothesis is established. According to the empirical results, the government's pollution charge punishment will crack down on the development of green technology innovation activities, while the R&D investment of enterprises promotes the development of green technology innovation activities, and they show that state-owned and collective enterprises have more enthusiasm to carry out green technology innovation activities.

Second, the impact of the environmental regulation policy on green technological innovation is affected by regional differences. That is, the environmental regulation policy has different effects on green technology innovation under different economic levels, education levels and energy consumption levels. According to the regression results based on regional differences, $treat_{i,t} \times post_{i,t} \times Hgdp_{i,t}$, $treat_{i,t} \times post_{i,t} \times Lgdp_{i,t}$, $treat_{i,t} \times post_{i,t} \times Hedu_{i,t}$, $treat_{i,t} \times post_{i,t} \times Ledu_{i,t}$, $treat_{i,t} \times post_{i,t} \times Menergy_{i,t}$ and $treat_{i,t} \times post_{i,t} \times Lenergy_{i,t}$ can increase $app\_invention_{i,t}$ statistically significantly with all the control variables, but the coefficients of $treat_{i,t} \times post_{i,t} \times Mgdp_{i,t}$, $treat_{i,t} \times post_{i,t} \times Medu_{i,t}$ and $treat_{i,t} \times post_{i,t} \times Henergy_{i,t}$ are not obvious. On the one hand, in regions with high and low levels of economic development and education, the promulgation and implementation of the new environmental law has a significant positive promoting effect on green technology innovation, while the policy effect in regions with medium levels of economic development and education is not significant. Theoretically, regions with a higher economic development level and higher education levels have richer social resources, and the effect of policy implementation should be more obvious. In fact, regions with low levels of economic development and education will receive more government attention and support than medium-level regions. This may be the main reason for the insignificant policy effect in the medium economic development and education level areas in the regression results. On the other hand, in regions with medium and low energy consumption levels, the promulgation and implementation of the new environmental law has a significant positive promoting effect on green technology innovation, while the policy effect in regions with high energy consumption levels is not significant. This is consistent with common sense. The environmental problems in regions with high energy consumption levels are relatively more serious. Environmental regulation policy has played a more regulatory role in limiting their pollution emissions, and their impact on innovation activities is not obvious compared with regions with medium and low energy consumption levels.

6.2. Suggestions

In view of the above findings, this study proposes several suggestions on how to improve the effect of the environmental regulation policy and green technology innovation, so as to provide a reference for government and enterprise decision-making.

On the one hand, for the government, there are four policy suggestions. First, the central government of China should take into account the impact on the optimal level of green technology innovation exerted by decision-making models of enterprises and local governments when formulating effective environmental regulation policies [54]. The government should fully consider the differences between regions and clarify the characteristics of performance and root causes of regional development differences, such as the actual situation of regional industrial structure characteristics and business development. Furthermore, the government should realize that the level of regional economic development, education and energy consumption can impact the level of the effects of government policies [55,56]. The government needs to pay attention to the policy effect results of those regions with medium levels of economic development and education, and to not reduce their attention, but rather they need to provide targeted guidance policies. Second, regional governments should promote interregional exchanges and cooperation while paying attention to their own development. Government departments can act as intermediaries to promote the transfer of green technology innovation among regions and make good use of the external characteristics of green technology innovation. For regions with weak innovation ability, the introduction of green technology should be strengthened, while for regions with strong innovation ability, the output of green technology innovation should be encouraged, so that the efficient transfer and diffusion of green technology innovation can be promoted [57]. Third, environmental problems are not short-term. Similarly, the solution to environmental problems should be viewed from a long-term perspective, rather than being solved overnight. The government should formulate environmental regulation policies that conform to the long-term development of regions and the long-term operation of enterprises, adjusting the regulatory means according to the development stage. It is necessary to avoid business difficulties that may be caused by too strict an environmental regulation, while also avoiding too much environmental pollution from enterprises caused by a weak environmental regulation policy [58–60]. For example, the government should appropriately reduce pollution tax pressure when the level of green technology innovation reaches a certain level in the future. Finally, the government needs to measure the balance between strengthening environmental protection and encouraging green innovation when choosing positive incentive policies and negative punitive policies. If the policy choice is more inclined to encourage green technology innovation, then the government can appropriately reduce the collection of unnecessary punitive emission fees. While encouraging enterprises' innovation activities, the government should also pay attention to the quality of those enterprises' innovation and formulate an evaluation system of the innovation development from multiple perspectives.

On the other hand, in terms of enterprises, first of all, the necessary R&D investment contributes to the improvement of enterprises' innovation abilities. Enterprises can appropriately increase R&D investment for green technology innovation when their own capital base is strong and their operation ability is strong, which is the most important way to enhance enterprises' green technology innovation abilities and to carry out green transformation. Second, the improvement of enterprise nationalization and collectivization contributes to the improvement of that green technology innovation ability [61,62]. Enterprises should learn to cooperate with the government in projects while maintaining their independent development. Compared with large enterprises, small-sized and medium-sized enterprises need more government support and project funding because of their weak capital base and operation ability. In this way, while responding to the development of national green innovation and the protection of the environment, enterprises can also improve their adaptability for survival and thus have a longer-term development.

6.3. Limitations and Future Study

This study had some limitations that should be addressed in future work. First, the number of indicators that measure regional green technology innovation quality was limited. Only one indicator was chosen for measuring green technology innovation: the number of green patent applications and acquisitions; however, green technology innovation activities are not only documented by patents. Therefore, the number of indicators representing the green technological and economic benefits of innovation activities should be increased in future studies, and the measurement of green innovation quality will be involved in further studies. Second, the definition of environmental regulation in this study was relatively simple, the measurement method adopted was relatively singular, and many aspects were not refined. In future research, the definition and measurement of environmental regulation will be more accurately refined and classified. Third, this study measured the impact of environmental regulation on regional green technology innovation from the three aspects of regional economy, education, and energy consumption through a regression analysis. In particular, innovation may bring social benefits to the region, such as the reduction of the unemployment rate, an improvement in people's income and consumption levels, and an increase in people's happiness, which will be addressed in future research. Future studies will extend the influence factors of the regional innovation quality, such as the characteristics of green technology innovation subjects and the degree of urbanization.

Author Contributions: Conceptualization, Q.L. and Y.Z.; methodology, Q.L., Y.Z., W.Y. and X.W.; software, Y.Z.; validation, Q.L., Y.Z. and W.Y.; formal analysis, Q.L., Y.Z. and W.Y.; investigation, Y.Z. and X.W.; resources, Y.Z. and X.W.; data curation, Y.Z. and X.W.; writing—original draft preparation, Q.L., Y.Z., X.W. and W.Y.; writing—review and editing, Q.L., Y.Z. and W.Y.; visualization, Q.L.; supervision, Q.L. and W.Y. All authors have read and agreed to the published version of the manuscript.

Funding: This research was funded and supported by the National Natural Science Foundation of China (71003070), Ministry of Education Key Funding Project (15YJC790060), General Project of Shanghai Philosophy and Social Science Planning (2021BGL014), and Special Project of The Development Research Center of Shanghai Municipal People Government (2018-YJ-L04). We gratefully acknowledge the above financial supports.

Institutional Review Board Statement: Not applicable.

Informed Consent Statement: Not applicable.

Data Availability Statement: The data presented in this study are all from the statistical data officially released by China and have been explained in Section 4.3.

Conflicts of Interest: The authors declare no conflict of interest. The funders had no role in the design of the study; in the collection, analyses, or interpretation of data; in the writing of the manuscript, or in the decision to publish the results.

Appendix A

The regression results of parallel trend assumption and time trend are in Table A1.

Table A1. The regression results of parallel trend assumption and time trend.

	app_invention
Before4	−0.8143 ***
	(0.2034)
Before3	−0.4575 **
	(0.2034)
Before2	−0.2160
	(0.2034)
Current	0.2172
	(0.2034)
After1	0.5551 ***
	(0.2034)
After2	0.7886 ***
	(0.2034)
After3	0.1082 ***
	(0.0673)
After4	0.8108 ***
	(0.2034)
Year effects	Yes
City effects	Yes
Observations	270
R-squared	0.3813

Note: Standard errors in parentheses; ** and *** represent 5%, and 1% significance levels, respectively.

References

1. Cai, W.; Ye, P. How Does Environmental Regulation Influence Enterprises' Total Factor Productivity? A Quasi-Natural Experiment Based on China's New Environmental Protection Law. *J. Clean. Prod.* **2020**, *276*, 124105. [CrossRef]
2. 2018 Xi Jinping Attended the National Conference on Ecological Environment Protection and Delivered an Important Speech Published by the Ministry of Ecology and Environment of China. Available online: https://www.mee.gov.cn/home/ztbd/gzhy/qgsthjbhdh/qgdh_tt/201807/t20180713_446605.shtml (accessed on 21 December 2021). (In Chinese)

3. 'Green Patent', Building Protection Barrier of Ecological Civilization Published by the China National Intellectual Property Administration. Available online: https://www.cnipa.gov.cn/art/2017/8/4/art_55_125994.html (accessed on 21 December 2021). (In Chinese)
4. Zhang, H.; Zhu, Z.; Fan, Y. The Impact of Environmental Regulation on the Coordinated Development of Environment and Economy in China. *Nat. Hazards* **2018**, *91*, 473–489. [CrossRef]
5. Zhang, Y.; Wang, J.; Xue, Y.; Yang, J. Impact of Environmental Regulations on Green Technological Innovative Behavior: An Empirical Study in China. *J. Clean. Prod.* **2018**, *188*, 763–773. [CrossRef]
6. Zhang, N.; Jiang, X.-F. The Effect of Environmental Policy on Chinese Firm's Green Productivity and Shadow Price: A Metafrontier Input Distance Function Approach. *Technol. Forecast. Soc. Chang.* **2019**, *144*, 129–136. [CrossRef]
7. Tang, K.; Qiu, Y.; Zhou, D. Does Command-and-Control Regulation Promote Green Innovation Performance? Evidence from China's Industrial Enterprises. *Sci. Total Environ.* **2020**, *712*, 136362. [CrossRef] [PubMed]
8. Su, X.; Zhou, S. Dual environmental regulation, government subsidy and enterprise innovation output. *China Popul. Resour. Environ.* **2019**, *29*, 31–39. (In Chinese)
9. Zhao, Y.; Zhu, F.; He, L. Definition, Classification and Evolution of Environmental Regulations. *China Popul. Resour. Environ.* **2009**, *19*, 85–90. (In Chinese)
10. Feng, Z.; Chen, W. Environmental Regulation, Green Innovation, and Industrial Green Development: An Empirical Analysis Based on the Spatial Durbin Model. *Sustainability* **2018**, *10*, 223. [CrossRef]
11. Lanoie, P.; Patry, M.; Lajeunesse, R. Environmental Regulation and Productivity: Testing the Porter Hypothesis. *J Prod Anal* **2008**, *30*, 121–128. [CrossRef]
12. Zhang, C.; Lu, Y.; Guo, L.; Yu, T. The Intensity of Environmental Regulation and Technological Progress of Production. *Econ. Res. J.* **2011**, *46*, 113–124. (In Chinese)
13. Wang, R.; Sun, T. Research on the Influence of Environmental Regulation on China's Regional Green Economy Efficiency Based on Super Efficiency DEA Model. *Ecol. Econ.* **2019**, *35*, 131–136. (In Chinese)
14. Fu, J.; Li, L. A Case Study on the Environmental Regulation, the Factor Endowment and the International Competitiveness in Industries. *Manag. World* **2010**, *10*, 87–98. (In Chinese) [CrossRef]
15. Li, L.; Tao, F. Selection of Optimal Environmental Regulation Intensity for Chinese Manufacturing Industry—Based on the Green TFP Perspective. *China Ind. Econ.* **2012**, *5*, 70–82. (In Chinese) [CrossRef]
16. Qian, Z.; Liu, X. Environmental Regulation and Green Economic Efficiency. *Stat. Res.* **2015**, *32*, 12–18. (In Chinese) [CrossRef]
17. Feng, M.; Li, X. Evaluating the Efficiency of Industrial Environmental Regulation in China:A Three-Stage Data Envelopment Analysis Approach. *J. Clean. Prod.* **2020**, *242*, 118535. [CrossRef]
18. Liu, Y.; Zhu, J.; Li, E.Y.; Meng, Z.; Song, Y. Environmental Regulation, Green Technological Innovation, and Eco-Efficiency: The Case of Yangtze River Economic Belt in China. *Technol. Forecast. Soc. Chang.* **2020**, *155*, 119993. [CrossRef]
19. Fang, Z.; Bai, H.; Bilan, Y. Evaluation Research of Green Innovation Efficiency in China's Heavy Polluting Industries. *Sustainability* **2020**, *12*, 146. [CrossRef]
20. Fan, F.; Lian, H.; Liu, X.; Wang, X. Can Environmental Regulation Promote Urban Green Innovation Efficiency? An Empirical Study Based on Chinese Cities. *J. Clean. Prod.* **2021**, *287*, 125060. [CrossRef]
21. Li, J.; Du, Y. Spatial Effect of Environmental Regulation on Green Innovation Efficiency: Evidence from Prefectural-Level Cities in China. *J. Clean. Prod.* **2021**, *286*, 125032. [CrossRef]
22. Tao, F.; Zhao, J.; Zhou, H. Does Environmental Regulation Improve the Quantity and Quality of Green Innovation——Evidence from the Target Responsibility System of Environmental Protection. *China Ind. Econ.* **2021**, *2*, 136–154. (In Chinese) [CrossRef]
23. China Green Patent Statistics Report Published by the China National Intellectual Property Administration. Available online: https://www.cnipa.gov.cn/col/col87/index.html?uid=669&pageNum=3 (accessed on 21 December 2021). (In Chinese)
24. Porter, M.E.; van der Linde, C. Toward a New Conception of the Environment-Competitiveness Relationship. *J. Econ. Perspect.* **1995**, *9*, 97–118. [CrossRef]
25. Lanjouw, J.O.; Mody, A. Innovation and the International Diffusion of Environmentally Responsive Technology. *Res. Policy* **1996**, *25*, 549–571. [CrossRef]
26. Domazlicky, B.R.; Weber, W.L. Does Environmental Protection Lead to Slower Productivity Growth in the Chemical Industry? *Environ. Resour. Econ.* **2004**, *28*, 301–324. [CrossRef]
27. Yang, C.-H.; Tseng, Y.-H.; Chen, C.-P. Environmental Regulations, Induced R&D, and Productivity: Evidence from Taiwan's Manufacturing Industries. *Resour. Energy Econ.* **2012**, *34*, 514–532. [CrossRef]
28. Mazzanti, M.; Zoboli, R. Environmental Efficiency and Labour Productivity: Trade-off or Joint Dynamics? A Theoretical Investigation and Empirical Evidence from Italy Using NAMEA. *Ecol. Econ.* **2009**, *68*, 1182–1194. [CrossRef]
29. Xing, X.; Liu, T.; Wang, J.; Shen, L.; Zhu, Y. Environmental Regulation, Environmental Commitment, Sustainability Exploration/Exploitation Innovation, and Firm Sustainable Development. *Sustainability* **2019**, *11*, 6001. [CrossRef]
30. Yuan, B.; Xiang, Q. Environmental Regulation, Industrial Innovation and Green Development of Chinese Manufacturing: Based on an Extended CDM Model. *J. Clean. Prod.* **2018**, *176*, 895–908. [CrossRef]
31. Chintrakarn, P. Environmental Regulation and U.S. States' Technical Inefficiency. *Econ. Lett.* **2008**, *100*, 363–365. [CrossRef]
32. Song, M.; Wang, S.; Zhang, H. Could Environmental Regulation and R&D Tax Incentives Affect Green Product Innovation? *J. Clean. Prod.* **2020**, *258*, 120849. [CrossRef]

33. Wang, X.; Sun, C.; Wang, S.; Zhang, Z.; Zou, W. Going Green or Going Away? A Spatial Empirical Examination of the Relationship between Environmental Regulations, Biased Technological Progress, and Green Total Factor Productivity. *Int. J. Environ. Res. Public Health* **2018**, *15*, 1917. [CrossRef]
34. Yuan, B.; Ren, S.; Chen, X. Can Environmental Regulation Promote the Coordinated Development of Economy and Environment in China's Manufacturing Industry?–A Panel Data Analysis of 28 Sub-Sectors. *J. Clean. Prod.* **2017**, *149*, 11–24. [CrossRef]
35. Yuan, B.; Zhang, Y. Flexible Environmental Policy, Technological Innovation and Sustainable Development of China's Industry: The Moderating Effect of Environment Regulatory Enforcement. *J. Clean. Prod.* **2020**, *243*, 118543. [CrossRef]
36. Wang, X.; Shao, Q. Non-Linear Effects of Heterogeneous Environmental Regulations on Green Growth in G20 Countries: Evidence from Panel Threshold Regression. *Sci. Total Environ.* **2019**, *660*, 1346–1354. [CrossRef]
37. Guo, Y.; Xia, X.; Zhang, S.; Zhang, D. Environmental Regulation, Government R&D Funding and Green Technology Innovation: Evidence from China Provincial Data. *Sustainability* **2018**, *10*, 940. [CrossRef]
38. Xie, R.; Yuan, Y.; Huang, J. Different Types of Environmental Regulations and Heterogeneous Influence on "Green" Productivity: Evidence from China. *Ecol. Econ.* **2017**, *132*, 104–112. [CrossRef]
39. Cai, W.; Li, Q. Dual effect of environmental regulation on enterprise's eco-technology innovation. *Sci. Res. Manag.* **2019**, *40*, 87–95. (In Chinese) [CrossRef]
40. Feng, Z.; Zeng, B.; Ming, Q. Environmental Regulation, Two-Way Foreign Direct Investment, and Green Innovation Efficiency in China's Manufacturing Industry. *Int. J. Environ. Res. Public Health* **2018**, *15*, 2292. [CrossRef]
41. Wang, H.; Zhang, Y. Trade Structure Upgrading, Environmental Regulation and Green Technology Innovation in Different Regions of China. *China Soft Sci.* **2020**, *2*, 174–181. (In Chinese)
42. Cai, X.; Zhu, B.; Zhang, H.; Li, L.; Xie, M. Can Direct Environmental Regulation Promote Green Technology Innovation in Heavily Polluting Industries? Evidence from Chinese Listed Companies. *Sci. Total Environ.* **2020**, *746*, 140810. [CrossRef]
43. Wang, Z.; Cao, Y.; Lin, S. The characteristics and heterogeneity of environmental regulation's impact on enterprises' green technology innovation—Based on green patent data of listed firms in China. *Stud. Sci. Sci.* **2021**, *39*, 909–919. (In Chinese) [CrossRef]
44. Cai, X.; Lu, Y.; Wu, M.; Yu, L. Does Environmental Regulation Drive Away Inbound Foreign Direct Investment? Evidence from a Quasi-Natural Experiment in China. *J. Dev. Econ.* **2016**, *123*, 73–85. [CrossRef]
45. Jiang, F.; Wang, Z.; Bai, J. The Dual Effect of Environmental Regulations' Impact on Innovation—An Empirical Study Based on Dynamic Panel Data of Jiangsu Manufacturing. *China Ind. Econ.* **2013**, *7*, 44–55. (In Chinese) [CrossRef]
46. Milani, S. The Impact of Environmental Policy Stringency on Industrial R&D Conditional on Pollution Intensity and Relocation Costs. *Env. Resour. Econ* **2017**, *68*, 595–620. [CrossRef]
47. Wang, B.; Zhang, M. Marketization, collaboration with academics and innovation performance of enterprises. *Stud. Sci. Sci.* **2015**, *33*, 748–757. (In Chinese) [CrossRef]
48. Chen, S.X. The Effect of a Fiscal Squeeze on Tax Enforcement: Evidence from a Natural Experiment in China. *J. Public Econ.* **2017**, *147*, 62–76. [CrossRef]
49. Cui, G.; Jiang, Y. The Influence of Environmental Regulation on the Behavior of Enterprise Environmental Governance:Based on a Quasi-Natural Experiment of New Environmental Protection Law. *Bus. Manag. J.* **2019**, *41*, 54–72. (In Chinese) [CrossRef]
50. 2020 China Patent Survey Report Published by the China National Intellectual Property Administration. Available online: https://www.cnipa.gov.cn/col/col88/index.html (accessed on 21 December 2021). (In Chinese)
51. Jiang, Z.; Wang, Z.; Li, Z. The Effect of Mandatory Environmental Regulation on Innovation Performance: Evidence from China. *J. Clean. Prod.* **2018**, *203*, 482–491. [CrossRef]
52. He, R.; Zhu, D.; Chen, X.; Cao, Y.; Chen, Y.; Wang, X. How the Trade Barrier Changes Environmental Costs of Agricultural Production: An Implication Derived from China's Demand for Soybean Caused by the US-China Trade War. *J. Clean. Prod.* **2019**, *227*, 578–588. [CrossRef]
53. Dai, J. Research on the Attribute of Policy Instruments in State-owned Enterprises. *Economist* **2013**, *8*, 65–70. (In Chinese) [CrossRef]
54. Deng, Y.; You, D.; Wang, J. Optimal Strategy for Enterprises' Green Technology Innovation from the Perspective of Political Competition. *J. Clean. Prod.* **2019**, *235*, 930–942. [CrossRef]
55. Gao, H.; Yang, W.; Wang, J.; Zheng, X. Analysis of the Effectiveness of Air Pollution Control Policies based on Historical Evaluation and Deep Learning Forecast: A Case Study of Chengdu-Chongqing Region in China. *Sustainability* **2021**, *13*, 206. [CrossRef]
56. Liu, H.; Liu, J.; Yang, W.; Chen, J.; Zhu, M. Analysis and Prediction of Land Use in Beijing-Tianjin-Hebei Region: A Study Based on the Improved Convolutional Neural Network Model. *Sustainability* **2020**, *12*, 3002. [CrossRef]
57. Yuan, G.; Yang, W. Study on optimization of economic dispatching of electric power system based on Hybrid Intelligent Algorithms (PSO and AFSA). *Energy* **2019**, *183*, 926–935. [CrossRef]
58. Yang, W.; Yang, Y. Research on Air Pollution Control in China: From the Perspective of Quadrilateral Evolutionary Games. *Sustainability* **2020**, *12*, 1756. [CrossRef]
59. Yang, Y.; Yang, W.; Chen, H.; Li, Y. China's energy whistleblowing and energy supervision policy: An evolutionary game perspective. *Energy* **2020**, *213*, 118774. [CrossRef]
60. Shen, X.; Yang, W.; Sun, S. Analysis of the Impact of China's Hierarchical Medical System and Online Appointment Diagnosis System on the Sustainable Development of Public Health: A Case Study of Shanghai. *Sustainability* **2019**, *11*, 6564. [CrossRef]

61. Li, Y.; Yang, W.; Shen, X.; Yuan, G.; Wang, J. Water Environment Management and Performance Evaluation in Central China: A Research Based on Comprehensive Evaluation System. *Water* **2019**, *11*, 2472. [CrossRef]
62. Yang, W.; Li, L. Energy Efficiency, Ownership Structure, and Sustainable Development: Evidence from China. *Sustainability* **2017**, *9*, 912. [CrossRef]

Article

Trends of Studies on Controlled Halogenated Gases under International Conventions during 1999–2018 Using Bibliometric Analysis: A Global Perspective

Jing Wang [1,†], Hui-Zhen Fu [2,†], Jiaqi Xu [1], Danqi Wu [1], Yue Yang [1], Xiaoyu Zhu [1] and Jing Wu [1,*]

1 The MOE Key Laboratory of Resource and Environmental System Optimization, College of Environmental Science and Engineering, North China Electric Power University, Beijing 102206, China; 120192232390@ncepu.edu.cn (J.W.); jx2466@columbia.edu (J.X.); wudanqi8592@163.com (D.W.); e0732769@u.nus.edu (Y.Y.); zhuxiaoyu@ncepu.edu.cn (X.Z.)
2 Department of Information Resources Management, School of Public Affairs, Zhejiang University, Hangzhou 310058, China; fuhuizhen@zju.edu.cn
* Correspondence: wujing.108@163.com; Tel.: +86-1061772891
† Jing Wang and Hui-Zhen Fu contributed equally to the article.

Abstract: A lot of research on international convention-controlled halogenated gases (CHGs) has been carried out. However, few bibliometric analyses and literature reviews exist in this field. Based on 734 articles extracted from the Science Citation Index (SCI) Expanded database of the Web of Science, we provided the visualisation for the performance of contributors and trends in research content by using VOSviewer and Science of Science (Sci2). The results showed that the United States was the most productive country, followed by the United Kingdom and China. The National Oceanic and Atmospheric Administration had the largest number of publications, followed by the Massachusetts Institute of Technology (MIT) and the University of Bristol. In terms of disciplines, environmental science and meteorological and atmospheric science have contributed the most. By using cluster analysis of all keywords, four key research topics of CHGs were identified and reviewed: (1) emissions calculation, (2) physicochemical analysis of halocarbons, (3) evaluation of replacements, and (4) environmental impact. The change in research substances is closely related to the phase-out schedule of the Montreal Protocol. In terms of environmental impact, global warming has always been the most important research hotspot, whereas research on ozone-depleting substances and biological toxicity shows a gradually rising trend.

Keywords: global research trend; SCI-Expanded database; scientometrics; halogenated gases; climate change; ozone depletion

Citation: Wang, J.; Fu, H.-Z.; Xu, J.; Wu, D.; Yang, Y.; Zhu, X.; Wu, J. Trends of Studies on Controlled Halogenated Gases under International Conventions during 1999–2018 Using Bibliometric Analysis: A Global Perspective. *Sustainability* **2022**, *14*, 806. https://doi.org/10.3390/su14020806

Academic Editors: Weixin Yang, Guanghui Yuan and Yunpeng Yang

Received: 9 November 2021
Accepted: 5 December 2021
Published: 12 January 2022

Publisher's Note: MDPI stays neutral with regard to jurisdictional claims in published maps and institutional affiliations.

Copyright: © 2022 by the authors. Licensee MDPI, Basel, Switzerland. This article is an open access article distributed under the terms and conditions of the Creative Commons Attribution (CC BY) license (https://creativecommons.org/licenses/by/4.0/).

1. Introduction

Halogenated gases deplete ozone and contribute to global warming and have received widespread attention. One of the major characteristics is their extremely high reactivity with electrons [1]. When they reach the stratosphere after being emitted from the Earth's surface, they absorb ultraviolet radiation and decompose, generating halogen radicals. Halogen radicals are involved in very effective catalytic chain reactions that deplete the ozone layer, causing a decrease in the ozone concentrations of the stratosphere. Ozone depletion allows more solar ultraviolet-B radiation (290–320 nm wavelength) to reach the surface [2], which can cause severe harm to animals, plants, and microorganisms [3–5]. In addition, these halogenated gases are potent greenhouse gases. The Synthesis Report (SYR) of the Intergovernmental Panel on Climate Change (IPCC) Fifth Assessment Report [6] highlighted that the cumulative radiative forcing of all halocarbons from 1750 to 2011 accounted for approximately 13% of the total radiative forcing of greenhouse gases. Due to the dual environmental impact of halogenated gases, the Montreal Protocol and its amendments included chlorofluorocarbons (CFCs), hydrochlorofluorocarbons (HCFCs), and carbon tetrachloride (CCl_4) as ozone-depleting substances that need to be regulated. Subsequently, hydrofluorocarbons (HFCs), perfluorocarbons (PFCs), sulfur hexafluoride (SF_6), and nitrogen trifluoride (NF_3) were listed by the Kyoto Protocol and the Paris Agreement as greenhouse gases that need to be regulated. The Kigali Amendment to the Montreal Protocol called for the phase-down of HFCs in 2016. Hence, the research object of this study is the above-mentioned international convention controlled halogenated gases (CHGs).

The international community attaches great importance to the impact of ozone depletion on the environment. The United Nations Environment Programme (UNEP) organised the Environmental Impact Assessment Committee for Ozone Depletion. Since 1988, research progress on the impact of ozone layer depletion on the environment has been announced to the world in the form of assessment reports every four years. Since the implementation of the Montreal Protocol, ozone depletion has been alleviated to a certain extent [7,8]. Stratospheric ozone is expected to return to the 1960 levels by the end of the 21st century [9]. However, recent scientific studies have found that some CHG concentration and emission trends are different from those expected. For example, the study of Montzka et al. [10] shows that although reported production has been close to zero since 2006, CFC-11 emissions have increased by 13 ± 5 gigagrams ($25 \pm 13\%$) per year since 2012. This discovery brought more attention to CHGs.

The greenhouse effect and global warming caused by CHGs have also received wide international attention, and the focus has increased more in recent years. IPCC has published reports on climate change since 1990. The reports show the growth trend of halogenated compounds over several decades and indicate that the atmospheric content of HFCs, PFCs, SF_6, etc. has increased rapidly since the 1990s. In April 2021, at the China-France-Germany Video Summit, China announced that it decided to accept the Kigali Amendment to strengthen the control of non-CO_2 greenhouse gases such as HFCs. The joint statement issued by China and the United States in response to the climate crisis also highlighted that the two countries will separately implement measures to gradually reduce the production and consumption of HFCs.

CHGs are concerned with two major scientific issues of global concern; however, there are few quantitative summaries and critical reviews in this field. Therefore, it is necessary to systematically summarise the literature and clarify the research hotspots and future research trends in this field.

Bibliometrics was first introduced by Pritchard (1969). It has been widely used as an effective and useful tool for evaluating scientific results and research topics in specific research fields [11,12]. The performance of national contributors, institutional contributors, and authors at different levels is an important factor in understanding a field [13,14]. These studies assume that the number of publications of a country in a specific scientific subfield reflects its commitment to the state of science and is a reasonable indicator of its contribution to research and development in that field. Collaboration has intensified in recent years owing to the rapid development of scientific communication [15]. Collaboration also leads to a higher citation impact in practically all science areas [16–18]. Therefore, the study of collaboration patterns between researchers and regions could provide important references for other scientific researchers and policy managers.

Another important concern in bibliometric studies is the identification of the research topics. The performance of contributors and their collaborations is not a complete indication of trends or future directions in the research field [19]. Information closer to the study itself, including source title, author keyword, keywords plus, and abstracts [20,21], should be introduced to study the research trend. The analysis combining the words in the title, author keywords, and keywords plus could minimise some limitations, such as the incomplete meaning of single words in the title, small sample size for author keywords, and indirect relationship between keywords plus and the research emphases [11]. These types of words are checked by time periods to show the trends and to minimise year-to-year fluctuations. Therefore, word cluster analysis combining author keywords, keywords plus, and title content words has proved to be a more effective and comprehensive bibliometric method, which has been successfully applied to reveal research trends and hotspots in the research fields of risk assessment [22], drinking water [14], and pluripotent stem cells [23]. The researchers express the opinion that the collaborative application of co-occurrence analysis and word cluster analysis can shed light not only on research trends, but also on the role of landmark works in the evolution of the research field.

In recent years, there has been increasing interest in visualising scientometrics using data mining and information retrieval to uncover possible collaborative behaviour patterns among contributors. VOSViewer [18] and Science of Science (Sci 2) [24] are newly developed tools that are interactive visualisation and exploration platforms for various networks and complex systems. These tools have been used to develop interactive superposed scientific maps based on the relationship between text words [25], co-author [26], and cooperation among research institutes [27].

Based on the relevant publications retrieved from the web of science database from 1990 to 2018, this study carried out a bibliometric measurement of the CHGs field. Through the quantitative analysis of the literature, the contributions of countries/regions, institutions, individuals, and disciplines are studied. More importantly, through co-citation analysis and word clustering analysis, the themes and hotspots in the field of CHGs are elaborated to provide a quick and in-depth understanding of the field.

2. Methodology

2.1. Data Collection

Data were obtained from the online version of the Science Citation Index (SCI) Expanded database of the Web of Science from Thomson Reuters in September 2019. ("chlorofluorocarbon*", "hydrochlorofluorocarbon*", "hydrofluorocarbon*", "hydrofluoroolefin*", " perfluorocarbon*", "sulphur hexafluoride", "nitrogen trifluoride", "halons", "carbon tetrachloride", "methyl bromide", "bromochloromethane", "dichloromethane", "chloroform", "trichloromethane", "perchloroethylene") AND ("global warming", "climate warming*", "ozone deplet*", "climate chang*", "climatic chang*", "Greenhouse gas", "radiation forc*") were searched in terms of topic within the publication years of 1999–2018. A total of 1116 publications met the inclusion criteria. Journal articles were selected for further analysis because they are the predominant article type, and the entire research objectives and results are also included in the article [20]. The "front page" was another filter condition [28]; therefore, only articles that contain search terms in the text of their "front page" (including article titles, abstracts, and keywords) were included. This resulted in 734 publications on CHGs over an 18-year period that were considered herein.

To obtain an overview of CHG research, the number of articles published annually from 1999 to 2018 is shown in Figure 1. The number of CHGs publications increased with several fluctuations from 31 in 1999 to 60 in 2018. An increasing number of journals published articles on CHGs. The average article length fluctuated slightly, with an overall average of 8.6 pages. In 1999, there were 40 references per paper, whereas in 2018, there were 48 references per paper—a slight increase over the past two decades.

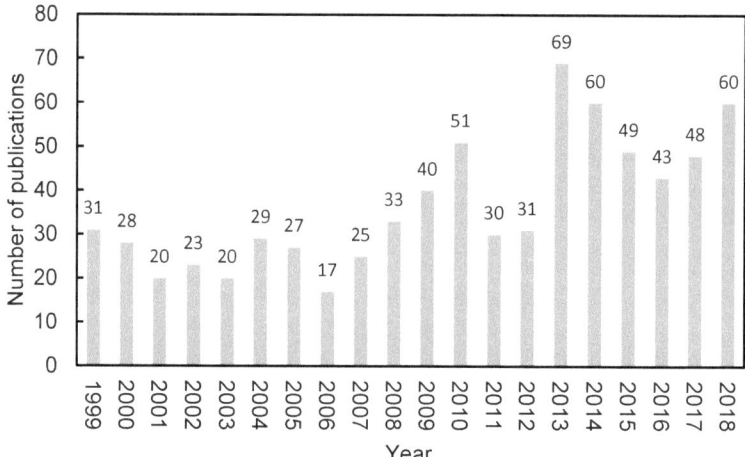

Figure 1. Annual number of publications on CHGs research during 1999–2018.

2.2. Methods

The downloaded content included author names, title, journal, abstract, contact addresses of the authors, year of publication, keywords, and keywords plus and Web of Science categories of the article. Records were downloaded into a spreadsheet; the country of origin of the collaborator, impact factor of the journals, and number of authors for additional coding [29] were incorporated manually. Articles from England, Scotland, Northern Ireland, and Wales were defined as coming from the United Kingdom (UK). Articles originating from Hong Kong were classified as originating from China. The contributions of different countries and institutions were set to include at least one author in the publication. The type of collaboration was determined by the researchers' address; if all the researchers were from the same country, the term "single-country article" was specified. The term "international collaborative articles" referred to articles written jointly by researchers in more than one country.

VOSviewer was developed by researchers at Leiden University in 2007. It is used to build a visual bibliometric network, such as researcher collaboration networks. VOSviewer also provides text mining functions, creating a visual co-occurrence network of important terms extracted from scientific literature [30]. In this study, VOSviewer was used to analyse the collaboration networks of contributors and co-occurrence networks. The Science of Science (Sci2) Tool is a modular toolset specifically designed to visualise scholarly datasets, study of science, network analysis, and supporting geospatial [24]. In this study, the Sci2 Tool was used to analyse the global geographic distribution of publications and the collaboration model of publications.

3. Performance of Contributors

3.1. Macro Contributors of Country/Territory

Cooperation between countries was discussed in depth using the Sci2 and Gephi tools. Figure 2 reveals the different patterns of the global geographic distribution of CHGs research publications. The shades of yellow to blue colour correspond to the total number of publications in the country from 1999 to 2018. The deeper the shade, the more papers the country publishes. The lines between any two countries represent a cooperative relationship between them. The thicker the line, the more intensive the international cooperation between the two countries.

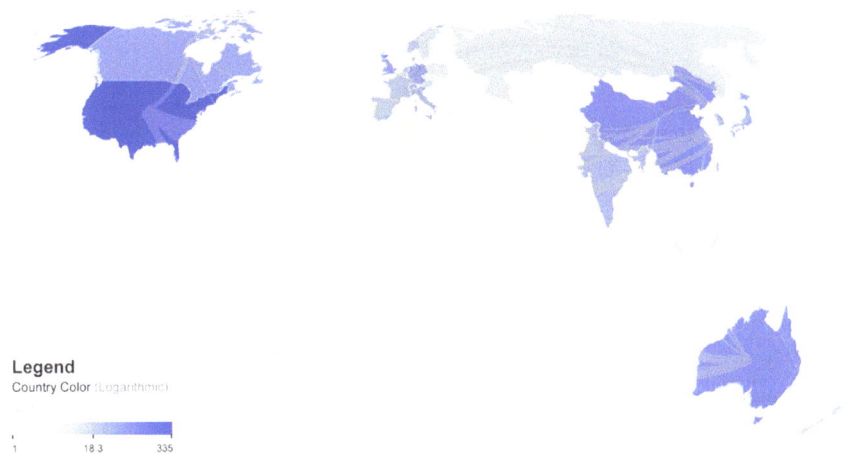

Figure 2. Global geographical distribution of publications and collaboration patterns of CHGs publications. Color represents the number of publications between 1999 and 2018; the darker the color, the greater the number of publications. Lines between any two countries represent a cooperative relationship between them. The thicker the line, the more intensive the international cooperation between the two countries.

Judging from the geographical distribution of publications, the number of countries/regions contributing articles to CHGs publications has increased significantly. Therefore, international cooperation has been greatly strengthened. The United States of America (USA) showed the highest contribution with 335 articles (46%), followed by the UK (125 articles; 17%) and Germany (75 articles).

To demonstrate international collaboration, Figure 2 shows the current partnerships among the 60 countries with high citation rates in the field. The map has 60 nodes and 263 undirected weighted edges, indicating 263 cooperative country pairs for the 60 countries. Nodes with more international cooperation articles are larger, whereas countries with more cooperation are connected through thicker edges. The United States is at the centre of the global network of cooperation. The US–UK collaboration was the strongest with 60 articles, followed by the Australia–UK collaboration with 35 articles. The United States is the most favoured national scientific partner in the field of electronics and electrical engineering, possibly due to its high level of research and its leading position. About half of the pairs (129 out of 263) had only one article. Other countries have not developed significant research networks between them, possibly because of the small number of publications. It was not surprising that countries with fewer publications dominate because this pattern has emerged in most scientific fields [31].

It should be noted that there is little research on CHGs in emerging countries. The rapid development and industrialization of those regions meant that large quantities of halogenated gases could be released from the region and cause damage to the ozone layer. It would be interesting to further stimulate more research in those regions and to foster international collaboration with more mature research teams.

3.2. Meso Contributors of Institution

The top 10 institutions in terms of productivity are listed in Table 1. Research institutes are concentrated in North America, Europe, and Asia. National Oceanic and Atmospheric Administration (NOAA) (65) published the most articles in North America, the University of Bristol (51) published the most articles in Europe, and the Peking University published the most articles in Asia (14). NOAA had the greatest number of publications with a total of 65 papers. At the second position is the Massachusetts Institute of Technology (MIT), with 52 publications, followed by the University of Bristol (51 publications), University of Colorado (50 publications), and University of California-San Diego (UCSD, 48 publications). In terms of citations, NOAA (3107 publications) was the most prolific institution, followed by MIT (2577), UCSD, National Aeronautics and Space Administration (NASA, 2489), the University of Bristol (2357), National Center for Atmospheric Research (NCAR) (2123), and Commonwealth Scientific and Industrial Research Organization (CSIRO) (2078).

Table 1. Top 10 research institutions in the field of CHGs.

Rank	The Name of Institution	Number of Publications	Citations/Publications	Number of Collaborators
1	NOAA	65	48	63
2	MIT	52	50	49
3	Univ Bristol	51	46	48
4	Univ Colorado	50	44	46
5	UCSD	48	52	52
6	NASA	47	53	44
7	NCAR	28	76	34
8	UNIV CALIF IRVINE	27	63	36
9	CEIRO	25	83	38
10	Ford Motor Company	20	882	10

The co-authorship institutional analysis network had a minimum threshold of five publications. The cooperative relationships among 87 institutions are shown in Figure 3. Each node in the figure represents an institution, the size of nodes represents the number of articles, the line between nodes represents the cooperation between institutions, and the thickness of the line represents the link strength between institutions. NOAA and the University of Colorado were the most strongly linked with 32 articles. In addition, the institutions with more cooperation are MIT, UCSD, and Univ Bristol. The cooperation between each of these organizations has reached more than 26. These phenomena are very reasonable. The work of NOAA Earth System Research Laboratories (ESRL) is dominated by its work in University of Colorado Boulder, so it is no surprise that there is a strong NOAA–Univ Colorado connection. MIT, UCSD, and Univ Bristol are research institutions of Advanced Global Atmospheric Gases Experiment (AGAGE), one of the most advanced, most systematic, and most contributing international observation networks for ozone-depleting substances (ODS) and fluorine-containing greenhouse gas observation technologies, and the results are shared among member institutions. This network is mainly sponsored by NASA's Atmospheric Composition Focus Area in Earth Science, and its research institutions also include CSIRO, Swiss Federal Laboratories for Materials Science and Technology (EMPA), University of Urbino, etc. It can be seen in Figure 3 that the AGAGE member units are all marked in blue, which shows that they have close cooperation and similar research.

3.3. Micro Contributors of Authors

The minimum number of publications for each author was set to five, and 82 authors were screened. Some of the 82 authors in the network were not connected to each other. To improve the visualisation, we eliminated unconnected authors, and finally presented 39 authors in the final network map of co-authorship authors. Due to this process, some authors do not appear in Figure 4.

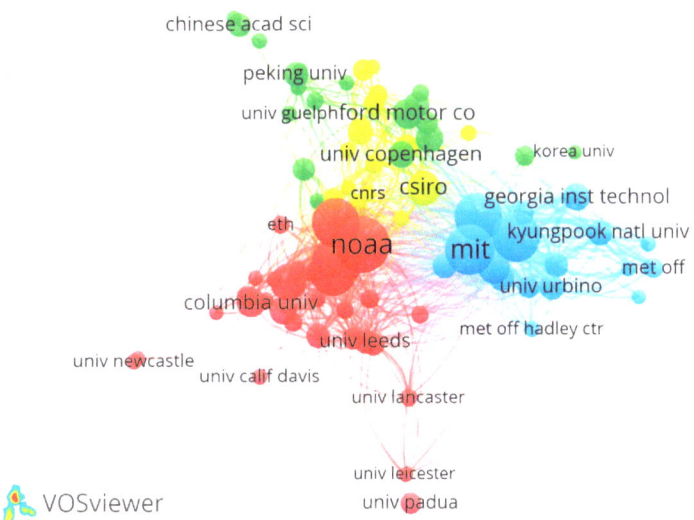

Figure 3. Cooperation relationship among different institution. Dots represent institutions, while dot size represents the number of published documents. Lines between dots indicate a connection between two institutions; thicker lines show a stronger connection and indicate a higher number of collaborated articles.

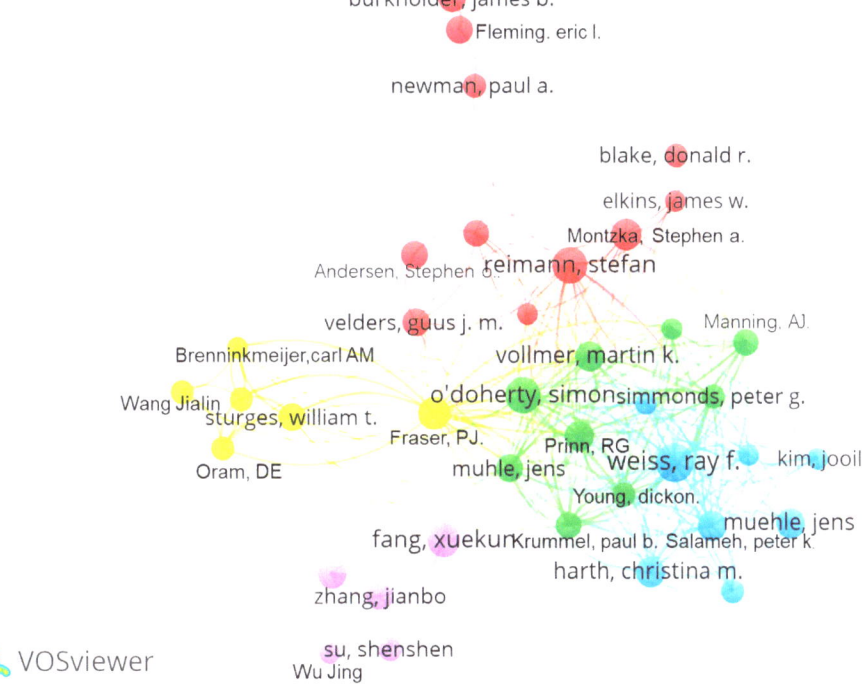

Figure 4. Cooperation relationship among different authors. Each node represents an author, the size of nodes represents the number of papers, the line between nodes represents the cooperation between authors, and the thickness of the line represents the link strength between authors.

McCulloch, A was the most prolific author, with 16 publications, followed by Weiss, RF with 15 publications. The top 10 authors are listed in Table 2. There were five clusters with different colours; authors in the same cluster usually suggested that they studied a similar field and had close cooperation with each other. There are two clusters of authors who collaborated more with others: blue cluster represented by Weiss, RF, Salameh, PK, Harth, CM, Jens Mühle, etc. (the main research direction was atmospheric observation); green cluster represented by Rigby, M, O'Doherty, S, and Martin K, whose main research direction was emission estimation. The authors in the red and yellow clusters are also doing observation and emission research, but there are slight differences. The research in yellow clustering is more inclined to aircraft-based observations, and more emphasis is placed on the emission of short-lived halocarbons. The main research direction of the author in purple clustering is to establish emission inventories based on production and consumption data.

Table 2. Top 10 authors ranked by publications.

Rank	The Name of Author	Number of Publications	Citations/Publications	Number of Collaborators
1	McCulloch, A	16	102	-
2	Weiss, RF	15	23	25
3	O'Doherty, S	13	8	27
4	Reimann, S	13	10	25
5	Fang, Xuekun	12	8	23
6	Fraser, PJ	11	12	28
7	Simmonds, PG	10	135	21
8	Montzka, SA	10	20	19
9	Shine, KP	10	61	-
10	Prinn, PG	9	4	27

4. Research Topics

4.1. Macro Topic of Category

To show the interdisciplinarity and distribution of disciplines, the interdisciplinary relationship among 90 disciplines in the current CHGs field was analysed (Figure 5). In total, 90 categories and 189 undirected weighted edges are present on the map, indicating 189 interdisciplinary pairs in these 90 categories. The node size is directly proportional to the number of articles published. Thicker edge connections represent more frequent interdisciplinary categories. Environmental sciences represent the largest macro-field with 650 articles, followed by physical sciences (456 articles), life sciences, and medicine (138 articles). Environmental sciences contributed the most (248 articles, 34%), followed by meteorology and atmospheric sciences (147 articles, 20%), and environmental engineering (71 articles, 9.6%). The interdisciplinarity of the environmental sciences, meteorology, and atmospheric sciences was the strongest with 100 articles, followed by the interdisciplinary pair of environmental sciences and environmental engineering with 64 articles, and the pair of the mechanical engineering and thermodynamics with 30 articles.

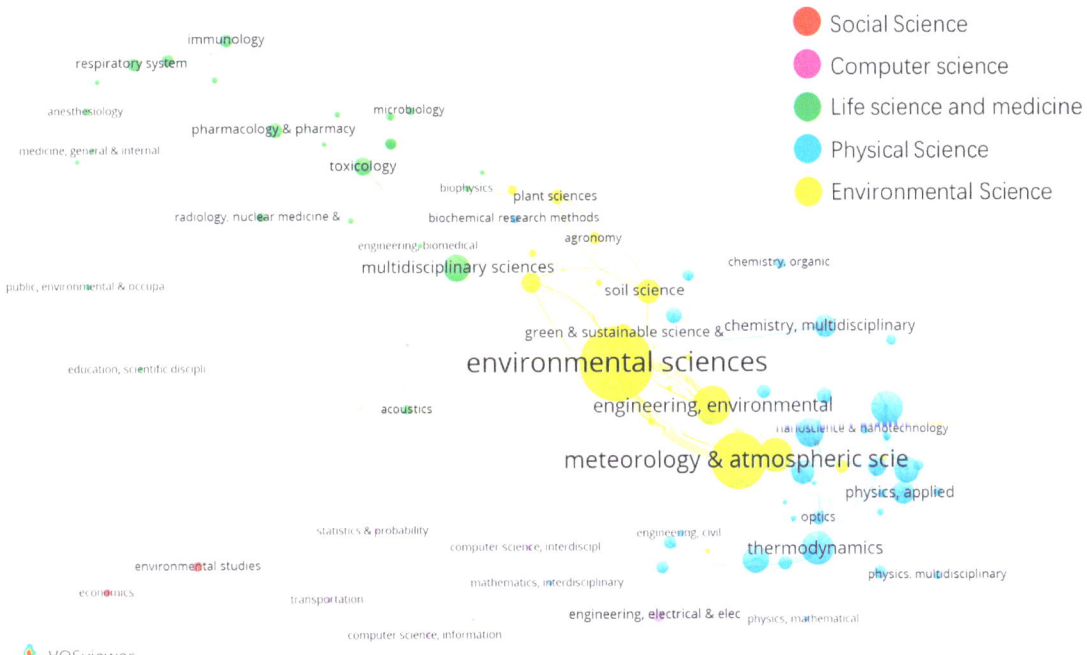

Figure 5. Current interdisciplinary relationships among categories of CHGs field. Nodes show web of science categories used to categorize CHGs research articles. The node size is directly proportional to the number of articles published. Thicker edge connections represent more frequent interdisciplinary categories.

4.2. Micro-Topic of Keywords

High-frequency keywords can reflect research hotspots. A total of 4526 co-occurrence keywords were extracted from 734 articles. The minimum occurrence of each keyword was set to eight times, and 181 co-occurrence keywords were finally presented. A keyword co-occurrence network map is shown in Figure 6. The top three keywords ranked by number of occurrences were as follows: emission ($n = 89$), chlorofluorocarbons ($n = 86$), and ozone ($n = 85$).

Nodes with the same colour belong to a cluster. The keywords were classified into four clusters. This study sorts out the topics of each cluster by reading the articles contained in each cluster to provide references for future research directions.

4.2.1. Cluster 1 (Green): Research on CHGs Emissions Calculation

Keywords: emission, halocarbons, in situ measurements, mixing ratios, global emissions, European emissions.

Two types of methods, bottom-up and top-down methods, are often used to calculate CHG emissions. The bottom-up methods include the mass balance method and emission factor method recommended by the IPCC National Greenhouse Gas Inventory Guidelines and are used based on the acquired chemical substance sales data or market data. The top-down methods include the tracer ratio correlation method, model inversion method, and box model method and are used based on the observed concentration data and the atmospheric behaviour of substances. Both types of emission calculation methods have advantages and disadvantages. Therefore, the 2019 Refinement to the 2006 IPCC Guidelines for National Greenhouse Gas Inventories proposed for the first time a method for retrieving greenhouse gas emissions based on atmospheric concentration to verify the bottom-up inventory results.

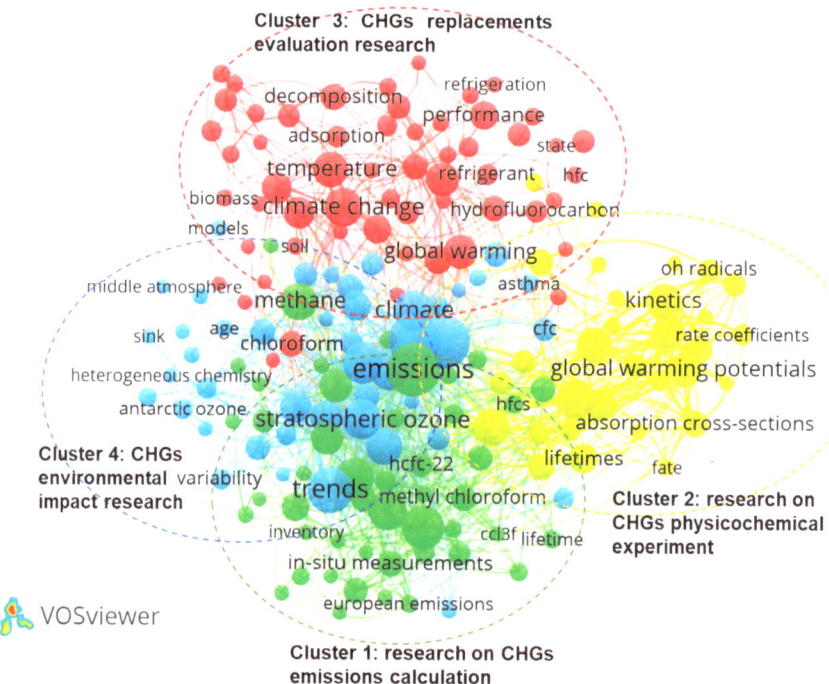

Figure 6. Cluster of CHGs co-occurrence keywords. The sizes of the nodes in Figure 6 represent the weights of the nodes. The larger the node, the larger the weight. The line between two keywords indicates that they appeared together. The thicker the line, the more co-occurrence they have.

Researchers often select appropriate emission calculation methods based on different research scales. The emission factor method and tracer ratio correlation method are mostly used to estimate emissions at the city or country scale [32–38]; the model inversion method is suitable for national and regional emission estimation [38–44]; and the box model is suitable for the estimation of emissions at the global scale.

4.2.2. Cluster 2 (Yellow): Physicochemical Analysis of CHGs

Keywords: atmospheric chemistry, kinetics, global warming potential, lifetime, rate constants, degradation, gas-phase reaction.

The research objective of physical and chemical experiments is to obtain the reaction rate, atmospheric lifespan, and radiation efficiency of the CHGs. These are important parameters for determining the global warming potential (GWP) and ozone-depleting substance potential (ODP) of CHGs. Some scholars have also used physical and chemical experiments to study the technologies for the removal of CHGs. In the study of reaction rate and atmospheric life, chemical experimental methods such as the relative rate method and absolute rate method are often used to determine the reaction rate of CHGs with OH radicals or chlorine atoms [45–53]. The reaction rate determines the atmospheric lifetime of a substance. In terms of radiation intensity efficiency, physical experimental methods, such as Fourier infrared spectroscopy, are often used to measure the infrared absorption cross-section of CHGs. In cases where the infrared (IR) spectrum cannot be measured, the radiation efficiency (RE) estimation based on computational chemistry is also very effective. In the study of halide molecule removal, chemical techniques such as thermal combustion or incineration are the most widely used [54–56]. These reactions often produce large quantities of complex chlorinated products; therefore, researchers are exploring combined processes or new alternatives, such as plasma-assisted technologies [57–60]. In addition, catalytic Decomposition, photooxidation, and biodegradation have also been studied as removal technologies [61–63].

4.2.3. Cluster 3 (Red): Evaluation of CHG Replacements

Keywords: global warming, climate change, refrigerant, temperature, system, performance, energy, dynamics, simulation.

Halogenated gases are widely used in refrigeration, foaming, and fumigation. However, due to its environmental impacts, such as ozone depletion and global warming, the search for replacements with little or no environmental impacts has gradually become a research hotspot. Performance tests are used to evaluate the feasibility of a substance as a replacement for halogenated gases. Refrigerant and synonyms appeared 41 times, higher than the words for other consumer applications, indicating that research on the replacement of refrigerants is the most popular research topic in this regard. The indicators that need to be examined for refrigerants replacement includes the cooling capacity, coefficient of performance, consumption, volumetric efficiency of the compressor, and safety [64–72]. For example, the latest experimental results show that R744, HFO-1234yf, etc. can replace the existing refrigerant in the use of refrigeration equipment. Research on the replacement of blowing agents is also a hot spot, which can be seen from the fact that blowing agents and their synonyms have appeared for 20 times. The substitution effect of blowing agents is usually evaluated from the aspects of foam opening rate, thermal insulation, and foam size stability [73]. For example, azeotropic mixtures such as HCFC-142b and HCFC-22 can be used instead of CFC-12 in the foam. The replacements for fumigants were evaluated in terms of soil fungal population, microbial biomass C (MBC), respiration, nitrification potential, and changes in enzyme activity after using various fumigants.

4.2.4. Cluster 4 (Blue): Environmental Impact Research on CHGs

Keywords: chlorofluorocarbons, ultraviolet radiation, stratospheric ozone, trends, recovery, destruction, toxic, asthma.

The harm caused by increased UV radiation, impact of climate change, and biological toxicity of CHGs are the main research directions in the field of environmental impacts of CHGs.

The depletion of stratospheric ozone caused by CHGs results in an increase in the UV radiation flux. Many scholars have studied the effects of increased UV radiation flux on human, plant, animal, and microbial growth [3,74–81]. The research on the impact of climate change mainly focuses on the surface temperature changes caused by greenhouse gas emissions, contribution of radiative forcing, and adverse impacts of global warming on microorganisms and plants [82–88]. The toxicology of halides has been well studied, and the methods of biotoxicity assessment often include exposure experiments on mice or follow-up studies on long-term exposures [89–95]. Studies have shown that human exposure to specific halides may increase the risk of immune-mediated hepatitis or lead to tumorigenesis and/or have severe toxic effects on reproductive systems.

5. Trends of Hotspots

Exploring research hotspots in different periods is very necessary to understand a re-search field. We have counted the numbers of each word appearing in title, author keywords, abstract, keywords plus in multiple consecutive time periods (1999–2003, 2004–2008, 2009–2013, 2014–2018), and then "Aggregate class" is identified. "Aggregate class" can represent a possible research hotspot, including important synonym words and phrases (supporting words), which is often summarized by the professional researchers in this field. Finally, an overview of research hotspots was revealed by analyzing the number of publications containing these supporting words [22]. In this study, two hot themes are obtained, including research substances and environmental impact.

5.1. Research Substances

The Aggregate class of Research substances consists of "chlorofluorocarbons", "CFCs", "CFC-11" and "CFC-12", "halons" and "CCl$_4$" and their substitutes, HCFCs, HFCs, and HFOs constitute the research hotspots of research substances. As shown in Figure 7 that most of the early studies focused on the first generation of ODS, and the research object of the first generation of ODS accounted for more than 50% of the literature from 1999 to 2004. With the control and elimination of the first generation of ODS by the Montreal Protocol, the proportion of literature with the first generation of ODS as the research object decreased gradually, accounting for 29% in 2018. In contrast, substitutes for HCFCs, HFCs, and HFOs have attracted more attention, and the number of articles has increased significantly. In 1999, the percentage of HCFCs literature was 10% and peaked in 2011, accounting for 21%. Compared with HFCs and HFOs, which are gradually being widely used, the proportion of research on HCFCs has decreased, with an average of 16% in 2012–2018. In 1999, HFCs literature accounted for 15% and HFOs literature accounted for 10%. By 2018, HFCs literature accounted for 35% and HFOs literature accounted for 21%.

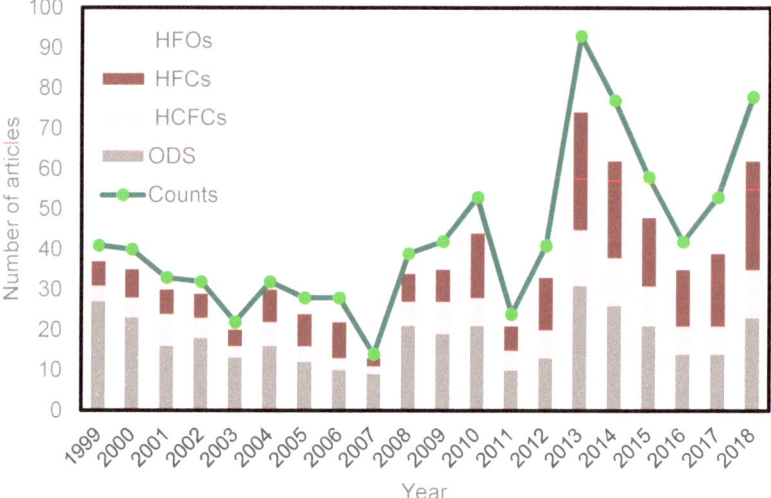

Figure 7. Number of articles published annually on different research substances.

5.2. Environmental Impact

The Aggregate class of Research substances consists of "ozone depletion", "global warming", "global warming potential" and "biological toxicity" and "inhalation toxicity" constitute the research hotspots of environmental impact. When the halogenated gases reach the stratosphere, they are exposed to ultraviolet radiation, and photolysis produces halogen radicals. Excessive halogen radicals accelerate the decomposition of ozone and destroy the balance between the generation and decomposition of the ozone layer. Furthermore, the absorption and reflection of infrared radiation by CHGs strengthened the greenhouse effect. CHGs are also biologically toxic. In this study, the annual literature on ozone depletion, global warming and biotoxicity was extracted to reveal hot research directions. Based on Figure 8, in 1999, the various environmental impacts of CHGs have been noticed. However, at this time, neither of them received much attention, with less than 20 papers published annually. As the impact of global warming became more apparent, research on the greenhouse effect of halides increased sharply, with the highest number of papers (58) being published in 2013. The number of publications on ozone depletion research and biological toxicity research is slowly increasing, and the trends of the two are very similar. Ozone depletion research reached its peak in 2014 (16 articles), and biological toxicity research reached its peak in 2013 (20 articles).

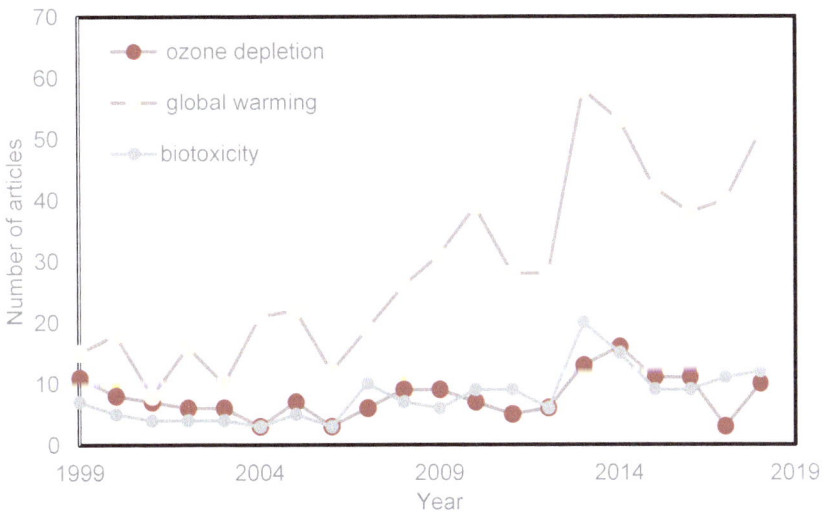

Figure 8. Number of articles published annually on different environmental impacts.

6. Conclusions

Based on 734 SCI publications on CHGs, this study provides an overview of the research on CHGs using two bibliometrics software VOSviewer and Sci2. The results showed that the United States was the most productive country, followed by UK and China. China has shown a strong growth momentum over the past decade. NOAA had the largest number of publications, followed by MIT and the University of Bristol. The most prolific authors were McCulloch, A, Ray F. Weiss, O'Doherty, S. and the authors who collaborated more with others mainly focused on the atmospheric concentrations and emissions of CHGs.

Using cluster analysis of all keywords and reading the articles in each cluster, four research hotspots of CHGs were identified and reviewed: (1) emissions calculation, (2) physicochemical analysis of halocarbons, (3) evaluation of replacements, and (4) environmental impact. Two types of methods are often used to estimate CHG emissions: top-down and bottom-up methods, and they can be used to verify each other's results. Physicochemical experiments are mainly carried out to study the removal process of CHGs and to obtain the key parameters, including reaction constants, atmospheric lifetime, and radiation efficiency, which can be used to calculate GWP and ODP. The purpose of replacement research is to find suitable substances to replace CHGs. The main evaluation index is the appraised index that includes cooling capacity, coefficient of performance, consumption, volumetric efficiency of the compressor, and safety. The environmental impact of CHGs is generally focused on ozone depletion, global warming, and biological toxicity.

The emerging topics and changes in research trends are closely related to the phase-out schedule of the Montreal Protocol. The original research topic was mainly CFCs and other ozone-depleting substances. With the phase-out of CFCs, research of HCFCs, as the transitional substitutes of ODSs, has gradually increased. Around the year when HCFCs began to freeze under the Montreal Protocol (2009 and 2010), research on HCFCs was gradually replaced by that on HFCs, HFOs, and other new substances. Global warming has always been the most concerning research hotspot, while research on ozone depletion shows a gradually rising trend. A total of 189 journals published articles on CHGs, referring to 90 disciplines, and the main disciplines were environmental science and physical science.

The purpose of this study is to provide an analysis of the publication knowledge related to CHGs. In addition, it provides guidance for researchers who want a comprehensive and quick understanding of the field.

Author Contributions: J.W. (Jing Wang): Conceptualization, Methodology, Investigation, Data curation, Writing-original draft, Validation. H.-Z.F.: Conceptualization, Methodology, Investigation, Data curation, Writing-original draft, Validation. J.X.: Investigation, Data curation, Writing-original draft. D.W.: Investigation, Data curation, Writing—original draft. Y.Y.: Investigation, Data curation, Writing-original draft. X.Z.: Supervision, Project administration. J.W. (Jing Wu): Conceptualization, Methodology, Supervision, Writing—reviewing and editing, Project administration, Validation. All authors have read and agreed to the published version of the manuscript.

Funding: This work was supported by the National Key R&D Program of China (Grant No. 2019YFC0214500), the National Natural Science Foundation of China (NSFC) (Grant No. 21976053 and No. 71804163), and Fundamental Research Funds for the Central Universities (Grant No. 2019MS042).

Institutional Review Board Statement: Not applicable.

Informed Consent Statement: Not applicable.

Data Availability Statement: The datasets analysed during the current study are available in the Science Citation Index (SCI) Expanded database of the Web of Science, https://www.webofscience.com/wos/woscc/basic-search (accessed on 24 September 2019).

Conflicts of Interest: The authors declare no conflict of interest.

References

1. Lu, Q.B. Cosmic-ray-driven reaction and greenhouse effect of halogenated molecules: Culprits for atmospheric ozone depletion and global climate change. *Int. J. Mod. Phys. B* **2013**, *27*, 1350073. [CrossRef]
2. Rowland, F.S. Stratospheric ozone depletion. *Philos. Trans. R. Soc. B-Biol. Sci.* **2006**, *361*, 769–790. [CrossRef] [PubMed]
3. Chakraborty, S.; Murray, G.M.; Magarey, P.A.; Yonow, T.; Emmett, R.W. Potential impact of climate change on plant diseases of economic significance to Australia. *Australas. Plant Pathol.* **1998**, *27*, 15–35. [CrossRef]
4. Manning, W.J.; Tiedemann, A.V. Climate change: Potential effects of increased atmospheric Carbon dioxide (CO_2), ozone (O_3), and ultraviolet-B (UV-B) radiation on plant diseases. *Environ. Pollut.* **1995**, *88*, 219–245. [CrossRef]
5. Paul, N.D. Stratospheric ozone depletion, UV-B radiation and crop disease. *Environ. Pollut.* **2000**, *108*, 343–355. [CrossRef]
6. IPCC. *Climate Change 2014: Synthesis Report. Contribution of Working Groups I, II and III to the Fifth Assessment Report of the Intergovernmental Panel on Climate Change*; Core Writing Team, Pachauri, R.K., Meyer, L.A., Eds.; IPCC: Geneva, Switzerland, 2014; 151p.
7. Maione, M.; Giostra, U.; Arduini, J.; Furlani, F.; Graziosi, F.; Lo Vullo, E.; Bonasoni, P. Ten years of continuous observations of stratospheric ozone depleting gases at Monte Cimone (Italy)—Comments on the effectiveness of the Montreal Protocol from a regional perspective. *Sci. Total Environ.* **2013**, *445*, 155–164. [CrossRef] [PubMed]
8. Wu, H.; Chen, H.; Wang, Y.; Ding, A.; Chen, J. The changing ambient mixing ratios of long-lived halocarbons under Montreal Protocol in China. *J. Clean. Prod.* **2018**, *188*, 774–785. [CrossRef]
9. Oman, L.D.; Plummer, D.A.; Waugh, D.W.; Austin, J.; Scinocca, J.F.; Douglass, A.R.; Salawitch, R.J.; Canty, T.; Akiyoshi, H.; Bekki, S.; et al. Multimodel assessment of the factors driving stratospheric ozone evolution over the 21st century. *J. Geophys. Res.-Atmos.* **2010**, *115*. [CrossRef]
10. Montzka, S.A.; Dutton, G.S.; Yu, P.; Ray, E.; Portmann, R.W.; Daniel, J.S.; Kuijpers, L.; Hall, B.D.; Mondeel, D.; Siso, C.; et al. An unexpected and persistent increase in global emissions of ozone-depleting CFC-11. *Nature* **2018**, *557*, 413. [CrossRef]
11. Fu, H.; Ho, Y. Independent research of China in Science Citation Index Expanded during 1980–2011. *J. Informetr.* **2013**, *7*, 210–222. [CrossRef]
12. Vega-Arce, M.; Salas, G.; Nunez-Ulloa, G.; Pinto-Cortez, C.; Fernandez, I.T.; Ho, Y. Research performance and trends in child sexual abuse research: A Science Citation Index Expanded-based analysis. *Scientometrics* **2019**, *121*, 1505–1525. [CrossRef]
13. Fu, H.; Ho, Y. Collaborative characteristics and networks of national, institutional and individual contributors using highly cited articles in environmental engineering in Science Citation Index Expanded. *Curr. Sci.* **2018**, *115*, 410–421. [CrossRef]
14. Fu, H.; Wang, M.; Ho, Y. Mapping of drinking water research: A bibliometric analysis of research output during 1992–2011. *Sci. Total Environ.* **2013**, *443*, 757–765. [CrossRef]
15. National Science Board. *Science and Engineering Indicators 2016*; NSB-2016-1; National Science Foundation: Arlington, VA, USA, 2016; p. 899.
16. Glnzel, W. National characteristics in international scientific co-authorship relations. *Scientometrics* **2001**, *51*, 69–115. [CrossRef]
17. Glnzel, W. Coauthorship Patterns and Trends in the Sciences (1980–1998): A Bibliometric Study with Implications for Database Indexing and Search Strategies. *Libr. Trends* **2002**, *5*, 461–473.
18. Persson, O.; Glanzel, W.; Danell, R. Inflationary bibliometric values: The role of scientific collaboration and the need for relative indicators in evaluative studies. *Scientometrics* **2004**, *60*, 421–432. [CrossRef]
19. Chiu, W.; Ho, Y. Bibliometric analysis of tsunami research. *Scientometrics* **2007**, *73*, 3–17. [CrossRef]

20. Fu, H.; Ho, Y.; Sui, Y.; Li, Z. A bibliometric analysis of solid waste research during the period 1993–2008. *Waste Manag.* **2010**, *30*, 2410–2417. [CrossRef]
21. Zhang, G.; Xie, S.; Ho, Y. A bibliometric analysis of world volatile organic compounds research trends. *Scientometrics* **2010**, *83*, 477–492. [CrossRef]
22. Mao, N.; Wang, M.; Ho, Y. A Bibliometric Study of the Trend in Articles Related to Risk Assessment Published in Science Citation Index. *Hum. Ecol. Risk Assess.* **2010**, *16*, 801–824. [CrossRef]
23. Lin, C.L.; Ho, Y. A Bibliometric Analysis of Publications on Pluripotent Stem Cell Research. *Cell J.* **2015**, *17*, 59–70. [CrossRef]
24. Sci2 Team. *Science of Science (Sci2) Tool*; Indiana University and SciTech Strategies: Bloomington, IN, USA, 2009. Available online: https://sci2.cns.iu.edu (accessed on 15 November 2019). [CrossRef]
25. Eck, N.; Waltman, L. Text mining and visualization using VOSviewer. *arXiv* **2011**, arXiv:1109.2058.
26. Clemente, D.H.; Hsuan, J.; Carvalho, M. The intersection between business model and modularity: An overview of the literature. *Braz. J. Oper. Prod. Manag.* **2019**, *16*, 387–397. [CrossRef]
27. Yao, X.; Yan, J.; Ginda, M.; Borner, K.; Saykin, A.J.; Shen, L. Mapping longitudinal scientific progress, collaboration and impact of the Alzheimer's disease neuroimaging initiative. *PLoS ONE* **2017**, *12*, e0186095. [CrossRef]
28. Fu, H.; Wang, M.; Ho, Y. The most frequently cited adsorption research articles in the Science Citation Index (Expanded). *J. Colloid Interface Sci.* **2012**, *379*, 148–156. [CrossRef]
29. Chuang, K.; Huang, Y.; Ho, Y. A bibliometric and citation analysis of stroke-related research in Taiwan. *Scientometrics* **2007**, *72*, 201–212. [CrossRef]
30. Van Eck, N.J.; Waltman, L. VOS: A new method for visualizing similarities between objects. In *Studies in Classification Data Analysis and Knowledge Organization*; Decker, R., Lenz, H.J., Eds.; Social Science Electronic Publishing: Leiden, The Netherlands, 2007; p. 299. [CrossRef]
31. Mela, G.S.; Cimmino, M.A.; Ugolini, D. Impact assessment of oncology research in the European Union. *Eur. J. Cancer* **1999**, *35*, 1182–1186. [CrossRef]
32. Kim, J.; Li, S.; Kim, K.R.; Stohl, A.; Muehle, J.; Kim, S.K.; Park, M.K.; Kang, D.J.; Lee, G.; Harth, C.M.; et al. Regional atmospheric emissions determined from measurements at Jeju Island, Korea: Halogenated compounds from China. *Geophys. Res. Lett.* **2010**, *37*. [CrossRef]
33. Li, S.; Kim, J.; Kim, K.; Muehle, J.; Kim, S.; Park, M.; Stohl, A.; Kang, D.; Arnold, T.; Harth, C.M.; et al. Emissions of Halogenated Compounds in East Asia Determined from Measurements at Jeju Island, Korea. *Environ. Sci. Technol.* **2011**, *45*, 5668–5675. [CrossRef]
34. Li, Y.; Zhang, Z.; An, M.; Gao, D.; Yi, L.; Hu, J. The estimated schedule and mitigation potential for hydrofluorocarbons phase-down in China. *Adv. Clim. Chang. Res.* **2019**, *10*, 174–180. [CrossRef]
35. McCulloch, A.; Lindley, A.A. Global emissions of HFC-23 estimated to year 2015. *Atmos. Environ.* **2007**, *41*, 1560–1566. [CrossRef]
36. Mcculloch, A.; Midgley, P.M.; Ashford, P. Releases of refrigerant gases (CFC-12, HCFC-22 and HFC-134a) to the atmosphere. *Atmos. Environ.* **2003**, *37*, 889–902. [CrossRef]
37. Shao, M.; Huang, D.; Gu, D.; Lu, S.; Chang, C.; Wang, J. Estimate of anthropogenic halocarbon emission based on measured ratio relative to CO in the Pearl River Delta region, China. *Atmos. Chem. Phys.* **2011**, *11*, 5011–5025. [CrossRef]
38. Wu, J.; Fang, X.; Martin, J.W.; Zhai, Z.; Su, S.; Hu, X.; Han, J.; Lu, S.; Wang, C.; Zhang, J.; et al. Estimated emissions of chlorofluorocarbons, hydrochlorofluorocarbons, and hydrofluorocarbons based on an interspecies correlation method in the Pearl River Delta region, China. *Sci. Total Environ.* **2014**, *470*, 829–834. [CrossRef]
39. Lunt, M.F.; Rigby, M.; Ganesan, A.L.; Manning, A.J.; Prinn, R.G.; O'Doherty, S.; Muehle, J.; Harth, C.M.; Salameh, P.K.; Arnold, T.; et al. Reconciling reported and unreported HFC emissions with atmospheric observations. *Proc. Natl. Acad. Sci. USA* **2015**, *112*, 5927–5931. [CrossRef]
40. O'Doherty, S.; Cunnold, D.M.; Manning, A.; Miller, B.R.; Wang, R.; Krummel, P.B.; Fraser, P.J.; Simmonds, P.G.; McCulloch, A.; Weiss, R.F.; et al. Rapid growth of hydrofluorocarbon 134a and hydrochlorofluorocarbons 141b, 142b, and 22 from Advanced Global Atmospheric Gases Experiment (AGAGE) observations at Cape Grim, Tasmania, and Mace Head, Ireland. *J. Geophys. Res.-Atmos.* **2004**, *109*. [CrossRef]
41. Oram, D.E.; Mani, F.S.; Laube, J.C.; Newland, M.J.; Reeves, C.E.; Sturges, W.T.; Penkett, S.A.; Brenninkmeijer, C.A.M.; Rockmann, T.; Fraser, P.J. Long-term tropospheric trend of octafluorocyclobutane (c-C4F8 or PFC-318). *Atmos. Chem. Phys.* **2012**, *12*, 261–269. [CrossRef]
42. Stohl, A.; Seibert, P.; Arduini, J.; Eckhardt, S.; Fraser, P.; Greally, B.R.; Lunder, C.; Maione, M.; Muehle, J.; O'Doherty, S.; et al. An analytical inversion method for determining regional and global emissions of greenhouse gases: Sensitivity studies and application to halocarbons. *Atmos. Chem. Phys.* **2009**, *9*, 1597–1620. [CrossRef]
43. Vollmer, M.K.; Zhou, L.X.; Greally, B.R.; Henne, S.; Yao, B.; Reimann, S.; Stordal, F.; Cunnold, D.M.; Zhang, X.C.; Maione, M.; et al. Emissions of ozone-depleting halocarbons from China. *Geophys. Res. Lett.* **2009**, *36*. [CrossRef]
44. Yao, B.; Fang, X.; Vollmer, M.K.; Reimann, S.; Chen, L.; Fang, S.; Prinn, R.G. China's Hydrofluorocarbon Emissions for 2011–2017 Inferred from Atmospheric Measurements. *Environ. Sci. Technol. Lett.* **2019**, *6*, 479–486. [CrossRef]
45. Andersen, L.L.; Osterstrom, F.F.; Andersen, M.P.S.; Nielsen, O.J.; Wallington, T.J. Atmospheric chemistry of cis-$CF_3CH=CHCl$ (HCFO-1233zd(Z)): Kinetics of the gas-phase reactions with Cl atoms, OH radicals, and O_3. *Chem. Phys. Lett.* **2015**, *639*, 289–293. [CrossRef]

46. Andersen, M.P.S.; Nielsen, O.J. Atmospheric chemistry of a cyclic hydro-fluoro-carbon: Kinetics and mechanisms of gas-phase reactions of 1-trifluoromethyl-1,2,2-trifluorocyclobutane with Cl atoms, OH radicals, and O_3. *Phys. Chem. Chem. Phys.* **2019**, *21*, 1497–1505. [CrossRef] [PubMed]
47. Gour, N.K.; Paul, S.; Deka, R.C. Atmospheric impact of Z- and E-isomers of $CF_3CH=CHC_2F_5$ molecule initiated by OH radicals: Reaction mechanisms, kinetics and global warming potential. *Int. J. Refrig. Rev. Int. Froid* **2019**, *101*, 167–177. [CrossRef]
48. Guo, Q.; Zhang, N.; Uchimaru, T.; Chen, L.; Quan, H.; Mizukado, J. Atmospheric chemistry of cyc-$CF_2CF_2CF_2CH=CH$-: Kinetics, products, and mechanism of gas-phase reaction with OH radicals, and atmospheric implications. *Atmos. Environ.* **2018**, *179*, 69–76. [CrossRef]
49. Osterstrom, F.F.; Andersen, S.T.; Solling, T.I.; Nielsen, O.J.; Andersen, M.P.S. Atmospheric chemistry of Z- and E-$CF_3CH=CHCF_3$. *Phys. Chem. Chem. Phys.* **2017**, *19*, 735–750. [CrossRef]
50. Osterstrom, F.F.; Nielsen, O.J.; Wallington, T.J. Atmospheric chemistry of $CF_3CF_2OCH_3$. *Chem. Phys. Lett.* **2016**, *653*, 149–154. [CrossRef]
51. Sondergaard, R.; Nielsen, O.J.; Hurley, M.D.; Wallington, T.J.; Singh, R. Atmospheric chemistry of trans-$CF_3CH=CHF$: Kinetics of the gas-phase reactions with Cl atoms, OH radicals, and O_3. *Chem. Phys. Lett.* **2007**, *443*, 199–204. [CrossRef]
52. Tovar, C.M.; Blanco, M.B.; Barnes, I.; Wiesen, P.; Teruel, M.A. Gas-phase reactivity study of a series of hydrofluoroolefins (HFOs) toward OH radicals and Cl atoms at atmospheric pressure and 298 K. *Atmos. Environ.* **2014**, *88*, 107–114. [CrossRef]
53. Zhang, N.; Chen, L.; Mizukado, J.; Quan, H.; Suda, H. Rate constants for the gas-phase reactions of (Z)-$CF3CH=CHF$ and (E)-$CF3CH=CHF$ with OH radicals at 253–328 K. *Chem. Phys. Lett.* **2015**, *621*, 78–84. [CrossRef]
54. Han, W.; Li, Y.; Tang, H.; Liu, H. Treatment of the potent greenhouse gas, CHF_3-An overview. *J. Fluor. Chem.* **2012**, *140*, 7–16. [CrossRef]
55. Minami, W.; Fujii, H.; Kim, H. Combustion decomposition treatment of Freon. *Kagaku Kogaku Ronbunshu* **2006**, *32*, 190–195. [CrossRef]
56. Qin, L.; Han, J.; Liu, L.; Yang, X.; Kim, H.; Yu, F. Highly efficient decomposition of HFC-134a by combustion oxidization method. *Fresenius Environ. Bull.* **2013**, *22*, 1919–1923.
57. Kang, H. Decomposition of Chlorofluorocarbon by Non-thermal Plasma. *J. Ind. Eng. Chem.* **2002**, *8*, 488–492.
58. Mok, Y.S.; Demidyuk, V.; Whitehead, J.C. Decomposition of hydrofluorocarbons in a dielectric-packed plasma reactor. *J. Phys. Chem. A* **2008**, *112*, 6586–6591. [CrossRef] [PubMed]
59. Narengerile; Saito, H.; Watanabe, T. Decomposition Mechanism of Fluorinated Compounds in Water Plasmas Generated under Atmospheric Pressure. *Plasma Chem. Plasma Process.* **2010**, *30*, 813–829. [CrossRef]
60. Watanabe, T.; Tsuru, T. Water plasma generation under atmospheric pressure for HFC destruction. *Thin Solid Film.* **2008**, *516*, 4391–4396. [CrossRef]
61. Nagata, H.; Takakura, T.; Kishida, M.; Mizuno, K.; Tamori, Y.; Wakabayashi, K. Oxidative decomposition of chlorofluorocarbon (CFC-115) in the presence of butane over gamma-alumina catalysts. *Chem. Lett.* **1993**, *22*, 1545–1546. [CrossRef]
62. Tennakone, K.; Wijayantha, K. Photocatalysis of CFC degradation by titanium dioxide. *Appl. Catal. B-Environ.* **2005**, *57*, 9–12. [CrossRef]
63. Minami, W.; Kim, H. Decomposition of halocarbons using TiO_2 photocatalyst. *Kagaku Kogaku Ronbunshu* **2006**, *32*, 310–313. [CrossRef]
64. Aral, M.C.; Suhermanto, M.; Hosoz, M. Performance evaluation of an automotive air conditioning and heat pump system using R1234yf and R134a. *Sci. Technol. Built Environ.* **2021**, *27*, 44–60. [CrossRef]
65. Bolaji, B.O. Theoretical analysis of the energy performance of three low global warming potential hydro-fluorocarbon refrigerants as R134a alternatives in refrigeration systems. *Proc. Inst. Mech. Eng. Part A J. Power Energy* **2014**, *228*, 56–63. [CrossRef]
66. Dalkilic, A.S.; Wongwises, S. A performance comparison of vapour-compression refrigeration system using various alternative refrigerants. *Int. Commun. Heat Mass Transf.* **2010**, *37*, 1340–1349. [CrossRef]
67. Daviran, S.; Kasaeian, A.; Golzari, S.; Mahian, O.; Nasirivatan, S.; Wongwises, S. A comparative study on the performance of HFO-1234yf and HFC-134a as an alternative in automotive air conditioning systems. *Appl. Therm. Eng.* **2017**, *110*, 1091–1100. [CrossRef]
68. Hamza, A.; Khan, T.A. Comparative Performance of Low-GWP Refrigerants as Substitutes for R134a in a Vapor Compression Refrigeration System. *Arab. J. Sci. Eng.* **2020**, *45*, 5697–5712. [CrossRef]
69. Heredia-Aricapa, Y.; Belman-Flores, J.M.; Mota-Babiloni, A.; Serrano-Arellano, J.; Garcia-Pabon, J.J. Overview of low GWP mixtures for the replacement of HFC refrigerants: R134a, R404A and R410A. *Int. J. Refrig.* **2020**, *111*, 113–123. [CrossRef]
70. Kedzierski, M.A.; Brown, J.S.; Koo, J. Performance ranking of refrigerants with low global warming potential. *Sci. Technol. Built Environ.* **2015**, *21*, 207–219. [CrossRef]
71. Mota-Babiloni, A.; Navarro-Esbri, J.; Barragan, A.; Moles, F.; Peris, B. Drop-in energy performance evaluation of R1234yf and R1234ze(E) in a vapor compression system as R134a replacements. *Appl. Therm. Eng.* **2014**, *71*, 259–265. [CrossRef]
72. Prabakaran, R.; Sidney, S.; Iyyappan, R.; Lal, D.M. Experimental studies on the performance of mobile air conditioning system using environmental friendly HFO-123yf as a refrigerant. *Proc. Inst. Mech. Eng. Part E J. Process. Mech. Eng.* **2021**, *235*, 731–742. [CrossRef]
73. Vachon, C. Evaluation of HFC-245fa as an alternative blowing agent for extruded low-density polyethylene. *Cell. Polym.* **2004**, *23*, 109–121. [CrossRef]

74. Beardall, J.; Raven, J.A. The potential effects of global climate change on microalgal photosynthesis, growth and ecology. *Phycologia* **2004**, *43*, 26–40. [CrossRef]
75. Beardall, J.; Stojkovic, S.; Gao, K. Interactive effects of nutrient supply and other environmental factors on the sensitivity of marine primary producers to ultraviolet radiation: Implications for the impacts of global change. *Aquat. Biol.* **2014**, *22*, 5–23. [CrossRef]
76. Croteau, M.C.; Davidson, M.A.; Lean, D.R.S.; Trudeau, V.L. Global Increases in Ultraviolet B Radiation: Potential Impacts on Amphibian Development and Metamorphosis. *Physiol. Biochem. Zool.* **2008**, *81*, 743–761. [CrossRef]
77. Durif, C.M.F.; Fields, D.M.; Browman, H.I.; Shema, S.D.; Enoae, J.R.; Skiftesvik, A.B.; Bjelland, R.; Sommaruga, R.; Arts, M.T. UV radiation changes algal stoichiometry but does not have cascading effects on a marine food chain. *J. Plankton Res.* **2015**, *37*, 1120–1136. [CrossRef]
78. Lago Londero, J.E.; Dos Santos, M.B.; Schuch, A.P. Impact of solar UV radiation on amphibians: Focus on genotoxic stress. *Mutat. Res. Genet. Toxicol. Environ. Mutagens* **2019**, *842*, 14–21. [CrossRef] [PubMed]
79. Llabres, M.; Agusti, S.; Fernandez, M.; Canepa, A.; Maurin, F.; Vidal, F.; Duarte, C.M. Impact of elevated UVB radiation on marine biota: A meta-analysis. *Glob. Ecol. Biogeogr.* **2013**, *22*, 131–144. [CrossRef]
80. Mitchell, T.; Alton, L.A.; White, C.R.; Franklin, C.E. Relations between Conspecific Density and Effects of Ultraviolet-B Radiation on Tadpole Size in the Striped Marsh Frog. *Conserv. Biol.* **2012**, *26*, 1112–1120. [CrossRef]
81. Schuch, A.P.; Dos Santos, M.B.; Lipinski, V.M.; Peres, L.V.; Dos Santos, C.P.; Cechin, S.Z.; Schuch, N.J.; Pinheiro, D.K.; Da Silva Loreto, E.L. Identification of influential events concerning the Antarctic ozone hole over southern Brazil and the biological effects induced by UVB and UVA radiation in an endemic treefrog species. *Ecotoxicol. Environ. Saf.* **2015**, *118*, 190–198. [CrossRef]
82. Aprea, C.; Greco, A.; Maiorino, A. An experimental evaluation of the greenhouse effect in the substitution of R134a with CO_2. *Energy* **2012**, *45*, 753–761. [CrossRef]
83. Convey, P.; Bindschadler, R.; di Prisco, G.; Fahrbach, E.; Gutt, J.; Hodgson, D.A.; Mayewski, P.A.; Summerhayes, C.P.; Turner, J. Antarctic climate change and the environment. *Antarct. Sci.* **2009**, *21*, 541–563. [CrossRef]
84. Newman, P.A.; Oman, L.D.; Douglass, A.R.; Fleming, E.L.; Frith, S.M.; Hurwitz, M.M.; Kawa, S.R.; Jackman, C.H.; Krotkov, N.A.; Nash, E.R.; et al. What would have happened to the ozone layer if chlorofluorocarbons (CFCs) had not been regulated? *Atmos. Chem. Phys.* **2009**, *9*, 2113–2128. [CrossRef]
85. Turner, J.; Barrand, N.E.; Bracegirdle, T.J.; Convey, P.; Hodgson, D.A.; Jarvis, M.; Jenkins, A.; Marshall, G.; Meredith, M.P.; Roscoe, H.; et al. Antarctic climate change and the environment: An update. *Polar Rec.* **2014**, *50*, 237–259. [CrossRef]
86. Velders, G.J.M.; Andersen, S.O.; Daniel, J.S.; Fahey, D.W.; McFarland, M. The importance of the Montreal Protocol in protecting climate. *Proc. Natl. Acad. Sci. USA* **2007**, *104*, 4814–4819. [CrossRef]
87. Velders, G.J.M.; Fahey, D.W.; Daniel, J.S.; McFarland, M.; Andersen, S.O. The large contribution of projected HFC emissions to future climate forcing. *Proc. Natl. Acad. Sci. USA* **2009**, *106*, 10949–10954. [CrossRef]
88. Zhang, H.; Wu, J.; Luc, P. A study of the radiative forcing and global warming potentials of hydrofluorocarbons. *J. Quant. Spectrosc. Radiat. Transf.* **2011**, *112*, 220–229. [CrossRef]
89. Ema, M.; Naya, M.; Yoshida, K.; Nagaosa, R. Reproductive and developmental toxicity of hydrofluorocarbons used as refrigerants. *Reprod. Toxicol.* **2010**, *29*, 125–131. [CrossRef] [PubMed]
90. Gaku, I.; Yu, X.; Junzoh, K.; Nobuyuki, A.; Toshihiko, K.; Hisakazu, I.; Eiji, S.; Tetsuya, Y.; Wang, H.; Xie, Z. Reproductive toxicity of 1-bromopropane, a newly introduced alternative to ozone layer depleting solvents, in male rats. *Toxicol. Sci.* **2000**, *54*, 416–423. [CrossRef]
91. Lee, J.; Lee, C.; Kim, C.H. Uncontrolled Occupational Exposure to 1,1-Dichloro-1-Fluoroethane (HCFC-141b) Is Associated with Acute Pulmonary Toxicity. *Chest* **2009**, *135*, 149–155. [CrossRef] [PubMed]
92. Rusch, G.M.; Tveit, A.; Muijser, H.; Tegelenbosch-Schouten, M.; Hoffman, G.M. The acute, genetic, developmental and inhalation toxicology of trans-1,3,3,3-tetrafluoropropene (HFO-1234ze). *Drug Chem. Toxicol.* **2012**, *36*, 170–180. [CrossRef] [PubMed]
93. Sabik, L.M.E.; Abbas, R.A.; Ismail, M.M.; El-Refaei, S. Cardiotoxicity of Freon among refrigeration services workers: Comparative cross-sectional study. *Environ. Health* **2009**, *8*, 31. [CrossRef] [PubMed]
94. Tsai, W.T. An overview of environmental hazards and exposure risk of hydrofluorocarbons (HFCs). *Chemosphere* **2005**, *61*, 1539–1547. [CrossRef] [PubMed]
95. Tveit, A.; Rusch, G.M.; Muijser, H.; van den Hoven, M.J.W.; Hoffman, G.M. The acute, genetic, developmental and inhalation toxicology of trans-1-chloro,3,3,3-trifluoropropene (HCFO-1233zd(E)). *Drug Chem. Toxicol.* **2013**, *37*, 83–92. [CrossRef] [PubMed]

Article

Modeling Spatial Distribution and Determinant of $PM_{2.5}$ at Micro-Level Using Geographically Weighted Regression (GWR) to Inform Sustainable Mobility Policies in Campus Based on Evidence from King Abdulaziz University, Jeddah, Saudi Arabia

Alok Tiwari * and Mohammed Aljoufie

Department of Urban and Regional Planning, King Abdulaziz University, Jeddah 21589, Saudi Arabia; maljufie@kau.edu.sa
* Correspondence: atwari@kau.edu.sa

Abstract: Air pollution is fatal. Fine particles, such as $PM_{2.5}$, in ambient air might be the cause of many physical and psychological disorders, including cognitive decline. This is why educational policymakers are adopting sustainable mobility, and other policy measures, to make their campuses carbon-neutral; however, car-dependent cities and their university campuses are still lagging behind in this area. This study attempts to model the spatial heterogeneity and determinants of $PM_{2.5}$ at the King Abdulaziz University campus in Jeddah, which is ranked first among the Saudi Arabian universities, as well as in the MENA region. We developed four OLS and GWR models of different peak and off-peak periods during weekdays in order to estimate the determinants of the $PM_{2.5}$ concentration. The number of cars, humidity, temperature, windspeed, distance from trees, and construction sites were the estimators in our analysis. Because of a lack of secondary data at a finer scale, we collected the samples of all dependent and independent variables at 51 locations on the KAU campus. Model selection was based on RSS, log-likelihood, adjusted R2, and AICc, and a modal comparison shows that the GWR variant of Model-2 outperformed the other models. The results of the GWR model demonstrate the geographical variability of the $PM_{2.5}$ concentration on the KAU campus, to which the volume of car traffic is the key contributor. Hence, we recommend using the results of this study to support the development of a car-free and zero-carbon campus at KAU; furthermore, this study could be exploited by other campuses in Saudi Arabia and the Gulf region.

Keywords: King Abdulaziz University; $PM_{2.5}$; GWR; zero-carbon campus; spatial heterogeneity

Citation: Tiwari, A.; Aljoufie, M. Modeling Spatial Distribution and Determinant of $PM_{2.5}$ at Micro-Level Using Geographically Weighted Regression (GWR) to Inform Sustainable Mobility Policies in Campus Based on Evidence from King Abdulaziz University, Jeddah, Saudi Arabia. *Sustainability* **2021**, *13*, 12043. https://doi.org/10.3390/su132112043

Academic Editors: Weixin Yang, Guanghui Yuan and Yunpeng Yang

Received: 18 September 2021
Accepted: 27 October 2021
Published: 31 October 2021

Publisher's Note: MDPI stays neutral with regard to jurisdictional claims in published maps and institutional affiliations.

Copyright: © 2021 by the authors. Licensee MDPI, Basel, Switzerland. This article is an open access article distributed under the terms and conditions of the Creative Commons Attribution (CC BY) license (https://creativecommons.org/licenses/by/4.0/).

1. Introduction

Contaminated air is an issue of serious concern worldwide and causes one out of every nine deaths. It is also frightening to consider that exposure to $PM_{2.5}$ has lowered the average life expectancy by one year, according to 2016 data [1]. The UNEP data further shows that fine particle pollution caused 17,795 deaths in 2019; this can be translated into 500 fatalities per 1 million people. There is no doubt that clean air is a boon for public health and an essential prerequisite for the healthy living of all human beings [2]. There are many root causes of air pollution, including several voluntary and nonvoluntary human activities, such as fossil-fuel-based vehicular transportation, manufacturing, and construction [3]. Furthermore, rapid urbanization and population growth have made the problem of air pollution even worse in both developed and developing countries [4–6]. Automobiles are the major source of fine particle air pollution—a study in Sao Paulo confirmed that a trucker's strike was associated with potential economic and health benefits [7]. A study in Mexico showed that people living in spatial proximity to high-traffic roads are highly exposed to traffic-related air pollution, which is a matter of grave concern for urban planners and public health professionals, while another study in Santiago, Chile, identified three distinct episodes of fine particle air pollution, which is helpful in the mitigation of $PM_{2.5}$ and PM_{10} pollutants in urban settings [8,9].

In the past few decades, several studies, including those of Englert [10] and Valavanidis et al. [11], have confirmed that air pollution damages the physical and neurological health of urban residents, increases economic costs, and aggravates the unpleasant impacts of climate change. According to the EPA [12], outdoor air pollution encompasses a diverse blend of chemical, physical, and biological substances; however, this study considers solely the concentration of $PM_{2.5}$ (fine particles with a diameter of less than 2.5 μm), as these fine particles pose serious threats to human health.

University campuses worldwide are faced with fine particle pollution, mainly $PM_{2.5}$, and this is negatively affecting the health of university attendants [13,14]. We chose King Abdulaziz University (KAU) campus for our study for two reasons: firstly, the university mimics a small city and, secondly, campuses are a hub of innovation and sustainable development. In addition, universities have realized that pollution and resource consumption have increased on their campuses and in classrooms in the past few decades. This has compelled university administrations to consider devoting their best efforts to achieving sustainability in their indoor and outdoor environments [15,16].

Although several studies have modeled the determinant factors of $PM_{2.5}$ levels and their spatial heterogeneity through GWR, there is a lack of studies performed at the micro level of the local geographical region, particularly at the level of a university campus. This gap is due to the unavailability of fine data at micro geographical scales, such as university campuses. Therefore, this study attempts to model the spatial distribution and determinants of $PM_{2.5}$ at the campus of King Abdulaziz University, utilizing primary data collected by the authors. This paper is further organized into the following sections: Literature Review, Material and Methods, Results, Discussions, and Conclusions.

2. Literature Review
2.1. Health Consequences of Particulate Matter ($PM_{2.5}$)

A number of epidemiological studies have been carried out to date demonstrating the impacts of deteriorated ambient air quality on human health, although the health effects of fine particulate matter depend on four factors: the source and composition of particulate matter, the time of exposure, its depth of travel inside the human body, and the age of the affected person. Persistent exposure to human-induced $PM_{2.5}$ is positively associated with a series of pulmonary and cardiovascular diseases. For instance, in children, chronic exposure to $PM_{2.5}$ poses risks for acute lung respiratory infections (ALRI), while in adults, it can generate chronic obstructive pulmonary disease (COPD), ischemic heart disease (IHD), cardiopulmonary disorders, lung cancer, and stroke [17–19]. Pope et al. [20] highlight that there is a 4–6% increased risk of cardiovascular and lung cancer mortality with each 10 mg/m^3 increase in fine particulate matter in the air.

Ambient air pollutants might not only harm physical health, but can also cause damage to people's nervous systems and cognitive abilities, leading to neurological disorders of different magnitudes, ranging from headaches and migraines, to strokes and various types of dementia [21]. In addition, a few studies have stated that the $PM_{2.5}$ in the air contains certain neurotoxicants that can either produce or accelerate neurodegenerative diseases related to cognitive decline, schizophrenia, and brain damage. These studies strongly suggest that persistent exposure to the fine particulate matters present in ambient air might harm the central nervous system [22–26]. A recent study by Ranzani et al. [27] reported that adult human lungs have the capacity to purify 10,000 liters of air daily; however, increasing levels of pollutants could reduce immunity, resulting in increased inflammation and poor bone health.

2.2. Campus Response to Challenges of Poor Ambient Air Quality

Educational administrators have launched several initiatives to tackle toxic particulate matter and the other pollutants present in ambient air. The low-carbon campus project at the Massachusetts Institute of Technology [28] is an endeavor of the MIT Office of Sustainability (MITOS) to establish a healthy and low-carbon campus by exploiting the competence and experience of prestigious MIT alumni to create the testbed for expandable solutions. Four core components were explored to accomplish the goal of carbon neutrality at the MIT campus, including mobility, building, climate, and energy. Under the mobility theme, the students, faculty, and staff members of MIT are encouraged to select flexible, affordable, and low-impact modes of transportation, including walking and bicycle riding, with only a small number of members commuting by car.

Similarly, the University of Leeds [29] launched a living lab for their air quality project in 2017 that uses the inhouse air quality and pollution to enhance the environment and health of the community's members. The living lab regularly helps to diminish emissions from vehicles and curb exposure to poor air quality. In a recent related project, the Institute for Transport Studies has been investigating pollution exposure to university staff, faculty, and students via commuter routes. Volunteers carry air quality monitoring devices while walking, using public transport, or driving; the results are used to make comparisons on the levels of pollution exposure between different transport modes and routes.

It is also noteworthy that smoking emits 10 times more air pollutants into the ambient air than a car. The smoking of one cigarette daily produces an equivalent $PM_{2.5}$ level of 22 $\mu g/m^3$ [30,31]. Considering this, universities, including KAU [32], have formulated strategies to create smoking-free campuses. In compliance with the Smoke Free Environments Act, 1990, of New Zeeland, which prohibits smoking in workplaces, Lincoln University [33] accepted the "Clean Air Policy", aimed at providing a healthy and safe smoke-free working and learning environment on their campus.

Moreover, universities in developing economies are the worst affected by $PM_{2.5}$ exposure and are struggling to transform their campuses towards carbon neutrality. According to Express Web Desk [34], the University of Hyderabad in India has recently introduced e-rickshaw services on weekdays from 8:00 a.m. to 6:00 p.m. This affordable (USD 0.14 per trip) and zero-emission commuting service might improve ambient air quality inside the university campus. Some other universities are trying to increase awareness about air pollution by disseminating and displaying information on air quality. The Central University of Columbia at Bogota has established an Air Quality Monitoring Network (Red de monitoreo de calidad del aire) that is equipped with low-cost sensors and Internet of Things technology [35]. The Times News Network [36] reported on Punjab University at Chandigarh (India) as an excellent example of this, as they have set up a Continuous Ambient Air Quality Monitoring Station (CAAQMS) that offers hourly data in real time. Air quality information is thus available to everyone through large electronic display panels. Xi'an Jiao Tong University in China is transforming its campus into a green energy-fueled smart campus, aiming to enhance ambient air quality for the academic community [37].

Furthermore, Monash University [38] has suggested that a future without change will be dismal; on the dangers of air pollution, they state, "We don't believe in a future where people can't go outside".

If the current trends of air pollution continue, then breathing in ambient oxygen might significantly risk the health of people. The Monash Climate and Air Quality Research Group (CARE) maintain that the fine-particulate matter in ambient air can significantly accelerate the risk of miscarriage among pregnant women, and the group also found a robust association between air pollution and autism. In their plan to combat air pollution, Monash set the goal of attaining net zero emissions on Australian campuses by 2030. In Canada, the University of Victoria [39] has developed a Campus Cycling Plan to make their campus bicycle-friendly, with the goal of maximizing cycling, walking, public transit use, and carpooling up to 70% by expanding facilities for cycling by 10%. To promote bicycle riding in and around the campus, the university plans to improve the cycling network, the safety of bike-users in shared spaces, bicycle parking, bicycle sharing, and end-of-trip amenities for users of all age groups.

The Surgeon General [40] of the USA has urged American universities to construct walkable campuses, suggesting that walking is a win–win strategy for community health, as increased physical activity offers significant health benefits. Bopp, Kaczynski, and Wittman [41] suggest that colleges and universities should become the ultimate locations for walking. Policies related to walkable campuses may not only inspire students, faculty members, and personnel to embrace active living, but may also encourage students to consider future roles as public health professionals, urban planners, urban designers, transport planners, and architects. Stevens [42] reported that the University of Kentucky has installed a large amount of signage on their campus, with QR codes in collaboration with the WALK [Your City] app, which helps university students approximate the time required for traveling by foot as an alternative to driving. This also helps university researchers in their investigations into how university attendees use information technology to plan their day. Scott et al. [43] state that Canadian universities are working to make their campuses car-free. The University of British Columbia [39] has been successfully implementing its Transport Strategic Plan (TSP) since 1999 (reviewed in 2005). UBC has a large cycling and pedestrian network and is aspiring towards sustainable campus transit by 2040.

2.3. $PM_{2.5}$ Modeling and Geographically Weighted Regression

Several studies before now have investigated the spatial heterogeneity and spatial dependence of $PM_{2.5}$ on the associated socioeconomic and environmental factors, using geographically weighted regression (GWR). GWR permits the exploration of spatially varying relationships [44]. Nearly all the studies have validated that GWR addresses the implicit spatial attributes of $PM_{2.5}$ data, and improves upon the outcomes offered by traditional OLS regression, which is nonspatial in nature [45].

Lin et al. [46] emphasize the urban green belt area, population density, and economic growth as the key factors affecting the concentration of $PM_{2.5}$ in Chinese cities. Guan et al. [47] stressed that China's foreign trade is responsible for most of the $PM_{2.5}$ pollution. According to Hao and Liu [48], motor vehicles and industrial activities are the factors of $PM_{2.5}$ exposure. Zhang et al. [49] deployed the enhanced vegetation index (EVI) with GWR, and concluded that meteorological parameters, together with fused aerosol optical depth (AOD) products, explain nearly 87% of the spatial variance in $PM_{2.5}$ concentrations. Similarly, Pateraki et al. [50] concluded that humidity and temperature fluctuations were strongly correlated with $PM_{2.5}$ concentration, while Onat and Stakeeva [51] affirmed that accelerated wind speed (>2m/s) might significantly lower the intensity of $PM_{2.5}$.

In recent years, researchers have frequently used GWR models to understand $PM_{2.5}$ exposure in various cities and regions. Through a generalized additive model (GAM), He and Lin [52] confirmed that the $PM_{2.5}$ concentration change in Nanjing was strongly correlated with air pressure, water vapor pressure, and temperature. The seasonal and daily variability in $PM_{2.5}$ levels was modeled by several spatial scientists in the Yangtze River delta region via GWR, while the spatiotemporal mapping of fine particle concentrations in mainland China was carried out by combining Bayesian maximum entropy (BME) with GWR [53,54].

Many types of GWR models have been effectively employed to quantify the spatiotemporal heterogeneity of $PM_{2.5}$ pollution in Chinese cities. Zhai et al. [55] developed an enhanced-subset regression model, which combines Principal Component Analysis (PCA) and GWR to predict the independent variables responsible for spatial variations in the levels of $PM_{2.5}$. Hajiloo, Hamzeh, and Gheysari [56] developed models to understand the impacts of metrological and environmental parameters on the intensity of $PM_{2.5}$ using satellite data and GWR analysis. Other GWR-based studies by Cheng et al. [57], Dong et al. [58], and Lou et al. [59] demonstrated the various determinant factors responsible for the geographical heterogeneity of $PM_{2.5}$.

A recent study by Gu et al. [60] suggested that $PM_{2.5}$ increases in Chinese cities are positively associated with people's income; growths in income in certain geographical areas have aggravated $PM_{2.5}$ emissions. Wang and Wang [61] observed that the density of the population, the proportions of industrial land uses, car ridership, and the amount of foreign direct investment (FDI) all contribute significantly to the level of $PM_{2.5}$, and show qualities of spatial heterogeneity. There were also significant variations in the levels of influence of these factors between different time periods and locations.

3. Material and Methods

3.1. Study Area

King Abdulaziz University, Jeddah, is a renowned public university in Saudi Arabia. The main campus of the university is situated between Prince Majid Road and Al-Haramin Expressway, spread over 6.35 square kilometers (Figure 1). It was set up in 1964 as a private university and, later, in 1974, became a public university by royal order. There are 77,000 full time students in the university at present, with 4059 academic faculty staff members, 4000 administrative staff members, and an additional 4200 support staff. The Vice Presidency for Projects KAU data show that the number of total daily trips through all six gates is approximately 56,000 during the academic semester.

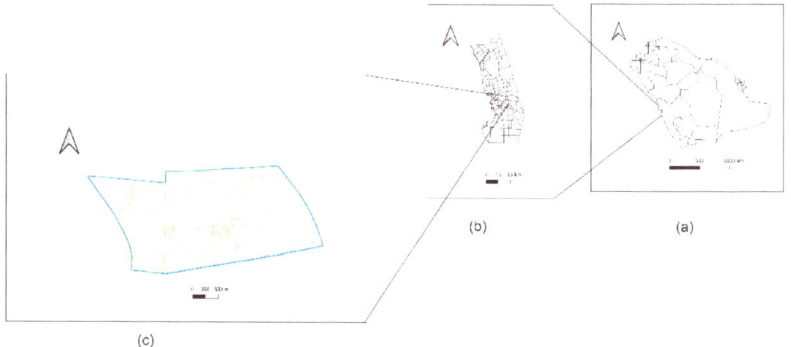

Figure 1. Location of study area: (**a**) Kingdom of Saudi Arabia; (**b**) Jeddah City; and (**c**) King Abdulaziz University campus.

The university has five main gates through which one can enter and exit the campus and sample locations near these gates capture the attributes of adjacent districts. In the north, where Gates no. 2, 3 and 4 are located, the Al-Sulaymaniyah district is residential. The Al-Jamia district is in the south; it is residential, with some commercial streets, and is accessed via Gate no.6. The main gate, or Gate no. 1 (The Eagle Gate), opens into the west towards Al-Fahya district, which is also residential with some commercial streets. Gate no. 5 opens onto a service road next to Al-Haramin highway (Abruq Ar Rughama district) in the east. More than 90% of the traffic around the campus is directed towards or away from the university.

3.2. Methods

We chose our independent variables from previous studies that are relevant to the KAU campus, including the number of cars, the windspeed, the temperature, the humidity, distance from trees, and distance from construction sites [46,56,61]. In addition to the dependent variable, $PM_{2.5}$, the data on the independent variables were collected at 51 locations inside the KAU campus, including the six entry and exit gates. The sampling period was from 29 September 2019 to 31 October 2019, with four distinct time points during weekdays (from Sunday to Thursday), inclusive of both peak and nonpeak traffic, and hourly data were recorded at 7 a.m., 9 a.m., 11 a.m., and 2 p.m. All of the independent variables were tested for multicollinearity. To ensure the representativeness of the sample locations, we employed the stratified random sampling technique in the QGIS environment, which has been found to be reliable in numerous ecological studies [62].

Multicollinearity may adversely affect the quality of estimators in a GWR model; hence, the detection of multicollinearity is a prerequisite when developing statistical models. The variance inflation factor (VIF) is a popular measure for detecting multicollinearity among independent variables [63]. We confirmed the absence of multicollinearity, as the VIF value for each predictor among the parameters was lower than 4 [64] (Table 1).

Table 1. VIF and descriptive statistics for the variables.

	Variables	Min.	Max.	Mean	Std. Dev.	VIF
Dependent Variables	$PM_{2.5}$ off-peak 7 am (Model-1)	17	37	24.1	4.8	-
	$PM_{2.5}$ peak 9 am (Model-2)	19	68	35.3	10.3	-
	$PM_{2.5}$ peak 11 am (Model-3)	18	65	33.3	9.7	-
	$PM_{2.5}$ peak 2 pm (Model-4)	21	70	36.6	9.2	-
Independent Variables	Number of cars	25	6989	1022.0	1452.4	2.9
	Windspeed	12	21	18.2	2.4	1.4
	Temperature	27	34	30.5	1.8	2.2
	Humidity	28	33	30.4	1.3	1.3
	Distance from construction sites	74	937	312.1	201.4	2.5
	Distance from trees	51	650	215.7	163.0	1.9

Thereafter, four OLS and GWR models were developed to assess the contributory factors for the $PM_{2.5}$ concentration on the KAU campus.

Ordinary least squares (OLS) is a linear least squares method for assessing the unknown parameters in a linear regression model. The OLS equation could be expressed as:

$$y_i = \sum_{j=0}^{m} \beta_i + \varepsilon_i, \quad i = 1, 2, \ldots, n \tag{1}$$

where m is the total number of predictors, while n denotes the number of observations, β_i shows the coefficient to be estimated, and ε_i is used for random errors.

A GWR model is a special type of regression model that considers geographically varying parameters. The GWR4 software was used to calibrate the GWR models [65]. A conventional GWR equation is:

$$y_i = \Sigma_k \, \beta_k \, (u_i, v_i) \, x_{ki} + \varepsilon_i \tag{2}$$

where y_i, x_{ki}, and ε_i are the dependent variable, the kth independent variable, and the Gaussian error at location i, respectively; (u_i, v_i) is the x–y coordinate of the ith location; and the coefficients, $\beta_k \, (u_i, v_i)$, are the varying conditionals for the location.

We developed four models to understand the spatial variability and the determinants of $PM_{2.5}$ at the KAU campus. The OLS and GWR alternatives of each model were compared to choose the best fit, as suggested by Grekousis [64].

To test the significance of the regression coefficient in our models, we used a t-test under the set of assumptions called the Gauss–Markov conditions [66]. The equation for the t-scores is:

$$t\beta 1 = \frac{\beta 1}{S_{\beta 1}} \tag{3}$$

where t is the t-score for the regression estimate, $\beta 1$, and S is the standard error. The null hypothesis for the t-test states that the t-score for the regression coefficient $\beta 1$ is 0. For better visualization of the local t-scores (and local R2), we drew Voronoi polygons around sample locations. In general, Voronoi or Thiessen polygons help in mapping the influence area of an individual data point [67].

AICc estimates prediction errors. By default, GWR 4 facilitates comparisons of the relative quality of statistical models for a given dataset.

4. Results

4.1. OLS and GWR Regression Results

As mentioned earlier, we developed four different geographically weighted regression models to understand the key predictors that explain the levels of $PM_{2.5}$ in the ambient air of the KAU campus.

The results of the OLS regression and GWR are summarized in Tables 2–6. The results in all four models are quite similar. However, the GWR models performed better than the OLS regression models.

The results of Model-1 reveal that the two environmental parameters of wind speed and temperature, along with the distances from trees and construction sites, were not significant (p-value > 0.5) in either the OLS or GWR models (Table 2). The OLS results indicate that the number of cars and the humidity were significant predictors (p-value ≤ 0.5) in Model-1. Additionally, the regression coefficient suggests that a 1 unit increase in the number of cars increases the $PM_{2.5}$ concentration by 0.915 units, and a 1 unit increase in humidity decreases the $PM_{2.5}$ concentration by 0.122 units. The GWR models show that humidity was a significant parameter at 29.41% of the locations with a negative sign (Figure 2), while the number of cars was a significant predictor at 100% of the locations with a positive sign (Figure 3a).

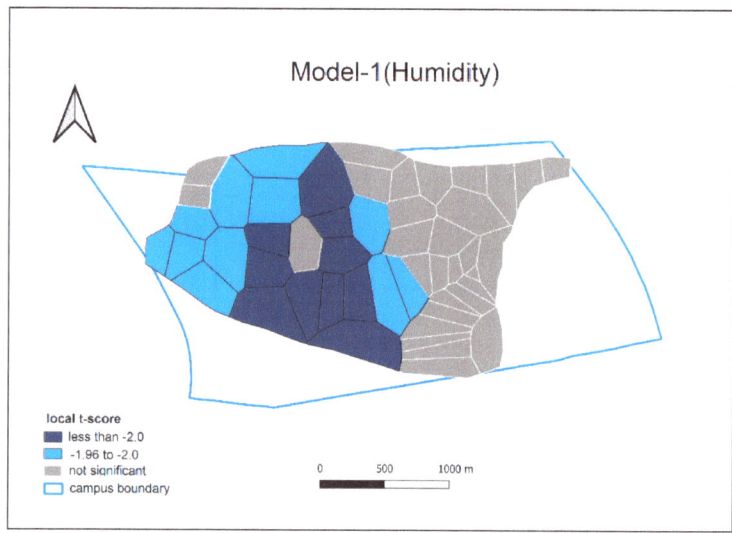

Figure 2. Local t-scores for humidity in Model-1 (GWR).

The results of Models -2, -3, and -4 were not very dissimilar; however, the number of cars was the sole significant predictor (p-value ≤ 0.5) in all three models, while other parameters were found to be statically insignificant (p-value >0.5) in both the OLS and GWR models (Tables 3–5).

In comparison to the OLS model, the benefit of the GWR models is that they facilitate the coefficient estimation of the predictor parameter for each geographical location. The spatial distribution of the local coefficient estimates from the GWR models are shown in Figures 2 and 3.

Table 2. OLS and GWR estimates (Model-1).

	OLS Model				GWR Model		
	Est.	SE	T (Est/SE)	p-Value	$p \leq 0.05$	+ (%)	− (%)
Intercept	0.001	0.051	0.019	1.000	0	0	0
Number of cars	0.915	0.058	15.809	0.000	100	100	0
Wind speed	−0.049	0.053	−0.920	0.357	0	0	0
Temperature	−0.084	0.056	−1.503	0.133	0	0	0
Humidity	−0.122	0.055	−2.227	0.026	29.41	0	100.0
Distance from construction	0.032	0.056	0.579	0.563	0	0	0
Distance from trees	−0.032	0.060	−0.530	0.596	0	0	0

Table 3. OLS and GWR estimates (Model-2).

	OLS Model				GWR Model		
	Est.	SE	T (Est/SE)	p-Value	$p \leq 0.05$	+ (%)	− (%)
Intercept	0.000	0.045	0.000	1.000	0	0	0
Number of cars	0.953	0.056	17.167	0.000	100	100	0
Wind speed	−0.037	0.048	−0.782	0.434	0	0	0
Temperature	0.010	0.051	0.195	0.846	0	0	0
Humidity	0.003	0.052	0.066	0.947	0	0	0
Distance from construction	−0.090	0.043	−2.097	0.036	0	0	0
Distance from trees	−0.027	0.046	−0.581	0.561	0	0	0

Table 4. OLS and GWR estimates (Model-3).

	OLS Model				GWR Model		
	Est.	SE	T (Est/SE)	p-Value	$p \leq 0.05$	+ (%)	− (%)
Intercept	0.000	0.069	0.000	1.000	0	0	0
Number of cars	0.855	0.076	11.285	0.000	100	100	0
Wind speed	−0.029	0.071	−0.405	0.685	0	0	0
Temperature	0.120	0.073	1.650	0.099	0	0	0
Humidity	−0.003	0.076	−0.040	0.968	0	0	0
Distance from construction	−0.78	0.070	−1.124	0.261	0	0	0
Distance from trees	−0.34	0.072	−0.475	0.635	0	0	0

Table 5. OLS and GWR estimates (Model-4).

	OLS Model				GWR Model		
	Est.	SE	T (Est/SE)	p-Value	$p \leq 0.05$	+ (%)	− (%)
Intercept	0.000	0.066	0.000	1.000	0	0	0
Number of cars	0.868	0.073	11.886	0.000	100	100	0
Wind speed	0.046	0.071	0.648	0.517	0	0	0
Temperature	−0.072	0.068	−1.061	0.288	0	0	0
Humidity	−0.089	0.068	−1.313	0.189	0	0	0
Distance from construction	0.067	0.065	1.040	0.298	0	0	0
Distance from trees	0.045	0.067	0.67	0.562	0	0	0

4.2. Model Comparison

We used the residual sum of squares (RSS), log-likelihood, adjusted R-squared, and corrected Akaike information criterion indices to estimate the suitability of the models. RSS, or the sum of the squared estimate of error (SEE), which is used to quantify the variance in a dataset, remained unexplained by the regression model [68]. Additionally, the corrected Akaike information criterion (AICc) is an estimator of estimation error, and thereby an indicator of the quality of the regression models for a given dataset. By default, GWR 4 facilitates comparisons between the relative quality of statistical models for a given dataset [69]. AICc facilitates a better model fit if the AICc value declines [64].

In effect, the lower values of RSS and AICc indicate the better fit of the model. Additionally, the log-likelihood function is a measure of how well a specific model fits the data. This function explains the suitability of a parameter for explaining an observed value [70]. Furthermore, R-squared is a coefficient of determination, and a modified version of R-squared is adjusted for the number of predictors in the model. This assesses the explanatory power of the linear regression models [71]. As regards the log-likelihood function-adjusted R-squared, comparatively higher values suggest a better fit.

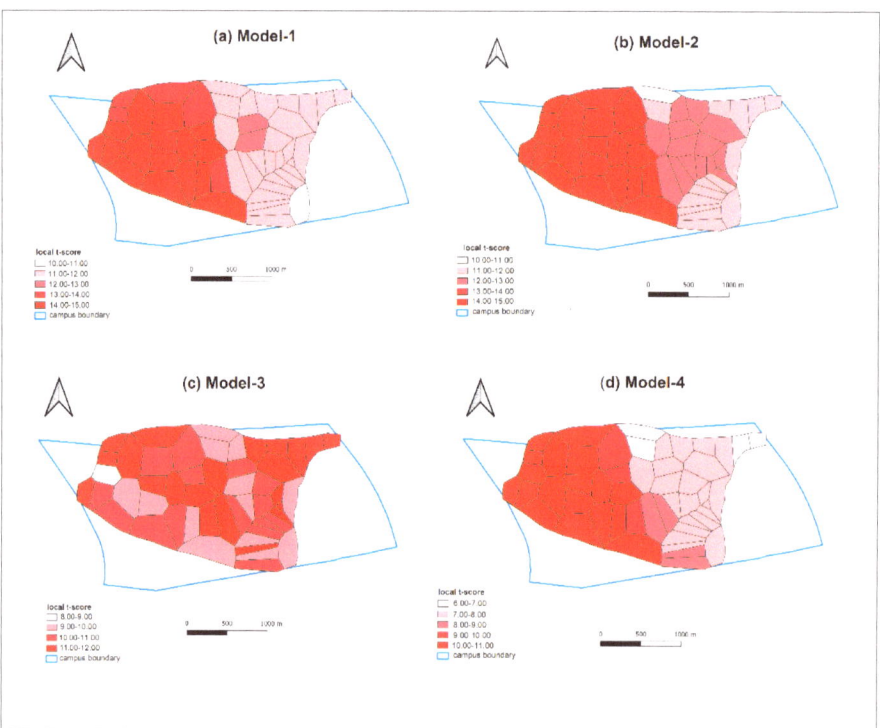

Figure 3. Local t-scores for numbers of cars in GWR models (**a**) Model-1, (**b**) Model-2, (**c**) Model-3, and (**d**) Model-4 (all t-scores are significant because of t-values > 1.96).

The model comparison demonstrates (in Table 6 and Figure 4) that the RSS values in all four models are lower for the GWR variants than for OLS. For GWR, the RSS value is lowest (4.526) in Model-2, and highest (8.640) in Model-4. Similarly, the AICc value in all four models is again lower for the GWR variants, and the lowest AICc (38.435) is present in the GWR variant of Model-2. Next, the log-likelihood values are higher in all four GWR models than in OLS, and the GWR variant of Model-2 has the highest log-likelihood value (−10.667). Furthermore, the adjusted R2 values are higher in all GWR models in comparison to the OLS models. The highest adjusted R2 value (0.897) was reported in the GWR variant of Model-2. In brief, Model-2 has a better fit in all four models.

Table 6. Goodness of fit comparison for all four OLS and GWR models.

	Model-1		Model-2		Model-3		Model-4	
	OLS	GWR	OLS	GWR	OLS	GWR	OLS	GWR
RSS	6.214	5.475	4.830	4.526	11.272	6.476	10.113	8.640
Log-likelihood	−18.688	−15.459	−12.263	−10.607	−33.874	−19.741	−31.107	−27.091
Adjusted R2	0.871	0.868	0.893	0.897	0.760	0.844	0.784	0.795
AICc	54.89	51.285	45.594	38.435	81.658	66.094	73.581	72.213

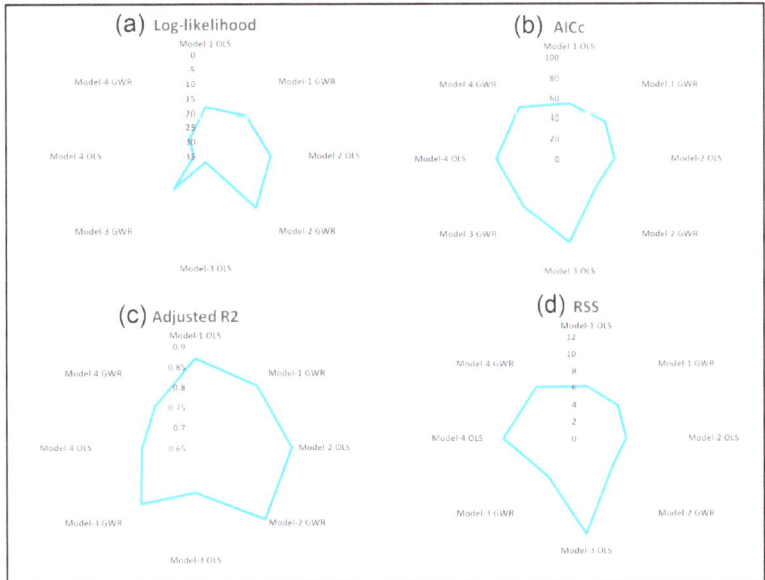

Figure 4. Model comparison of (a) log-likelihood; (b) AICc; (c) adjusted R2; and (d) RSS.

5. Discussion

This study investigates the effects of various independent variables on the $PM_{2.5}$ concentration at the KAU campus through OLS and GWR models. We showed that car traffic is the single most significant factor contributing to the presence of fine particles in the ambient air of KAU at the ground level.

This study has certain limitations related to sampling. We collected samples from 51 locations at four distinct times of the day for one month; however, extending the time of data collection to one year or more may capture seasonal variations in $PM_{2.5}$ concentrations, resulting in more sophisticated models.

An earlier study by Khodeir et al. [72] confirmed that the main source of $PM_{2.5}$ at the KAU campus is dense vehicular traffic; however, this study does not provide any evidence regarding the spatial variability of $PM_{2.5}$ within the campus. A higher concentration of $PM_{2.5}$ during peak hours (68–70 µg/m³) is a frightening result.

As mentioned earlier, prolonged exposure to fine particles might result in a higher prevalence of diseases, and even premature casualties. In general, ambient air quality at the KAU campus is the worst for pedestrians during peak hours. This makes a robust case for policymakers to make the campus car-free, as curbing sources of fine particle emissions might reduce deadly health consequences [73]. Our study reinforces the findings of Ruben Garnica-Monroy et al. [8], which suggest that the exposure of a population to roads with heavy traffic exposes it to higher health risks and necessitates changes in urban planning and public health policies.

Health consequences associated with fine particles have been well-researched in the global north and the global south, but there are very few studies in the Middle East region, particularly in Saudi Arabia. An important study by Nayebare et al. [73] noted that the $PM_{2.5}$ concentration in Jeddah City is increasing the risk of cardiopulmonary and respiratory morbidity.

Through GWR models, we have specified the locations on the KAU campus that are at the greatest risk of $PM_{2.5}$ exposure because of car use for on-campus mobility.

We recommend addressing the $PM_{2.5}$ concentration on the KAU campus through policies promoting walkability and bicycle riding, which will help to discourage car traffic, similar to other universities in the global north that are trying to make their campuses car-free and carbon-neutral.

6. Conclusions

Higher concentrations of $PM_{2.5}$ at the ground level might accelerate many physical and cognitive diseases. In this study, we assembled data on dependent and independent variables at 51 locations on the KAU campus, and developed OLS- and GWR-based regression models to explain the spatial variability and determinants of $PM_{2.5}$ exposure at the micro geographical scale. Unlike other studies, we used primary data to develop four OLS and GWR models, suggesting that car traffic is a significant factor in $PM_{2.5}$ concentration on the KAU campus, while environmental factors and other activities (construction) are not significant at this scale. We compared all four models based on RSS, log-likelihood, adjusted R-squared, and AICc, and concluded that all the GWR estimates are better than the traditional OLS estimates, with Model-2 (attributing peak hour traffic in the morning) representing the best fit. On the basis of our findings, we recommend adopting sustainable mobility policies on the KAU campus in order to improve the general health of KAU students and staff.

Author Contributions: Data curation, and formal analysis by M.A.; methodology, writing, review and editing by A.T. All authors have read and agreed to the published version of the manuscript.

Funding: This project was funded by the Deanship of Scientific Research (DSR), King Abdulaziz University, Jeddah, under grant No. (DF-830-137-1441). The authors, therefore, gratefully acknowledge DSR technical and financial support.

Institutional Review Board Statement: Not applicable.

Informed Consent Statement: Not applicable.

Data Availability Statement: Data are available on request.

Conflicts of Interest: The authors declare no conflict of interest.

References

1. UNEP. Pollution Action Note—Data You Need to Know. Available online: https://www.unep.org/interactive/air-pollution-note/ (accessed on 15 August 2021).
2. Stauffer, A.; Perroud, S. Clean air policy challenges in Europe: Achieving air standards that prevent disease. *Eur. J. Public Health* **2020**, *30*. [CrossRef]
3. Klompmaker, J.O.; Hoek, G.; Bloemsma, L.D.; Marra, M.; Wijga, A.H.; van den Brink, C.; Brunekreef, B.; Lebret, E.; Gehring, U.; Janssen, N.A.H. Surrounding green, air pollution, traffic noise exposure and non-accidental and cause-specific mortality. *Environ. Int.* **2020**, *134*, 105341. [CrossRef] [PubMed]
4. An, X.; Hou, Q.; Li, N.; Zhai, S. Assessment of human exposure level to PM10 in China. *Atmos. Environ.* **2013**, *70*, 376–386. [CrossRef]
5. Fang, T.; Guo, H.; Verma, V.; Peltier, R.E.; Weber, R.J. PM 2.5 water-soluble elements in the southeastern United States: Automated analytical method development, spatiotemporal distributions, source apportionment, and implications for heath studies. *Atmos. Chem. Phys.* **2015**, *15*, 11667–11682. [CrossRef]
6. Fattore, E.; Paiano, V.; Borgini, A.; Tittarelli, A.; Bertoldi, M.; Crosignani, P.; Fanelli, R. Human health risk in relation to air quality in two municipalities in an industrialized area of Northern Italy. *Environ. Res.* **2011**, *111*, 1321–1327. [CrossRef] [PubMed]

7. Debone, D.; Leirião, L.F.L.; Miraglia, S.G.E.K. Air quality and health impact assessment of a truckers' strike in Sao Paulo state, Brazil: A case study. *Urban Clim.* **2020**, *34*, 100687. [CrossRef]
8. Garnica-Monroy, R.; Garibay-Bravo, V.; Gonzalez-Gonzalez, A.; Martínez Salgado, H.; Hernández-Reyes, M. Spatial Analysis of Exposure to Traffic-Related Air Pollution in Mexico: Implications for Urban Planning to Improve Public Health. *Appl. Spat. Anal. Policy* **2021**. [CrossRef]
9. Toro, R.; Kvakić, M.; Klaić, Z.B.; Koračin, D. Exploring atmospheric stagnation during a severe particulate matter air pollution episode over complex terrain in Santiago, Chile. *Environ. Pollut.* **2019**, *244*, 705–714. [CrossRef]
10. Englert, N. Fine particles and human health—A review of epidemiological studies. *Toxicol. Lett.* **2004**, *149*, 235–242. [CrossRef]
11. Valavanidis, A.; Fiotakis, K.; Vlachogianni, T. Airborne Particulate Matter and Human Health: Toxicological Assessment and Importance of Size and Composition of Particles for Oxidative Damage and Carcinogenic Mechanisms. *J. Environ. Sci. Health Part C* **2008**, *26*, 339–362. [CrossRef]
12. EPA. How Does PM Affect Human Health? Available online: https://www3.epa.gov/region1/airquality/pm-human-health.html (accessed on 9 September 2021).
13. Gao, P.; Lei, T.; Jia, L.; Song, Y.; Lin, N.; Du, Y.; Feng, Y.; Zhang, Z.; Cui, F. Exposure and health risk assessment of $PM_{2.5}$-bound trace metals during winter in university campus in Northeast China. *Sci. Total Environ.* **2017**, *576*, 628–636. [CrossRef]
14. Chen, P.; Bi, X.; Zhang, J.; Wu, J.; Feng, Y. Assessment of heavy metal pollution characteristics and human health risk of exposure to ambient PM2.5 in Tianjin, China. *Particuology* **2015**, *20*, 104–109. [CrossRef]
15. Mascarelli, A.L. How green is your campus? *Nature* **2009**, *461*, 154–155. [CrossRef] [PubMed]
16. Stafford, S.L. How green is your campus? An Analysis of the factors that drives universities to embrace Sustainability. *Contemp. Econ. Policy* **2011**, *29*, 337–356. [CrossRef]
17. Anenberg, S.C.; Horowitz, L.W.; Tong, D.Q.; West, J.J. An Estimate of the Global Burden of Anthropogenic Ozone and Fine Particulate Matter on Premature Human Mortality Using Atmospheric Modeling. *Environ. Health Perspect.* **2010**, *118*, 1189–1195. [CrossRef] [PubMed]
18. Cao, Q.; Rui, G.; Liang, Y. Study on PM2.5 pollution and the mortality due to lung cancer in China based on geographic weighted regression model. *BMC Public Health* **2018**, *18*, 925. [CrossRef]
19. Chowdhury, S.; Dey, S. Cause-specific premature death from ambient $PM_{2.5}$ exposure in India: Estimate adjusted for baseline mortality. *Environ. Int.* **2016**, *91*, 283–290. [CrossRef]
20. Pope III, C.A. Lung Cancer, Cardiopulmonary Mortality, and Long-term Exposure to Fine Particulate Air Pollution. *JAMA* **2002**, *287*, 1132. [CrossRef]
21. Bandyopadhyay, A. Neurological Disorders from Ambient (Urban) Air Pollution Emphasizing UFPM and $PM_{2.5}$. *Curr. Pollut. Rep.* **2016**, *2*, 203–211. [CrossRef]
22. Ailshire, J.A.; Crimmins, E.M. Fine Particulate Matter Air Pollution and Cognitive Function Among Older US Adults. *Am. J. Epidemiol.* **2014**, *180*, 359–366. [CrossRef] [PubMed]
23. Costa, L.G.; Cole, T.B.; Coburn, J.; Chang, Y.-C.; Dao, K.; Roque, P. Neurotoxicants Are in the Air: Convergence of Human, Animal, and In Vitro Studies on the Effects of Air Pollution on the Brain. *Biomed Res. Int.* **2014**, *2014*, 1–8. [CrossRef]
24. Fu, P.; Yung, K.K.L. The association between $PM_{2.5}$ exposure and neurological disorders. In *Air Pollution, Climate, and Health*; Elsevier: Amsterdam, The Netherlands, 2021; pp. 229–245.
25. Kilian, J.; Kitazawa, M. The emerging risk of exposure to air pollution on cognitive decline and Alzheimer's disease—Evidence from epidemiological and animal studies. *Biomed. J.* **2018**, *41*, 141–162. [CrossRef]
26. Zhou, Y.-M.; Fan, Y.-N.; Yao, C.-Y.; Xu, C.; Liu, X.-L.; Li, X.; Xie, W.-J.; Chen, Z.; Jia, X.-Y.; Xia, T.-T.; et al. Association between short-term ambient air pollution and outpatient visits of anxiety: A hospital-based study in northwestern China. *Environ. Res.* **2021**, *197*, 111071. [CrossRef]
27. Ranzani, O.T.; Milà, C.; Kulkarni, B.; Kinra, S.; Tonne, C. Association of Ambient and Household Air Pollution with Bone Mineral Content Among Adults in Peri-urban South India. *JAMA Netw. Open* **2020**, *3*, e1918504. [CrossRef]
28. Massachusetts Institute of Technology. Designing the Zero-Carbon Campus of the Future. Available online: https://sustainability.mit.edu/topic/zero-carbon-campus#!introduction (accessed on 22 July 2021).
29. University of Leeds. Living Lab for Air Quality. Available online: https://sustainability.leeds.ac.uk/the-living-lab/airquality/ (accessed on 6 July 2021).
30. BBC. Smoking More Toxic than Car Fumes. Available online: http://news.bbc.co.uk/2/hi/health/3590578.stm (accessed on 8 July 2021).
31. Muller, R.A.; Muller, E.A. Air Pollution and Cigarette Equivalence. Available online: http://berkeleyearth.org/air-pollution-and-cigarette-equivalence/ (accessed on 15 July 2021).
32. King Abdulaziz University. King Abdulaziz University: Competition to Design "A Smoking-free University" Logo. Available online: https://www.kau.edu.sa/Content-0-EN-36146 (accessed on 5 July 2021).
33. Lincoln University. Clean Air Policy. Available online: https://www.lincoln.ac.nz/assets/PoliciesAndProcedures/Clean-Air-Policy-.pdf (accessed on 9 July 2021).
34. Express Web Desk. Hyderabad University Launches E-Rickshaws for Campus. Available online: https://indianexpress.com/article/cities/hyderabad/hyderabad-university-launches-e-rickshaws-for-campus-6162327/ (accessed on 12 July 2021).

35. Columbia Central University. Red de Monitoreo de Calidad del Aire. Available online: https://www.ucentral.edu.co/noticentral/red-monitoreo-calidad-del-aire (accessed on 21 October 2021).
36. Times News Network. Chandigarh: Air Quality Improves to 'Moderate'. Available online: https://timesofindia.indiatimes.com/city/chandigarh/air-quality-improves-to-moderate/articleshow/71929952.cms (accessed on 5 July 2021).
37. Grundfos. Green Energy Fuels Smart Campus in Western China. Available online: https://www.grundfos.com/solutions/learn/cases/green-energy-fuels-smart-campus-in-western-china (accessed on 22 August 2021).
38. Monash University Air Pollution. Available online: https://www.monash.edu/future-without-change/air-pollution (accessed on 22 July 2021).
39. University of Victoria. *Campus Cycling Plan*; University of Victoria: Victoria, BC, Canada, 2019.
40. Surgeon General Step It Up! The Surgeon General's Call to Action to Promote Walking and Walkable Communities. Available online: https://www.hhs.gov/surgeongeneral/reports-and-publications/physical-activity-nutrition/index.html (accessed on 18 July 2021).
41. Bopp, M.; Kaczynski, A.; Wittman, P. Active Commuting Patterns at a Large, Midwestern College Campus. *J. Am. Coll. Health* **2011**, *59*, 605–611. [CrossRef] [PubMed]
42. Stevens, A. How Colleges Are Stepping Up Campus Walkability. Available online: https://www.citylab.com/life/2015/12/how-colleges-are-stepping-up-campus-walkability/419220/ (accessed on 15 July 2021).
43. Scott, A.; Nwadike, N.; Seibel, L.; Dosch, G.; Uchendu, N. Creating a Car Free Campus. Available online: http://umanitoba.ca/campus/sustainability/media/Creating_a_Car_Free_Campus.pdf (accessed on 11 July 2021).
44. Páez, A.; Wheeler, D.C. Geographically Weighted Regression. In *International Encyclopedia of Human Geography*; Elsevier: Amsterdam, The Netherlands, 2009; pp. 407–414.
45. Sheehan, K.R.; Strager, M.P.; Welsh, S.A. Advantages of Geographically Weighted Regression for Modeling Benthic Substrate in Two Greater Yellowstone Ecosystem Streams. *Environ. Model. Assess.* **2013**, *18*, 209–219. [CrossRef]
46. Lin, G.; Fu, J.; Jiang, D.; Hu, W.; Dong, D.; Huang, Y.; Zhao, M. Spatio-Temporal Variation of PM2.5 Concentrations and Their Relationship with Geographic and Socioeconomic Factors in China. *Int. J. Environ. Res. Public Health* **2013**, *11*, 173–186. [CrossRef] [PubMed]
47. Guan, D.; Su, X.; Zhang, Q.; Peters, G.P.; Liu, Z.; Lei, Y.; He, K. The socioeconomic drivers of China's primary $PM_{2.5}$ emissions. *Environ. Res. Lett.* **2014**, *9*. [CrossRef]
48. Hao, Y.; Liu, Y.-M. The influential factors of urban $PM_{2.5}$ concentrations in China: A spatial econometric analysis. *J. Clean. Prod.* **2016**, *112*, 1443–1453. [CrossRef]
49. Pateraki, S.; Asimakopoulos, D.N.; Flocas, H.A.; Maggos, T.; Vasilakos, C. The role of meteorology on different sized aerosol fractions (PM_{10}, $PM_{2.5}$, $PM_{2.5-10}$). *Sci. Total Environ.* **2012**, *419*, 124–135. [CrossRef]
50. Onat, B.; Stakeeva, B. Personal exposure of commuters in public transport to $PM_{2.5}$ and fine particle counts. *Atmos. Pollut. Res.* **2013**, *4*, 329–335. [CrossRef]
51. He, X.; Lin, Z.-S. Interactive Effects of the Influencing Factors on the Changes of $PM_{2.5}$ Concentration Based on GAM Model. *Huan Jing Ke Xue Huanjing Kexue* **2017**, *38*, 22–32. [CrossRef]
52. Jiang, M.; Sun, W.; Yang, G.; Zhang, D. Modelling Seasonal GWR of Daily $PM_{2.5}$ with Proper Auxiliary Variables for the Yangtze River Delta. *Remote Sens.* **2017**, *9*, 346. [CrossRef]
53. Xiao, L.; Lang, Y.; Christakos, G. High-resolution spatiotemporal mapping of $PM_{2.5}$ concentrations at Mainland China using a combined BME-GWR technique. *Atmos. Environ.* **2018**, *173*, 295–305. [CrossRef]
54. Zhai, L.; Li, S.; Zou, B.; Sang, H.; Fang, X.; Xu, S. An improved geographically weighted regression model for $PM_{2.5}$ concentration estimation in large areas. *Atmos. Environ.* **2018**, *181*, 145–154. [CrossRef]
55. Hajiloo, F.; Hamzeh, S.; Gheysari, M. Impact assessment of meteorological and environmental parameters on $PM_{2.5}$ concentrations using remote sensing data and GWR analysis (case study of Tehran). *Environ. Sci. Pollut. Res.* **2019**, *26*, 24331–24345. [CrossRef] [PubMed]
56. Cheng, Z.; Luo, L.; Wang, S.; Wang, Y.; Sharma, S.; Shimadera, H.; Wang, X.; Bressi, M.; de Miranda, R.M.; Jiang, J.; et al. Status and characteristics of ambient $PM_{2.5}$ pollution in global megacities. *Environ. Int.* **2016**, *89–90*, 212–221. [CrossRef] [PubMed]
57. Dong, F.; Wang, Y.; Zheng, L.; Li, J.; Xie, S. Can industrial agglomeration promote pollution agglomeration? Evidence from China. *J. Clean. Prod.* **2020**, *246*, 118960. [CrossRef]
58. Lou, C.-R.; Liu, H.-Y.; Li, Y.-F.; Li, Y.-L. Socioeconomic Drivers of $PM_{2.5}$ in the Accumulation Phase of Air Pollution Episodes in the Yangtze River Delta of China. *Int. J. Environ. Res. Public Health* **2016**, *13*, 928. [CrossRef]
59. Gu, K.; Zhou, Y.; Sun, H.; Dong, F.; Zhao, L. Spatial distribution and determinants of $PM_{2.5}$ in China's cities: Fresh evidence from IDW and GWR. *Environ. Monit. Assess.* **2021**, *193*, 15. [CrossRef] [PubMed]
60. Wang, M.; Wang, H. Spatial Distribution Patterns and Influencing Factors of $PM_{2.5}$ Pollution in the Yangtze River Delta: Empirical Analysis Based on a GWR Model. *Asia-Pacific J. Atmos. Sci.* **2021**, *57*, 63–75. [CrossRef]
61. Barabesi, L.; Fattorini, L. Random versus stratified location of transects or points in distance sampling: Theoretical results and practical considerations. *Environ. Ecol. Stat.* **2013**, *20*, 215–236. [CrossRef]
62. Mansfield, E.R.; Helms, B.P. Detecting Multicollinearity. *Am. Stat.* **1982**, *36*, 158–160. [CrossRef]
63. Grekousis, G. *Spatial Analysis Methods and Practice*; Cambridge University Press: Cambridge, UK, 2020; ISBN 9781108614528.

64. Oshan, T.; Li, Z.; Kang, W.; Wolf, L.; Fotheringham, A. mgwr: A Python Implementation of Multiscale Geographically Weighted Regression for Investigating Process Spatial Heterogeneity and Scale. *ISPRS Int. J. Geo Inf.* **2019**, *8*, 269. [CrossRef]
65. Allen, M.P. The t test for the simple regression coefficient. In *Understanding Regression Analysis*; Springer: Boston, MA, USA, 1997; pp. 66–70.
66. Harrison, B. Tips for quicker focused evaluation. In *Data Room Management for Mergers and Acquisitions in the Oil and Gas Industry*; Elsevier: Amsterdam, The Netherlands, 2020; pp. 145–167.
67. Morgan, J.A.; Tatar, J.F. Calculation of the Residual Sum of Squares for all Possible Regressions. *Technometrics* **1972**, *14*, 317–325. [CrossRef]
68. Hurvich, C.M.; Tsai, C.-L. A corrected akaike information criterion for vector autoregressive model selection. *J. Time Ser. Anal.* **1993**, *14*, 271–279. [CrossRef]
69. Ishiguro, M.; Sakamoto, Y.; Kitagawa, G. Bootstrapping Log Likelihood and EIC, an Extension of AIC. *Ann. Inst. Stat. Math.* **1997**, *49*, 411–434. [CrossRef]
70. Miles, J. R -Squared, Adjusted R -Squared. In *Encyclopedia of Statistics in Behavioral Science*; John Wiley & Sons, Ltd: Chichester, UK, 2005.
71. Khodeir, M.; Shamy, M.; Alghamdi, M.; Zhong, M.; Sun, H.; Costa, M.; Chen, L.-C.; Maciejczyk, P. Source apportionment and elemental composition of $PM_{2.5}$ and PM_{10} in Jeddah City, Saudi Arabia. *Atmos. Pollut. Res.* **2012**, *3*, 331–340. [CrossRef]
72. Schwartz, J.; Laden, F.; Zanobetti, A. The concentration-response relation between $PM_{2.5}$ and daily deaths. *Environ. Health Perspect.* **2002**, *110*, 1025–1029. [CrossRef] [PubMed]
73. Nayebare, S.R.; Aburizaiza, O.S.; Siddique, A.; Carpenter, D.O.; Arden Pope, C.; Mirza, H.M.; Zeb, J.; Aburiziza, A.J.; Khwaja, H.A. Fine particles exposure and cardiopulmonary morbidity in Jeddah: A time-series analysis. *Sci. Total Environ.* **2019**, *647*, 1314–1322. [CrossRef]

Article

Temporal and Spatial Analysis of PM$_{2.5}$ and O$_3$ Pollution Characteristics and Transmission in Central Liaoning Urban Agglomeration from 2015 to 2020

Ju Wang *, Yue Zhong, Zhuoqiong Li and Chunsheng Fang

College of New Energy and Environment, Jilin University, Changchun 130012, China; zhongyue20@mails.jlu.edu.cn (Y.Z.); zhuoqiong21@mails.jlu.edu.cn (Z.L.); fangcs@jlu.edu.cn (C.F.)
* Correspondence: wangju@jlu.edu.cn

Citation: Wang, J.; Zhong, Y.; Li, Z.; Fang, C. Temporal and Spatial Analysis of PM$_{2.5}$ and O$_3$ Pollution Characteristics and Transmission in Central Liaoning Urban Agglomeration from 2015 to 2020. *Sustainability* **2022**, *14*, 511. https://doi.org/10.3390/su14010511

Academic Editors: Weixin Yang, Hone-Jay Chu, Guanghui Yuan and Yunpeng Yang

Received: 22 October 2021
Accepted: 27 December 2021
Published: 4 January 2022

Publisher's Note: MDPI stays neutral with regard to jurisdictional claims in published maps and institutional affiliations.

Copyright: © 2022 by the authors. Licensee MDPI, Basel, Switzerland. This article is an open access article distributed under the terms and conditions of the Creative Commons Attribution (CC BY) license (https://creativecommons.org/licenses/by/4.0/).

Abstract: The central Liaoning urban agglomeration is an important heavy industry development base in China, and also an important part of the economy in northeast China. The atmospheric environmental problems caused by the development of heavy industry are particularly prominent. Trajectory clustering, potential source contribution (PSCF), and concentration weighted trajectory (CWT) analysis are used to discuss the temporal and spatial pollution characteristics of PM$_{2.5}$ and ozone concentrations and reveal the regional atmospheric transmission pattern in central Liaoning urban agglomeration from 2015 to 2020. The results show that: (1) PM$_{2.5}$ in the central Liaoning urban agglomeration showed a decreasing trend from 2015 to 2020. The concentration of PM$_{2.5}$ is the lowest in 2018. Except for Benxi (34.7 µg/m^3), the concentrations of PM$_{2.5}$ in other cities do not meet the standard in 2020. The ozone concentration in Anshan, Liaoyang, and Shenyang reached the peaks in 2017, which are 68.76 µg/m^3, 66.27 µg/m^3, and 63.46 µg/m^3 respectively. PM$_{2.5}$ pollution is the highest in winter and the lowest in summer. The daily variation distribution of PM$_{2.5}$ concentration showed a bimodal pattern. Ozone pollution is the most serious in summer, with the concentration of ozone reaching 131.14 µg/m^3 in Shenyang. Fushun is affected by Shenyang intercity pollution, and the ozone concentration is high. (2) In terms of spatial distribution, the high values of PM$_{2.5}$ are concentrated in monitoring stations in urban areas. On the contrary, the concentration of ozone in suburban stations is higher. The high concentration of ozone in the northeast of Anshan, Liaoyang, Shenyang to Tieling, and Fushun extended in a band distribution. (3) Through cluster analysis, it is found that PM$_{2.5}$ and ozone in Shenyang are mainly affected by short-distance transport airflow. In winter, the weighted PSCF high-value area of PM$_{2.5}$ presents as a potential contribution source zone of the northeast trend with wide coverage, in which the contribution value of the weighted CWT in the middle of Heilongjiang is the highest. The main potential source areas of ozone mass concentration in spring and summer are coastal cities and the Bohai Sea and the Yellow Sea. We conclude that the regional transmission of pollutants is an important factor of pollution, so we should pay attention to the supply of industrial sources and marine sources of marine pollution in the surrounding areas of cities, and strengthen the joint prevention and control of air pollution among regions. The research results of this article provide a useful reference for the central Liaoning urban agglomeration to improve air quality.

Keywords: PM$_{2.5}$; O$_3$; transmission pathways; backward trajectory; PSCF; CWT

1. Introduction

With the rapid development of the social economy, air pollution has become an increasingly serious environmental problem. Epidemiological studies have found that long-term exposure to air pollution will increase the risk of disease. For example, fine particulate matter (PM$_{2.5}$) is significantly associated with an increase in the prevalence of diabetes; short-term exposure to high concentrations of PM$_{2.5}$ and ozone(O$_3$) will increase cardiovascular disease, respiratory disease, and non-risk of accidental death [1–3]. Since 2013, regional atmospheric environmental problems mainly characterized by $PM_{2.5}$ pollution have attracted widespread attention in China. Therefore, China has formulated and issued a series of policies to alleviate air pollution. From 2015 to 2020, $PM_{2.5}$ pollution in China has been significantly reduced, while ozone pollution has become increasingly serious [4–9].

Central Liaoning Urban Agglomeration (CLUA) is located in the Northeast Plain. Due to the concentrated urban distribution, dense population, and industrial structure dominated by heavy industry, air pollution is serious. In 2020, the proportion of $PM_{2.5}$ and O_3 as the primary pollutants in Liaoning Province in the days exceeding the standard was 62.9% and 33.8% respectively. The annual average concentration of $PM_{2.5}$ was 38 µg/m^3 which exceeds the secondary standard of ambient air quality [10]. The air quality of Central Liaoning Urban Agglomeration may be affected by regional anthropogenic emissions. In winter, the transboundary pollution of $PM_{2.5}$ is extremely significant. Some studies also found that there was regional transmission of pollutants in Central Liaoning Urban Agglomeration and Harbin-Changchun urban agglomeration [11,12]. Straw burning and coal-burning heating during winter resulted in heavy $PM_{2.5}$ pollution, and the interannual change of $PM_{2.5}$ in Northeast China showed an obvious upward trend from 1998 to 2016. MDA8 (maximum daily 8 h average) O_3 increased and the number of days exceeding the standard continued to increase which can be attributed to the superimposed effects of atmospheric long-distance transport and anthropogenic emissions [13–17].

Many studies have mentioned the occurrence of air pollution in various urban agglomerations, such as the Beijing-Tianjin-Hebei region, Harbin-Changchun region, Yangtze River Delta, etc., while there are few studies on Northeast China and Central Liaoning Urban Agglomerations [18–20]. This study collected $PM_{2.5}$ and O_3 concentration data in CLUA (Anshan, Benxi, Fushun, Liaoyang, Shenyang, and Tieling) from 2015 to 2020 and analyzed its temporal and spatial characteristics to determine the level of pollutants. Specifically, the aim was to (1) study the long-term temporal and spatial changes of the mass concentrations of $PM_{2.5}$ and O_3 and (2), through the back trajectory HYSPLIT model and cluster analysis, discuss the regional transportation of $PM_{2.5}$ and ozone in Shenyang, which is the center of Central Liaoning Urban Agglomeration, in order to provide a useful reference for CLUA to improve air quality.

2. Data Sources and Methods

2.1. Data Sources

The geographical location of the study area is shown in Figure 1. This study used in situ data from a total of 38 air quality monitoring stations (Table 1) in six cities in central Liaoning urban agglomeration from 2015 to 2020 (http://www.cnemc.cn/ accessed on 14 May 2021). Among them, O_3 data used ozone eight-hour moving average (O_3-8 h), and $PM_{2.5}$ used hourly monitoring data. Ozone is measured by UV spectrophotometry (Thermo Scientific Model 49iQ Ozone Analyzer). $PM_{2.5}$ is passed β X-ray absorption method and light scattering method were used for real-time determination (Thermo Scientific Model 5030i Sharp Particulate Monitor). The season division refers to the meteorological industry-standard "Climate Season Division" (QX/T152-2012) and the annual difference is adjusted. The results are: spring (1 April–31 May), summer (1 June–31 August), autumn (1 September–31 October), winter (1 November–31 March of the following year).

2.2. Research Method

The Geographical Information System (GIS)-TrajStat software is used for backward trajectory clustering, potential source contribution analysis, and concentration weighted trajectory analysis, and grids were divided by $1° \times 1°$ within the trajectory range [21,22].

2.2.1. Cluster Analysis

The Hybrid Single Particle Lagrangian Integrated Trajectory (HYSPLIT) model (http://ready.arl.noaa.gov/HYSPLIT.php accessed on 12 July 2021) developed by the National Oceanic and Atmospheric Administration (NOAA) was used to simulate 72 h backward trajectories at 100 m altitude for Shenyang to analyze atmospheric pollutant transportation [23,24].

In the study, to understand the regional transport processes effects, we used $1° \times 1°$ Global Data Assimilation System (GDAS) data from National Centers for Environmental Prediction (NCEP). We also calculated the 72 h backward airflow trajectory at 0:00, 6:00, 12:00, and 18:00 every day from 2015 to 2020. Euclidean classification method is used to analyze the transport path of regional air mass through data statistics of various trajectories after clustering, which can distinguish trajectories with similar directions but large difference in length [25]. Then we could better distinguish the contribution of long-distance and short-distance transmissions from different directions.

Table 1. The details of 38 air quality monitoring stations.

Cities	Air Quality Monitoring Sites			
	Name	Abbr.	Longitude (°E)	Latitude (°N)
Anshan	MingDa New District	MD	123.1289	41.0228
	QianShan Mountain	QS	123.0156	41.0831
	ShenGouSi	SG	123.044	41.1196
	TaiPing	TP	123.0485	41.1442
	TieXi District Industrial Park	TD	122.9481	41.0833
	Tiexi Sandao Street	TSS	122.9642	41.0971
	TaiYang Cheng	TYC	123.011	41.0931
Benxi	Cai Tun	CT	123.7308	41.3047
	Da Yu	DY	123.8436	41.3283
	Dong Ming	DM	123.7669	41.2864
	Wei Ning	WN	123.8142	41.3472
	Xi Lake	XL	123.7528	41.3369
	Xinli Tun	XT	123.7989	41.2692
Fushun	DaHuoFang Reservoir	DHF	124.0878	41.8864
	DongZhou District	DZ	124.0383	41.8625
	ShenFuXinCheng	SF	123.7117	41.8417
	ShunCheng District	AC	123.9169	41.8828
	WangHua District	WH	123.81	41.8469
	XinFu District	XF	123.9	41.8594
Liaoyang	BinHe Road	BH	123.1761	41.2736
	HongWei District	HW	123.2	41.1953
	TieXi District industrial park	TXD	123.1417	41.2894
	XinHua Yuan	XY	123.15	41.2553
Shenyang	CangHai Road	CH	123.284	41.7694
	LingDong Street	LD	123.428	41.8472
	DongLing Road	DL	123.542	41.8336
	JingShen Street	JS	123.3783	41.9228
	TaiYuan Street	TY	123.3997	41.7972
	YuNong Road	YN	123.5953	41.9086
	WenHua Road	WH	123.41	41.765
	XiaoHeYan Road	XHY	123.478	41.7775
	SenLin Road	SL	123.6836	41.9339
	Eastern of HunNan Road	HN	123.535	41.7561
	ShenLiaoXi Road	SLX	123.2444	41.7347
Tieling	Western of HuiGong Street	HG	123.8139	42.3022
	Northern of JinShaJiang Road	JSJ	123.7153	42.2217
	ShuiShang Park	SP	123.8469	42.292
	Eastern of YinZhou Road	YZE	123.8489	42.2864

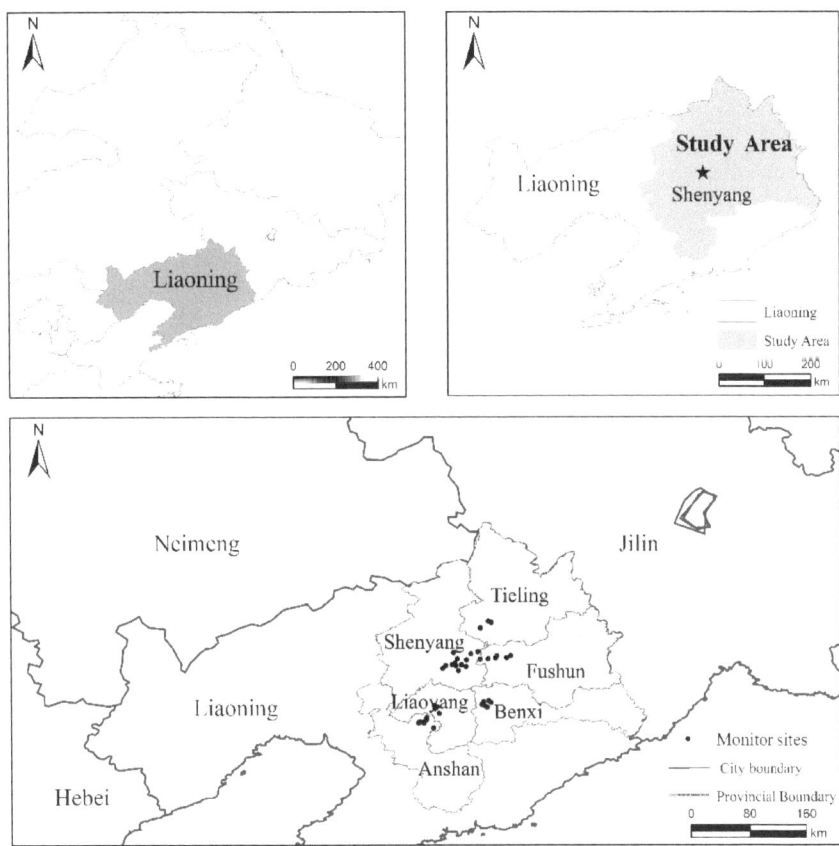

Figure 1. Air quality monitoring station in CLUA.

2.2.2. PSCF Analysis

Potential source contribution analysis (PSCF) is also called residence time analysis (RT). The ratio of the length of track l in the grid (i, j) to the length of the whole track is multiplied by the backward track 72 h, and the result is the residence time of pollutants in the grid [26]. PSCF identifies the possible pollution emission source areas by combining the air mass clustering trajectory and pollutant concentration value. PSCF function is a conditional probability function that pollutants carried by air masses passing through a unit area exceed the set pollutant threshold [27]. The PSCF value calculated by this function is the ratio of the number of pollution tracks as m_{ij} passing through the grid (i, j) in all track ranges to the number of tracks as n_{ij} passing through the grid [28]. The calculation method is shown in Equation (1).

$$PSCF_{ij} = \frac{m_{ij}}{n_{ij}} \quad (1)$$

In this paper, potential source contribution analysis (PSCF) is used to further identify the potential sources of air pollution in Shenyang [29,30], setting $PM_{2.5}$ concentration threshold of 35 μg/m^3 and O_3 concentration threshold of 100 μg/m^3 for all pollution trajectories, to determine the location of potential pollution sources affecting the atmospheric environment in Shenyang. However, in order to reduce the uncertainty of PSCF value, the weight factor W_{ij} is introduced and called WPSCF [31], such as Equations (2) and (3).

$$W_{ij} = \begin{cases} 1.00 & (80 < n_{ij}) \\ 0.70 & (25 < n_{ij} \leq 80) \\ 0.42 & (15 < n_{ij} \leq 25) \\ 0.17 & (n_{ij} \leq 15) \end{cases} \quad (2)$$

$$WPSCF_{ij} = \frac{m_{ij}}{n_{ij}} W_{ij} \quad (3)$$

where m_{ij} is the number of pollution tracks passing through the grid (i, j); n_{ij} is the total number of tracks passed.

2.2.3. CWT Analysis

The values calculated by the PSCF method are the same when the pollutant concentrations of the trajectories are only slightly higher or much higher than the standard [32,33]. As a result, it may be difficult to distinguish between moderate and strong sources by PSC method. Therefore, in this paper, the concentration-weighted trajectory (CWT) analysis method is used to calculate the average weighted concentration of each trajectory, which reveals the contribution of different regions to the air pollution in Shenyang by setting the mesh precision of CWT to be the same as that of PSCF [34,35]. The same weight factor W_{ij} from the PSCF method is introduced to distinguish the source intensity of potential sources, which is called the weighted average concentration value (WCWT value), as shown in Equation (4).

$$C_{ij} = \frac{W_{ij}}{\sum_{l=1}^{M} \tau_{ijl}} \sum_{l=1}^{M} C_l \tau_{ijl} \quad (4)$$

where C_{ij} is the average weight concentration in the grid (i,j), l is the index of the track, C_l is the pollutant concentration measured when track l arrives, M is the total number of tracks, and τ_{ijl} is the time that the trajectory l stays in the mesh (i,j). The higher the value of C_{ij}, the greater the contribution of the trajectory to the air pollution in Shenyang.

3. Results and Discussion

3.1. Temporal Variation

The annual average concentrations and change trends of $PM_{2.5}$ and O_3 in six cities of CLUA are investigated as shown in Figure 2. $PM_{2.5}$ basically showed a trend of declining from 2015 to 2020. The average concentration of $PM_{2.5}$ in Fushun decreased by 18% from 2015 to 2016, and increased slightly in 2017 and 2019, but was still lower than 53 µg/m^3 in 2015. China began to implement the environmental protection tax law in 2018 to strengthen the management of air pollution punishment, which explains why the $PM_{2.5}$ concentration in each city was the lowest in 2018. In addition, straw burning was prohibited in the study area, and Liaoning Province began to completely ban small coal-fired boilers of 10 tons and below in 2016 to improve air quality. The average concentration of $PM_{2.5}$ in Anshan City decreased from 67 µg/m^3 to 41.6 µg/m^3, which was 66.4% lower in 2020 than in 2015. In 2020, the lowest $PM_{2.5}$ value appeared in Benxi City, which was 34.7 µg/m^3. Except for Benxi, $PM_{2.5}$ in other cities (Anshan, Fushun, Liaoyang, Shenyang, Tieling) did not meet China Class II Environmental Air Quality Standard (CAAQS) limited to 35 µg/m^3. This shows that the government needs to strengthen particulate matter control in these cities. Unlike $PM_{2.5}$, the annual average concentration of ozone does not change much in various regions, and the trend of change is slightly different. This can be attributed to the fact that ozone is affected by more meteorological conditions and chemical reactions [36,37], so its regional characteristics are less obvious. The average concentration of O_3 in Shenyang increased from 2015 to 2017 and decreased significantly after reaching peak value in 2017. Among them, the ozone concentration in Anshan, Liaoyang, and Shenyang reached the peak concentrations in 2017, which were 68.76 µg/m^3, 66.27 µg/m^3, and 63.46 µg/m^3 respectively.

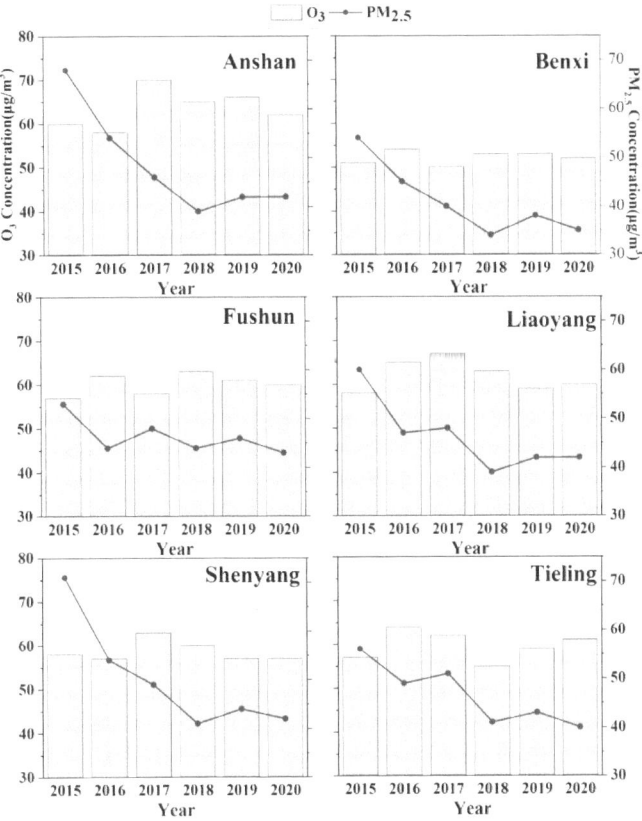

Figure 2. Inter-annual changes of PM$_{2.5}$ and O$_3$ years in CLUA from 2015 to 2020.

Figure 3 shows the monthly variation of the 24-h multi-year average concentration of PM$_{2.5}$ in CLUA from 2015 to 2020. It can be found that PM$_{2.5}$ pollution is the most serious at the beginning and the end of the year (January to March and October to December), and the PM$_{2.5}$ pollution is lower from May to September. In Northeast China, there is a large area of open burning of crop residues, which is one of the reasons for PM$_{2.5}$ pollution in autumn and winter. In addition, PM$_{2.5}$ pollution peaked in the heating period. Due to the long heating period, PM$_{2.5}$ pollution continued until the next spring. It can be seen that the frequency of low PM$_{2.5}$ concentration in Benxi area is higher, which also indicates that the PM$_{2.5}$ pollution in Benxi is relatively lower than in other areas, because Benxi is rainy in summer and autumn, and the wind direction is southeast, which is conducive to the diffusion of pollutants. In general, the PM$_{2.5}$ concentration in several cities resulted in high values at night and low values during the day. In this latter case, low-concentration PM$_{2.5}$ were registered from 14:00 to 17:00, while the peak of PM$_{2.5}$ mainly were registered from 7:00 to 9:00 and from 18:00 to 23:00. The first peak is related to atmospheric stability and the increase of human activities during this period [38]. This can be attributed to the intensification of atmospheric turbulence and the gradual decrease of PM$_{2.5}$ concentration with the increase of temperature in the morning. The evening peak in the cities appears after sunset, so the PM$_{2.5}$ emission increases. In addition, the surface radiation cooling reduces the height of the boundary layer, and the atmosphere tends to be stable, leading PM$_{2.5}$ concentration to continue to rise [39].

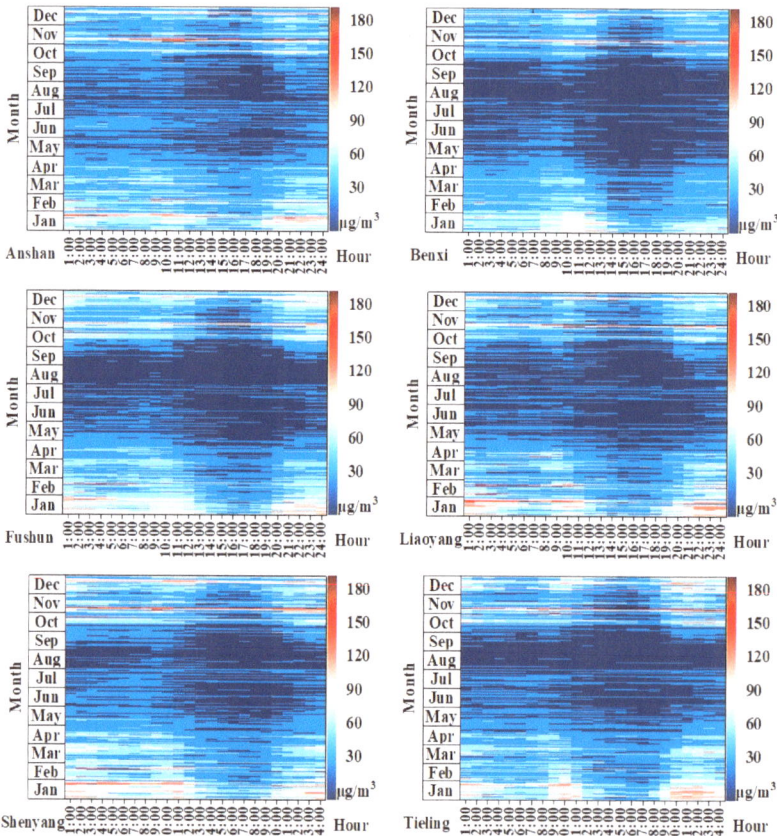

Figure 3. Hourly variation of multi-year average concentration of PM$_{2.5}$ in different months in CLUA from 2015 to 2020.

Figures 4 and 5 are box charts of daily average concentrations of PM$_{2.5}$ and O$_3$ in different seasons in CLUA from 2015 to 2020 respectively. The upper and lower frames of the box represent 75% and 25% quantiles respectively. The data points next to the box in the figures correspond to the daily average concentration of pollutants, and a normal distribution curve is added according to the concentration value. It is obvious that PM$_{2.5}$ pollution trends are the same in all cities. PM$_{2.5}$ pollution is the most serious in winter, followed by spring and autumn, and significantly lower in summer than in other seasons. In summer, the concentration of PM$_{2.5}$ in the Anshan area is the highest, 32.02 μg/m^3 (<35 μg/m^3). However, the concentration of PM$_{2.5}$ in other cities does not exceed 30 μg/m^3. It may be related to the fact that Anshan is a heavy industrial city and the main pollution source is a fixed source. Compared with other seasons, the rainy weather in summer has an obvious effect on the wet deposition of pollutants, so the PM$_{2.5}$ concentration is relatively lower [40]. In spring and autumn, PM$_{2.5}$ pollution is the most serious in Shenyang, followed by Tieling and Anshan. The concentration of PM$_{2.5}$ in Benxi is the lowest. In winter, PM$_{2.5}$ pollution in Shenyang is the most serious (69.57 μg/m^3), which is 2.46 times higher than that in summer. The concentration of PM$_{2.5}$ in Benxi City is the lowest, 57.9 μg/m^3. This is consistent with other results obtained using MODIS inversion data [41,42]. The high concentration in winter is mainly due to the high population density in the built-up area and the low temperature in winter in the north, so the amount of coal combustion increases. In addition, inversion is more likely to occur in winter, which is not conducive to the diffusion of pollutants [43,44].

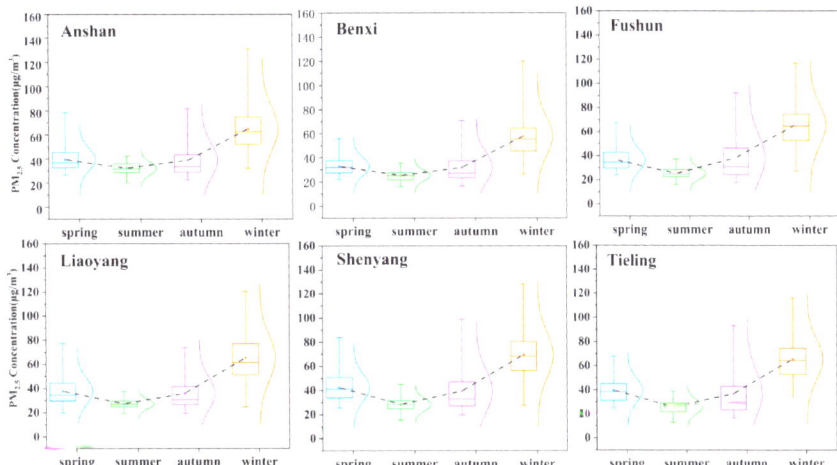

Figure 4. Seasonal variation of PM$_{2.5}$ in CLUA from 2015 to 2020.

Being affected by the seasonal variation of meteorological conditions, the ozone concentration is opposite to that of PM$_{2.5}$. In spring and summer, it is significantly higher than that in autumn and winter. The seasonal pattern of ozone in central Liaoning urban agglomeration is the same as that in other regions, with the overall pattern as summer > spring > autumn > winter [45–47]. In summer, the O$_3$ concentration in Shenyang is 131.14 µg/m^3, and the O$_3$ concentration in Benxi is 95.3 µg/m^3. Some studies have shown that direct emission of surface ozone is different from that of other air pollutants. O$_3$ is mainly generated by nitrogen oxides (NO$_x$) and volatile organic compounds (VOCs) through a series of complex photochemical reactions [48,49]. The high temperature and high chemical reaction rate in spring and summer make the ozone concentration much higher than that in other seasons. In addition, Benxi is different from other regions, and ozone concentration in summer is lower than that in spring. The ozone concentration in Fushun is higher in six cities. Huang et al. mentioned that there is intercity pollution between Fushun and Shenyang. Therefore, the ozone concentration in the monitoring stations near Shenyang is higher (Figure 6).

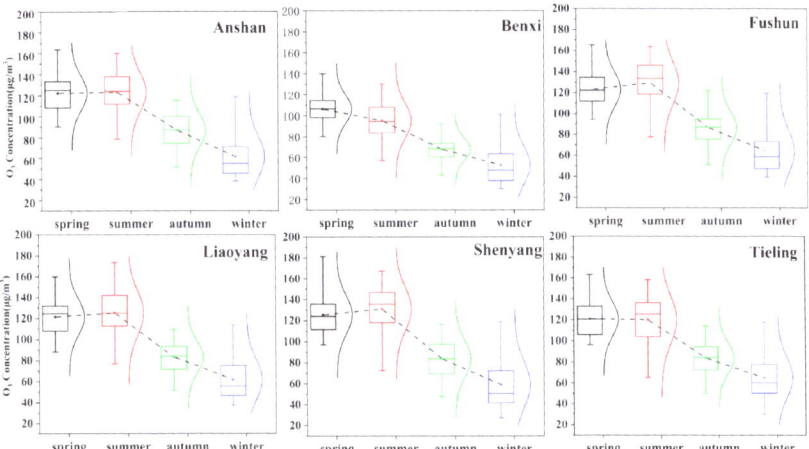

Figure 5. Seasonal variation of O$_3$ in CLUA from 2015 to 2020.

3.2. Spatial Analysis

The spatial sites distribution of O_3 and $PM_{2.5}$ in CLUA is shown in Figures 6 and 7. $PM_{2.5}$ and O_3 of all monitoring stations in Shenyang showed an obvious downward trend from 2015 to 2020. In 2015, the high concentrations of $PM_{2.5}$ were mainly located in LD, TY, SLX in Shenyang, TD, and TSS in Anshan. On the whole, since 2016, the concentration values of monitoring stations near the junction of Shenyang and Fushun have been higher. The stations with high concentrations of $PM_{2.5}$ are mainly concentrated in Shenyang urban areas, TD, SF, and WH. These stations are mostly concentrated in urban areas with high population, which leads to a higher concentration of pollutants caused by motor vehicles and industrial discharge.

The spatial distribution trend of ozone is basically the opposite of $PM_{2.5}$. The concentration of O_3 in rural areas is higher, such as XT, DY, DM, HW, and JSJ, etc. As a secondary pollutant, the formation of surface ozone is mainly related to its precursors (NO_x and VOCs). The source analysis of atmospheric VOCs in some areas shows that the contribution rate of motor vehicle sources is high, which is the main source of urban atmospheric VOCs [50,51]. The ozone pollution occurring in the urban areas is transported to the suburbs with certain meteorological conditions [52]. The higher vegetation coverage is conducive to photochemical reaction, resulting in higher ozone concentration in the suburbs [53]. In addition, it can be found that the O_3 concentration is higher at the stations near the junction of Liaoyang and Anshan. From 2015 to 2020, the O_3 concentration in western cities are lower than that in eastern cities. The high ozone concentration is distributed in a belt from Anshan, Liaoyang and Shenyang to Tieling and Fushun in the northeast. While the ozone concentration of Benxi in the east is always low, which is consistent with the results in Figure 2. Figure 8 shows that the ozone concentration in CLUA increased significantly from 2015 to 2017, and began to decrease gradually in 2018. It shows that while dealing with climate change, controlling pollutant emission reduction plays an active role in ozone mass concentration control.

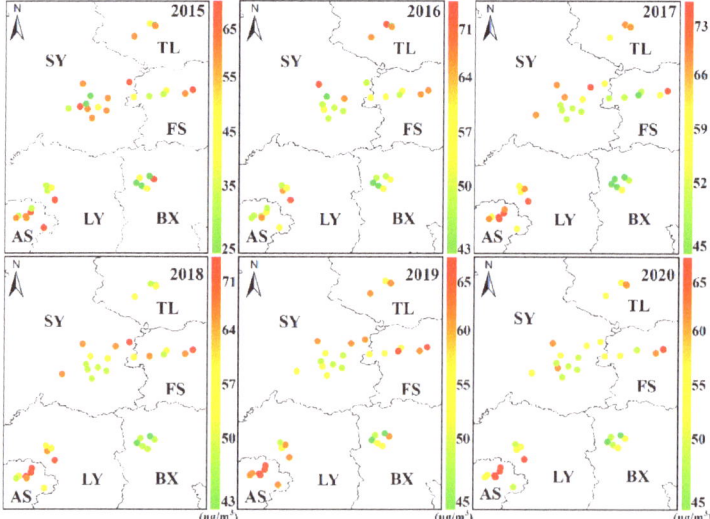

Figure 6. Spatial site distribution of annual average mass concentration of MDA8 O_3 in CLUA $PM_{2.5}$ annual average mass concentration of CLUA from 2015 to 2020 (AS: Anshan; BX: Benxi; FS: Fushun; LY: Liaoyang; SY: Shenyang; TL: Tieling).

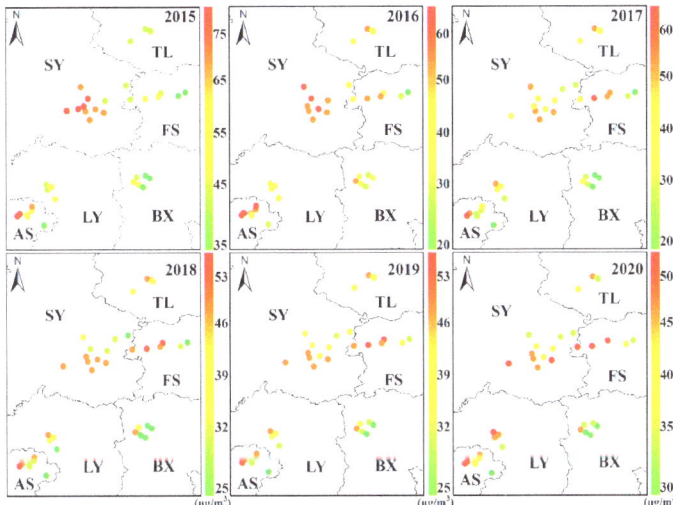

Figure 7. Spatial site distribution of PM$_{2.5}$ annual average mass concentration in CLUA from 2015 to 2020 (AS: Anshan; BX: Benxi; FS: Fushun; LY: Liaoyang; SY: Shenyang; TL: Tieling).

3.3. Transmission Path Characteristics in Shenyang Region

As the Shenyang area is the geographical and economic center of CLUA, its PM$_{2.5}$ and O$_3$ pollution is severe. Therefore, this study selects Shenyang city for backward trajectory cluster analysis and potential source area analysis. Cluster analysis of the backward airflow trajectories transported to Shenyang at 0:00, 6:00, 12:00, and 18:00 every day from 2015 to 2020 (Figure 9). There are six categories of backward trajectories in spring. The air flows in spring are mainly south (trajectory 1 and trajectory 5) and northwest (trajectory 3 and trajectory 6), which account for 67.79% of the airflows in spring. There are four categories of backward trajectories in summer. In summer, affected by the marine airflows, the southerly airflows (trajectory 1) and southeast airflows (trajectory 4) are predominant, accounting for 68.43% of the airflows in summer. There are seven categories of backward trajectories in autumn and six categories of backward trajectories in winter. In autumn and winter, the airflows are mostly northwest and the transmission distance is longer with faster speed, which may be related to the propagation of the East Asian winter monsoon. Except in summer, the airflows from the Beijing-Tianjin-Hebei region have all turned back significantly.

Based on the cluster analysis results of airflow backward trajectories in each season, the PM$_{2.5}$ and ozone concentration data in Shenyang are combined (Table 2) to quantitatively analyze the impact of various trajectories on PM$_{2.5}$ and ozone in Shenyang.

Consistent with the results discussed above, the PM$_{2.5}$ concentration corresponding to each air flow in autumn and winter is higher than that in spring and summer. In winter, the PM$_{2.5}$ concentration corresponding to the airflow (trajectory 3) from the junction of Liaoning and Hebei, Bohai Bay and southwest Liaoning is the highest, 106.02 µg/m^3. The second is the air flow from Northeast Inner Mongolia, northwest Jilin and northeast Liaoning (trajectory 6), and the corresponding PM$_{2.5}$ concentration is 105.23 µg/m^3. The corresponding concentration of ozone in the southerly flow (trajectory 1) from the Yellow Sea in spring is 142.41 µg/m^3, followed by the air flow (trajectory 5) is 136.5 µg/m^3.

The long-distance transport of northwest air flow from Hunshandak Sandy Land in Northwest Inner Mongolia and the Gobi Desert in central Mongolia is the main transport path affecting the PM$_{2.5}$ concentration in Shenyang in four seasons. In addition, southwest air flows through densely populated areas such as Beijing, Tianjin, and Hebei. The Bohai Sea has heavy shipping emissions, making it another main transmission path affecting PM$_{2.5}$ pollution in Shenyang.

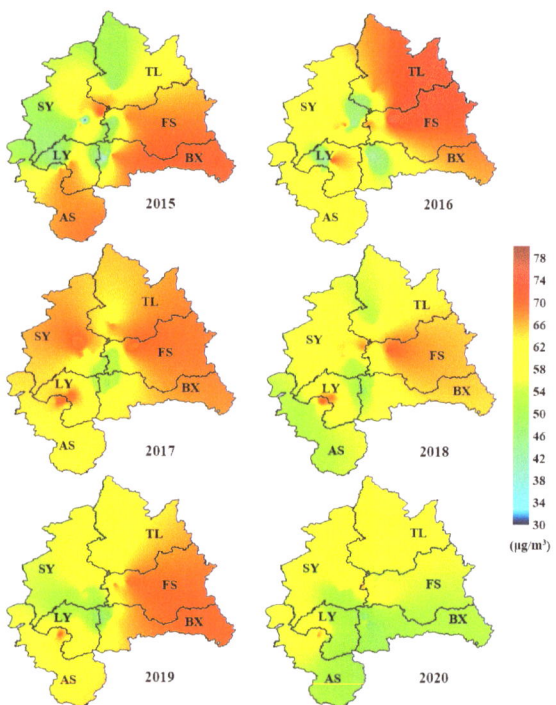

Figure 8. Spatial distribution of annual average mass concentration of MDA8 O$_3$ in CLUA from 2015 to 2020 (AS: Anshan; BX: Benxi; FS: Fushun; LY: Liaoyang; SY: Shenyang; TL: Tieling).

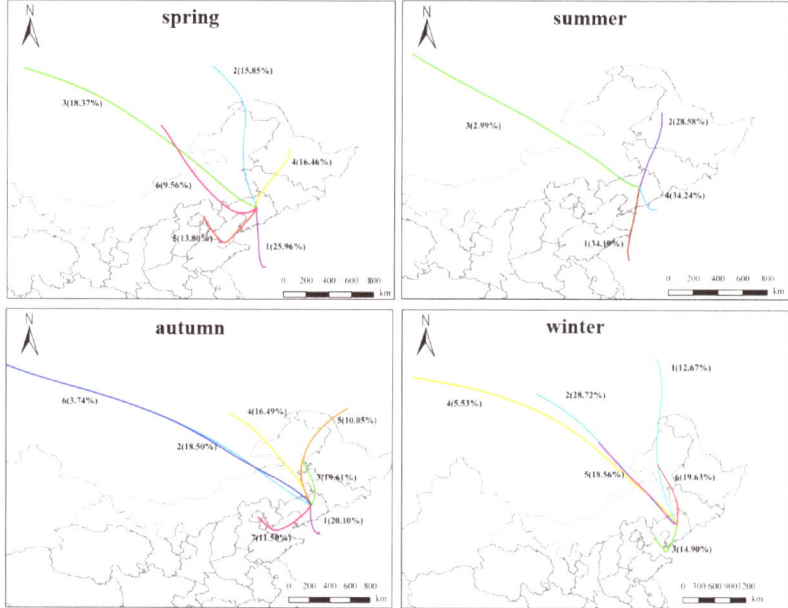

Figure 9. Clustering analysis of backward trajectories of each season in Shenyang from 2015 to 2020 (the trajectories are marked with number and frequency).

Table 2. Statistical results of the mass concentration of all kinds of air flow in the four seasons of Shenyang.

Season	Air Mass Type	PM$_{2.5}$	Stdev	Number	Ozone	Stdev	Number
spring	1	60.51	24.70	202	142.41	32.84	189
	2	60.64	28.59	65	111.46	7.57	14
	3	59.25	23.82	100	118.87	15.45	26
	4	69.00	34.10	124	128.95	21.41	50
	5	67.13	28.36	138	136.50	26.83	114
	6	68.01	31.71	79	126.39	20.51	31
summer	1	50.13	13.82	286	137.58	31.06	301
	2	48.56	11.93	87	126.94	21.51	113
	3	58.19	23.81	9	125.66	22.24	8
	4	49.32	13.65	224	139.83	32.71	230
autumn	1	72.59	38.98	144	119.96	15.78	11
	2	57.67	20.13	76	126.97	19.32	5
	3	70.91	30.63	117	127.91	20.74	37
	4	72.21	59.55	93	113.84	13.10	6
	5	63.65	48.37	22	0.00	0.00	0
	6	58.04	20.42	5	125.80	0.00	1
	7	59.25	25.67	150	121.29	21.59	44
winter	1	78.52	57.48	543	112.41	15.37	4
	2	90.24	65.41	506	109.51	5.96	10
	3	106.02	85.85	614	114.84	11.12	16
	4	66.15	33.99	45	0.00	0.00	0
	5	76.45	41.05	222	111.22	5.83	4
	6	105.23	54.45	489	120.49	19.19	50

3.4. Characteristics of Potential Source Areas in Shenyang

We conduct the potential source contribution factor (WPSCF) analysis and the concentration weight trajectory (WCWT) analysis based on the backward trajectory of each season in Shenyang from 2015 to 2020, in order to fully reflect the long-term impact characteristics and contribution of potential source regions on the mass concentration of PM$_{2.5}$ and O$_3$ in Shenyang. The results are shown in Figures 10–13. The larger the calculated value of WPSCF, the greater the impact of the area on the mass concentration of PM$_{2.5}$ and O$_3$ in Shenyang. The higher value of WCWT, the greater the contribution of the grid area to the pollution of PM$_{2.5}$ and O$_3$ in Shenyang.

The WPSCF value of PM$_{2.5}$ is the lowest in summer, and it can be found that WPSCF and WCWT have consistent spatial distribution characteristics in central Shandong. It can be seen that the regions with a higher contribution to the PM$_{2.5}$ concentration in Shenyang in spring were concentrated in the central Shandong Province and the northwestern parts of Jiangsu Province, with WPSCF value higher than 0.7 and corresponding WCWT value higher than 50 μg/m^3 (Figures 10 and 12). In summer, WPSCF value is higher near Zaozhuang and Jinan in Shandong Province, and WCWT value is 45~50 μg/m^3. The regions with relatively high WPSCF in autumn are mainly concentrated in Beijing-Tianjin-Hebei Urban Agglomeration and northwestern part of the Shandong Province, and the WCWT value is higher than 80 μg/m^3. In winter, due to the heavy PM$_{2.5}$ pollution, the WPSCF value of PM$_{2.5}$ is high and wide, indicating that the PM$_{2.5}$ pollution in Shenyang has certain regional characteristics. The high value of WPSCF mainly occurs in Beijing Tianjin Hebei Urban Agglomeration, Bohai Sea area, Lianyungang City, Shandong Province and Jiangsu Province, the northeast of Liaoning, the middle of Jilin, and the middle of Heilongjiang, showing a wide coverage of northeast trending potential contribution source zone (WPSCF > 0.8). It is worth noting that in the WCWT distribution, the central Heilongjiang shows more than 240 PM$_{2.5}$. It also mentioned that in winter, Liaoning

Province is subject to long-distance transportation from Heilongjiang Province, resulting in serious haze pollution [54].

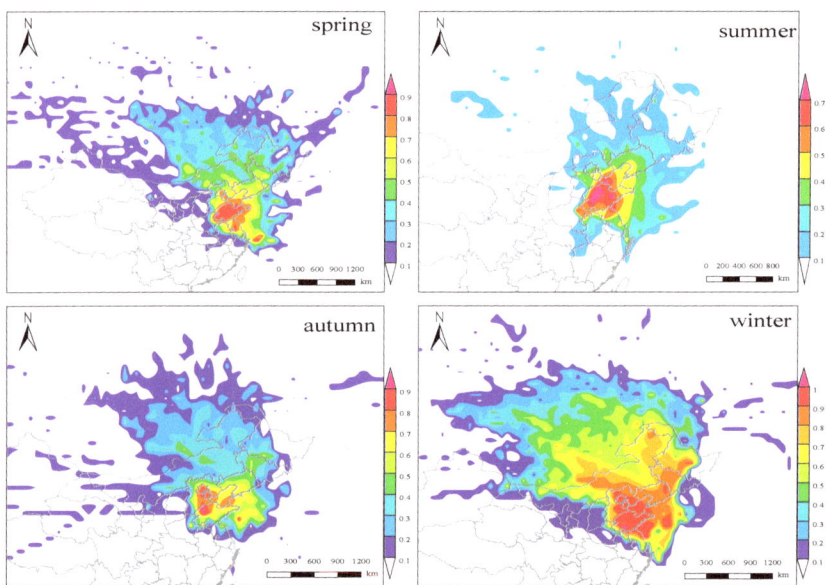

Figure 10. WPSCF distribution of $PM_{2.5}$ in Shenyang from 2015 to 2020.

It can be seen from Figure 11 that there are many regions with higher contributions of the O_3 concentration in Shenyang in spring and summer. They are mainly concentrated in Bohai Bay, Shandong Province, Jiangsu Province, and the nearby Yellow Sea. The WPSCF value is higher than 0.6 and the corresponding WCWT value is more than 110 μg/m³. Among them, Jiangsu Province is an economically developed region in China, with intensive secondary industry and serious pollutant emission. The high WPSCF values in the Bohai Sea and the Yellow Sea may be the pollutants in their adjacent areas and are transmitted to the nearby sea areas through the sea land winds. Then they are transported to the northeast along the Bohai Bay and the Yellow Sea bay, sinking in Shenyang. However, as mentioned above, compared with $PM_{2.5}$, ozone is unstable and the formation conditions are complex, so it is more difficult to determine the potential source area of ozone. PSCF and CWT methods show that the high ozone content in Shenyang mainly comes from the transmission in the Yellow Sea, Bohai Sea, and its adjacent areas.

Combined with the analysis of PSCF and CWT, it can be found that the main potential sources areas affecting the $PM_{2.5}$ mass concentration of in Shenyang in autumn and winter are Beijing-Tianjin-Hebei Urban Agglomeration, Shandong Province, Jiangsu Province, and nearby sea areas. This shows that the atmospheric circulation has an important impact on the regional transmission of the city. In addition, the main potential source areas of O_3 mass concentration in spring and summer are coastal cities and the Bohai Sea and Yellow Sea. This is also consistent with the backward air flow trajectory with heavy pollution in each season, and the influence of long-distance transmission of pollution concentration is small. Through potential source analysis, we should pay attention to the industrial source supply in the surrounding areas of cities and the marine source provided by marine pollution, and strengthen joint prevention and control of air pollution among regions.

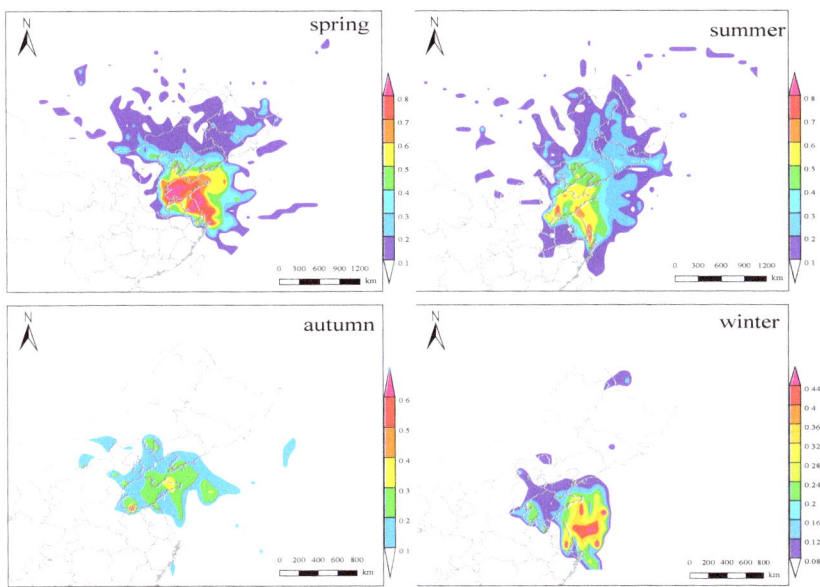

Figure 11. WPSCF distribution of O_3 in Shenyang from 2015 to 2020.

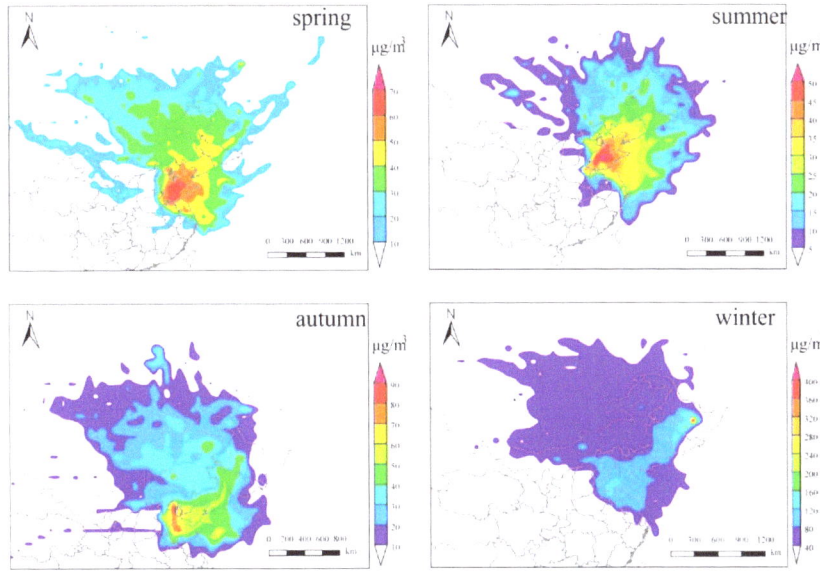

Figure 12. WCWT distribution of $PM_{2.5}$ in Shenyang from 2015 to 2020.

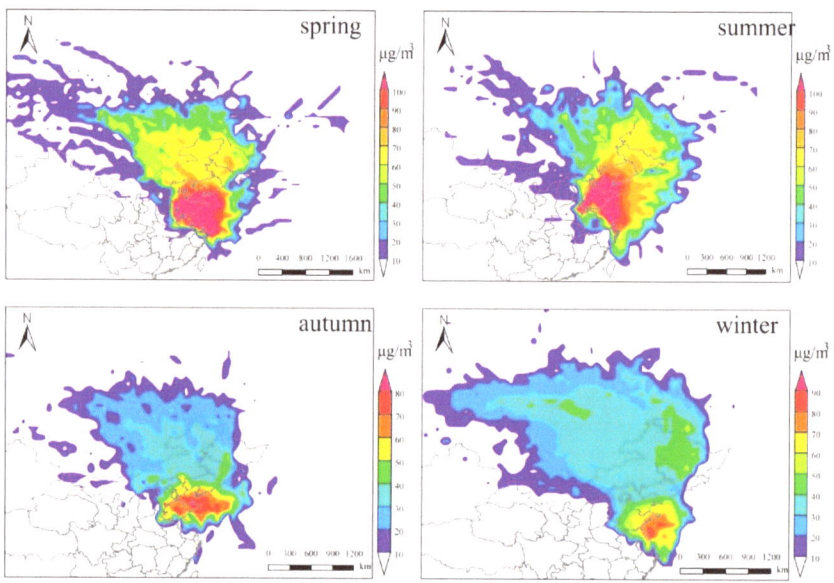

Figure 13. WCWT distribution of O_3 in Shenyang from 2015 to 2020.

4. Conclusions

The $PM_{2.5}$ concentration has fluctuated from 2016 to 2020 in Fushun, and has declined year by year from 2015 to 2020 in other areas of the CLUA. The lowest $PM_{2.5}$ level occurred in 2018. Except for Benxi, $PM_{2.5}$ in other cities (Anshan, Fushun, Liaoyang, Shenyang, Tieling) did not meet the China Class II Environmental Air Quality Standard (CAAQS) limit of 35 µg/m^3. The annual mean concentration of ozone has little change in different regions, and the variation trend is different. The ozone concentration in Anshan, Liaoyang, and Shenyang reached the peak value in 2017, which was 68.76 µg/m^3, 66.27 µg/m^3, and 63.46 µg/m^3, respectively. $PM_{2.5}$ pollution is the most severe at the beginning and the end of each year (January to March and October to December). The concentration of $PM_{2.5}$ in all cities showed the characteristics of high at night and low during the day. Affected by human activities and atmospheric movements, there are two peaks at 7:00~9:00 and 18:00~23:00. The seasonal pattern of $PM_{2.5}$ concentration was winter > spring > autumn > summer. In winter, $PM_{2.5}$ pollution in Shenyang is 69.57 µg/m^3, which is 2.46 times of that in summer. The seasonal pattern of ozone concentration is summer > spring > autumn > winter. In summer, The O_3 concentration in Shenyang is the highest, 131.14 µg/m^3, 1.37 times of that in Benxi. The ozone concentration in Fushun area is affected by the intercity pollution in Shenyang.

In terms of the spatial distribution of $PM_{2.5}$ and O_3 concentrations, the concentrations of $PM_{2.5}$ in the western cities (Shenyang, Liaoyang, and Anshan) present higher than that in the eastern ones (Tieling, Fushun, and Benxi), and the higher values are concentrated in urban monitoring stations. On the contrary, the concentration of O_3 is higher in rural areas, which is related to its precursors. The high concentration of ozone in Anshan, Liaoyang, and Shenyang extends to Tieling and Fushun in the northeast. The ozone concentration of Benxi city in the east is always low. Moreover, the ozone concentration in the central Liaoning urban agglomeration began to decrease gradually in 2018.

Through backward trajectory cluster analysis, it is found that the main transmission paths affecting Shenyang are southerly short-distance and northwest long-distance airflows in spring, southerly short-distance airflow in summer, southerly and northerly airflows in autumn, and northwestern long-distance airflow in winter. In winter, southwest airflows and northeast airflows have the highest $PM_{2.5}$ concentration, which is 106.02 µg/m^3 and

105.23 µg/m^3 respectively. The ozone concentration corresponding to the southerly airflows in spring and summer is the highest, which is 142.41 µg/m^3 and 139.83 µg/m^3, respectively.

Through the WPSCF and WCWT analysis, it is found that the main potential source areas affecting the mass concentration of PM$_{2.5}$ in Shenyang in autumn and winter are Beijing-Tianjin-Hebei urban agglomeration, Shandong Province, Jiangsu Province, and nearby sea areas, showing a potential contribution source belt with a wide coverage of northeast trend. In addition, the main potential source areas of ozone mass concentration in spring and summer are mainly coastal cities and the Bohai sea and Yellow Sea. Through the analysis of potential sources, we should pay attention to the industrial source supply in the surrounding areas and the marine source provided by marine pollution. We will strengthen joint prevention and control of air pollution between regions.

Author Contributions: Conceptualization, Y.Z.; methodology, J.W.; software, Y.Z.; validation, J.W. and C.F.; formal analysis, Z.L.; investigation, Y.Z.; resources, C.F.; data curation, C.F.; writing—original draft preparation, Y.Z.; writing—review and editing, J.W., Y.Z. and C.F.; visualization, C.F.; supervision, C.F.; project administration, J.W. and C.F. All authors have read and agreed to the published version of the manuscript.

Funding: This research received no external funding.

Acknowledgments: The authors would like to thank the group members of Laboratory 537 and 142 of Jilin University.

Conflicts of Interest: The authors declare no conflict of interest.

References

1. Yang, B.Y.; Qian, Z.; Li, S.; Markevych, I.; Bloom, M.S. Ambient air pollution in relation to diabetes and glucose-homoeostasis markers in China: A cross-sectional study with findings from the 33 Communities Chinese Health Study. *Lancet Planet. Health* **2018**, *2*, e113. [CrossRef]
2. Xia, S.Y.; Huang, D.S.; Jia, H.; Zhao, Y.; Li, N.; Mao, M.Q.; Lin, H.; Li, Y.X.; He, W.; Zhao, L. Relationship between atmospheric pollutants and risk of death caused by cardiovascular and respiratory diseases and malignant tumors in Shenyang, China, from 2013 to 2016: An ecological research. *Chin. Med. J.* **2019**, *132*, 2269–2277. [CrossRef] [PubMed]
3. Lei, R.Q.; Zhu, F.R.; Cheng, H.; Liu, J.; Shen, C.W.; Zhang, C.; Xu, Y.C.; Xiao, C.C.; Li, X.R.; Zhang, J.Q.; et al. Short-term effect of PM2.5/O-3 on non-accidental and respiratory deaths in highly polluted area of China. *Atmos. Pollut. Res.* **2019**, *10*, 1412–1419. [CrossRef]
4. Guan, Y.; Xiao, Y.; Wang, Y.M.; Zhang, N.N.; Chu, C.J. Assessing the health impacts attributable to PM2.5 and ozone pollution in 338 Chinese cities from 2015 to 2020. *Environ. Pollut.* **2021**, *287*, 117623. [CrossRef]
5. Dong, D.X.; Xu, B.Y.; Shen, N.; He, Q. The adverse impact of air pollution on China's economic growth. *Sustainability* **2021**, *13*, 9056. [CrossRef]
6. Wu, Z.Y.; Zhang, Y.Q.; Zhang, L.M.; Huang, M.J.; Zhong, L.J.; Chen, D.H.; Wang, X.M. Trends of outdoor air pollution and the impact on premature mortality in the Pearl River Delta region of southern China during 2006–2015. *Sci. Total Environ.* **2019**, *690*, 248–260. [CrossRef] [PubMed]
7. Zhang, N.N.; Ma, F.; Qin, C.B.; Li, Y.F. Spatiotemporal trends in PM2.5 levels from 2013 to 2017 and regional demarcations for joint prevention and control of atmospheric pollution in China. *Chemosphere* **2018**, *210*, 1176–1184. [CrossRef] [PubMed]
8. Chen, Z.; Wang, J.N.; Ma, G.X.; Zhang, Y.S. China tackles the health effects of air pollution. *Lancet* **2013**, *382*, 1959–1960. [CrossRef]
9. Guo, H.; Cheng, T.H.; Gu, X.F.; Wang, Y.; Chen, H.; Bao, F.W.; Shi, S.Y.; Xu, B.R.; Wang, W.N.; Zuo, X.; et al. Assessment of PM2.5 concentrations and exposure throughout China using ground observations. *Sci. Total Environ.* **2017**, *601*, 1024–1030. [CrossRef]
10. Liaoning Provincial Environmental Protection Bureau. *Bulletin on the Ecological Environment of Liaoning Province in 2020*; Liaoning Provincial Environmental Protection Bureau: Shenyang, China, 2020.
11. Liu, S.C.; Xing, J.; Wang, S.X.; Ding, D.A.; Chen, L.; Hao, J.M. Revealing the impacts of transboundary pollution on PM2.5-related deaths in China. *Environ. Int.* **2020**, *134*, 105323. [CrossRef]
12. Li, X.L.; Hu, X.M.; Shi, S.Y.; Shen, L.D.; Luan, L.; Ma, Y.J. Spatiotemporal variations and regional transport of air pollutants in two urban agglomerations in northeast China plain. *Chin. Geogr. Sci.* **2019**, *29*, 917–933. [CrossRef]
13. Nan, Y.; Zhang, Q.Q.; Zhang, B.H. Analysis on the influencing factors of long-term change of grid PM$_{2.5}$ in typical regions of China based on gam model. *Environ. Sci.* **2020**, *41*, 499–509.
14. Chen, W.W.; Zhang, S.C.; Tong, Q.S.; Zhang, X.L.; Zhao, H.M.; Ma, S.Q.; Xiu, A.J.; He, Y.X. Regional characteristics and causes of haze events in Northeast China. *Chin. Geogr. Sci.* **2018**, *28*, 836–850. [CrossRef]
15. Gao, C.; Xiu, A.; Zhang, X.; Chen, W.; Liu, Y.; Zhao, H.; Zhang, S. Spatiotemporal characteristics of ozone pollution and policy implications in Northeast China. *Atmos. Pollut. Res.* **2020**, *11*, 357–369. [CrossRef]

16. Liang, Z.H.; Ju, T.Z.; Dong, H.P.; Geng, T.Y.; Duan, J.L.; Huang, R.R. Study on the variation characteristics of tropospheric ozone in Northeast China. *Environ. Monit. Assess.* **2021**, *193*, 282. [CrossRef] [PubMed]
17. Wang, H.; Ding, K.; Huang, X.; Wang, W.; Ding, A. Insight into ozone profile climatology over northeast China from aircraft measurement and numerical simulation. *Sci. Total Environ.* **2021**, *785*, 147308. [CrossRef]
18. Luo, M.; Ji, Y.Y.; Ren, Y.Q.; Gao, F.H.; Zhang, H.; Zhang, L.H.; Yu, Y.Q.; Li, H. Characteristics and health risk assessment of PM2.5-Bound PAHs during heavy air pollution episodes in winter in urban area of Beijing, China. *Atmosphere* **2021**, *12*, 323. [CrossRef]
19. Li, B.; Shi, X.F.; Liu, Y.P.; Lu, L.; Wang, G.L.; Thapa, S.; Sun, X.Z.; Fu, D.L.; Wang, K.; Qi, H. Long-term characteristics of criteria air pollutants in megacities of Harbin-Changchun megalopolis, Northeast China: Spatiotemporal variations, source analysis, and meteorological effects. *Environ. Pollut.* **2020**, *267*, 10. [CrossRef] [PubMed]
20. Dai, H.B.; Zhu, J.; Liao, H.; Li, J.D.; Liang, M.X.; Yang, Y.; Yue, X. Co-occurrence of ozone and PM2.5 pollution in the Yangtze River Delta over 2013–2019: Spatiotemporal distribution and meteorological conditions. *Atmos. Res.* **2021**, *249*, 9. [CrossRef]
21. Wang, Y.Q.; Zhang, X.Y.; Draxler, R.R. TrajStat: GIS-based software that uses various trajectory statistical analysis methods to identify potential sources from long-term air pollution measurement data. *Environ. Model. Softw.* **2009**, *24*, 938–939. [CrossRef]
22. Wang, X.Q.; Zhang, T.S.; Xiang, Y.; Lv, L.H.; Fan, G.Q.; Ou, J.P. Investigation of atmospheric ozone during summer and autumn in Guangdong Province with a lidar network. *Sci. Total Environ.* **2021**, *751*, 141740. [CrossRef]
23. Stein, A.F.; Draxler, R.R.; Rolph, G.D.; Stunder, B.J.B.; Cohen, M.D.; Ngan, F. Noaa's hysplit atmospheric transport and dispersion modeling system. *bull. Amer. Meteorol. Soc.* **2015**, *96*, 2059–2077. [CrossRef]
24. Stohl, A.; Forster, C.; Frank, A.; Seibert, P.; Wotawa, G. Technical note: The Lagrangian particle dispersion model FLEXPART version 6.2. *Atmos. Chem. Phys.* **2005**, *5*, 2461–2474. [CrossRef]
25. Zhang, Y.R.; Zhang, H.L.; Deng, J.J.; Du, W.J.; Hong, Y.W.; Xu, L.L.; Qiu, Y.Q.; Hong, Z.Y.; Wu, X.; Ma, Q.L.; et al. Source regions and transport pathways of $PM_{2.5}$ at a regional background site in East China. *Atmos. Environ.* **2017**, *167*, 202–211. [CrossRef]
26. Cesari, R.; Paradisi, P.; Allegrini, P. Source identification by a statistical analysis of backward trajectories based on peak pollution events. *Int. J. Environ. Pollut.* **2014**, *55*, 94–103. [CrossRef]
27. Sparks, D.N. Euclidean cluster analysis. *R. Stat. Soc. Ser. C Appl. Stat.* **1973**, *22*, 126–130. [CrossRef]
28. Hong, Q.Q.; Liu, C.; Hu, Q.H.; Xing, C.Z.; Tan, W.; Liu, H.R.; Huang, Y.; Zhu, Y.; Zhang, J.S.; Geng, T.Z.; et al. Evolution of the vertical structure of air pollutants during winter heavy pollution episodes: The role of regional transport and potential sources. *Atmos. Res.* **2019**, *228*, 206–222. [CrossRef]
29. Zhang, Z.Y.; Wong, M.S.; Lee, K.H. Estimation of potential source regions of PM2.5 in Beijing using backward trajectories. *Atmos. Pollut. Res.* **2015**, *6*, 173–177. [CrossRef]
30. Ashbaugh, L.L.; Malm, W.C.; Sadeh, W.Z. A residence time probability analysis of sulfur concentrations at grand Canyon National Park. *Atmos. Environ.* **1985**, *19*, 1263–1270. [CrossRef]
31. Hsu, Y.K.; Holsen, T.M.; Hopke, P.K. Locating and quantifying PCB sources in Chicago: Receptor modeling and field sampling. *Environ. Sci. Technol.* **2003**, *37*, 681–690. [CrossRef] [PubMed]
32. Wang, Y.Q.; Zhang, X.Y.; Arimoto, R. The contribution from distant dust sources to the atmospheric particulate matter loadings at XiAn, China during spring. *Sci. Total Environ.* **2006**, *368*, 875–883. [CrossRef]
33. Jeong, U.; Kim, J.; Lee, H.; Jung, J.; Kim, Y.J.; Song, C.H.; Koo, J.H. Estimation of the contributions of long range transported aerosol in East Asia to carbonaceous aerosol and PM concentrations in Seoul, Korea using highly time resolved measurements: A PSCF model approach. *J. Environ. Monit.* **2011**, *13*, 1905–1918. [CrossRef]
34. Kabashnikov, V.P.; Chaikovsky, A.P.; Kucsera, T.L.; Metelskaya, N.S. Estimated accuracy of three common trajectory statistical methods. *Atmos. Environ.* **2011**, *45*, 5425–5430. [CrossRef]
35. Dimitriou, K. The dependence of PM size distribution from meteorology and local-regional contributions, in Valencia (Spain)—A CWT model approach. *Aerosol Air Qual. Res.* **2015**, *15*, 1979–1989. [CrossRef]
36. Zhang, Y.; Wang, W.; Wu, S.Y.; Wang, K.; Minoura, H.; Wang, Z.F. Impacts of updated emission inventories on source apportionment of fine particle and ozone over the southeastern US. *Atmos. Environ.* **2014**, *88*, 133–154. [CrossRef]
37. Otero, N.; Sillmann, J.; Schnell, J.L.; Rust, H.W.; Butler, T. Synoptic and meteorological drivers of extreme ozone concentrations over Europe. *Environ. Res. Lett.* **2016**, *11*, 13. [CrossRef]
38. Fan, H.; Zhao, C.F.; Yang, Y.K.; Yang, X.C. Spatio-temporal variations of the PM2.5/PM10 ratios and its application to air pollution type classification in China. *Front. Environ. Sci.* **2021**, *9*, 281. [CrossRef]
39. Pan, L.; Xu, J.M.; Tie, X.X.; Mao, X.Q.; Gao, W.; Chang, L.Y. Long-term measurements of planetary boundary layer height and interactions with PM2.5 in Shanghai, China. *Atmos. Pollut. Res.* **2019**, *10*, 989–996. [CrossRef]
40. Chen, Z.; Chen, D.; Zhao, C.; Kwan, M.P.; Cai, J.; Zhuang, Y.; Zhao, B.; Wang, X.; Chen, B.; Yang, J.; et al. Influence of meteorological conditions on PM2.5 concentrations across China: A review of methodology and mechanism. *Environ. Int.* **2020**, *139*, 105558. [CrossRef] [PubMed]
41. Zhao, H.; Che, H.; Zhang, X.; Ma, Y.; Wang, Y.; Wang, X.; Liu, C.; Hou, B.; Che, H. Aerosol optical properties over urban and industrial region of Northeast China by using ground-based sun-photometer measurement. *Atmos. Environ.* **2013**, *75*, 270–278. [CrossRef]

42. Zhao, H.; Gui, K.; Ma, Y.; Wang, Y.; Wang, Y.; Wang, H.; Zheng, Y.; Li, L.; Zhang, L.; Che, H.; et al. Climatological variations in aerosol optical depth and aerosol type identification in Liaoning of Northeast China based on MODIS data from 2002 to 2019. *Sci. Total Environ.* **2021**, *781*, 146810. [CrossRef]
43. Xu, Z.; Huang, X.; Nie, W.; Chi, X.; Xu, Z.; Zheng, L.; Sun, P.; Ding, A. Influence of synoptic condition and holiday effects on VOCs and ozone production in the Yangtze River Delta region, China. *Atmos. Environ.* **2017**, *168*, 112–124. [CrossRef]
44. Chen, J.; Shen, H.; Li, T.; Peng, X.; Cheng, H.; Ma, A.C. Temporal and spatial features of the correlation between $PM_{2.5}$ and O_3 concentrations in China. *Int. J. Environ. Res. Public Health* **2019**, *16*, 4824. [CrossRef] [PubMed]
45. Tui, Y.; Qiu, J.; Wang, J.; Fang, C. Analysis of spatio-temporal variation characteristics of main air pollutants in Shijiazhuang city. *Sustainability* **2021**, *13*, 941. [CrossRef]
46. Chan, K.L.; Wang, S.S.; Liu, C.; Zhou, B.; Wenig, M.O.; Saiz-Lopez, A. On the summertime air quality and related photochemical processes in the megacity Shanghai, China. *Sci. Total Environ.* **2017**, *580*, 974–983. [CrossRef]
47. Qian, Y.; Xu, B.; Xia, L.J.; Chen, Y.L.; Deng, L.C.; Wang, H.; Zhang, G. Characteristics of ozone pollution and relationships with meteorological factors in Jiangxi province. *Environ. Sci.* **2021**, *42*, 2190–2201. [CrossRef]
48. Huang, D.; Li, Q.L.; Wang, X.X.; Li, G.X.; Sun, L.Q.; He, B.; Zhang, L.; Zhang, C.S. Characteristics and trends of ambient ozone and nitrogen oxides at urban, suburban, and rural sites from 2011 to 2017 in Shenzhen, China. *Sustainability* **2018**, *10*, 530. [CrossRef]
49. Wang, T.; Xue, L.K.; Brimblecombe, P.; Lam, Y.F.; Li, L.; Zhang, L. Ozone pollution in China: A review of concentrations, meteorological influences, chemical precursors, and effects. *Sci. Total Environ.* **2017**, *575*, 1582–1596. [CrossRef]
50. Gu, Y.; Liu, B.S.; Li, Y.F.; Zhang, Y.F.; Bi, X.H.; Wu, J.H.; Song, C.B.; Dai, Q.L.; Han, Y.; Ren, G.; et al. Multi-scale volatile organic compound (VOC) source apportionment in Tianjin, China, using a receptor model coupled with 1-hr resolution data. *Environ. Pollut.* **2020**, *265*, 23. [CrossRef]
51. Li, C.; Liu, Y.; Cheng, B.; Zhang, Y.; Liu, X.; Qu, Y.; An, J.; Kong, L.; Zhang, Y.; Zhang, C.; et al. A comprehensive investigation on volatile organic compounds (VOCs) in 2018 in Beijing, China: Characteristics, sources and behaviours in response to O_3 formation. *Sci. Total Environ.* **2022**, *806*, 150247. [CrossRef] [PubMed]
52. Liu, H.; Zhang, M.; Han, X. A review of surface ozone source apportionment in China. *Atmos. Ocean. Sci. Lett.* **2020**, *13*, 470–484. [CrossRef]
53. Zhang, H.; Qiu, Z.; Sun, D.; Wang, S.; He, Y. Seasonal and interannual variability of satellite-derived chlorophyll-a (2000–2012) in the Bohai Sea, China. *Remote Sens.* **2017**, *9*, 582. [CrossRef]
54. Yin, S.; Wang, X.; Zhang, X.; Zhang, Z.; Xiao, Y.; Tani, H.; Sun, Z. Exploring the effects of crop residue burning on local haze pollution in Northeast China using ground and satellite data. *Atmos. Environ.* **2019**, *199*, 189–201. [CrossRef]

Article

Haze Pollution Levels, Spatial Spillover Influence, and Impacts of the Digital Economy: Empirical Evidence from China

Jie Zhou [1,2], Hanlin Lan [1,*], Cheng Zhao [1,*] and Jianping Zhou [1]

1. School of Economics, Zhejiang University of Technology, Hangzhou 310023, China; jali.zm@163.com (J.Z.); zjp126222@126.com (J.Z.)
2. School of Marxism, Zhejiang Chinese Medical University, Hangzhou 310053, China
* Correspondence: lhl@zjut.edu.cn (H.L.); zhaoc@zjut.edu.cn (C.Z.); Tel.: +86-13858187688 (H.L.); +86-13868054742 (C.Z.)

Abstract: With the development of digital technologies such as the Internet and digital industries such as e-commerce, the digital economy has become a new form of economic and social development, which has brought forth a new perspective for environmental governance, energy conservation, and emission reduction. Based on data from 30 Chinese provinces from 2011 to 2018, this study applies the space and threshold models to empirically examine the digital economy's influence on haze pollution and its spatial spillover. Furthermore, it investigates the spatial diffusion effect of regional digital economic development and haze pollution by constructing a spatial weight matrix. Subsequently, an instrumental variable robustness test is performed. Results indicate the following: (1) Haze pollution has spatial spillover effects and high emission aggregation characteristics, with haze pollution in neighbouring provinces significantly aggravating pollution levels in the focal province. (2) China's digital economy has positively impacted haze pollution, with digital economic development having a significant effect (i.e., most prominent in eastern China) on reducing haze pollution. (3) Changing the energy structure and supporting innovation can restrain haze pollution, and the digital economy can reduce the path mechanism of haze pollution through the mediating effect of an advanced industrial structure. It shows a non-linear characteristic that the influence of haze reduction continues to weaken. Thus, policymakers should include the digital economy as a mechanism for ecologically sustainable development in haze pollution control.

Keywords: haze pollution; digital economy; industrial structure; spatial spillover

Citation: Zhou, J.; Lan, H.; Zhao, C.; Zhou, J. Haze Pollution Levels, Spatial Spillover Influence, and Impacts of the Digital Economy: Empirical Evidence from China. *Sustainability* **2021**, *13*, 9076. https://doi.org/10.3390/su13169076

Academic Editors: Weixin Yang, Guanghui Yuan and Yunpeng Yang

Received: 6 July 2021
Accepted: 3 August 2021
Published: 13 August 2021

Publisher's Note: MDPI stays neutral with regard to jurisdictional claims in published maps and institutional affiliations.

Copyright: © 2021 by the authors. Licensee MDPI, Basel, Switzerland. This article is an open access article distributed under the terms and conditions of the Creative Commons Attribution (CC BY) license (https://creativecommons.org/licenses/by/4.0/).

1. Introduction

Since China's reform and opening up, factor cost advantages have enabled the nation to achieve rapid economic development. However, this long-term and extensive economic development model has caused severe environmental pollution. As haze effects are wide-ranging, long-lasting, and difficult to treat, this form of air pollution has attracted extensive attention from many researchers. Many studies show that severe haze pollution greatly harms people's physical and mental health and reduces life expectancy, and the resulting welfare cost hinders sustainable economic development [1–4]. Thus, haze pollution detracts from improvements to health, living standards, and quality of economic development, making its effective control a priority.

Scholars have studied the influence of haze on different aspects, such as the economy [5,6], population [7–11], and energy [12–15]. The existing research has comprehensively explored the mechanism of haze pollution. However, technological and industrial revolutions, global warming, water pollution, air pollution [16,17], and other environmental problems have occurred frequently. Thus, cloud computing, 5G, artificial intelligence, big data, and other digital technologies attempt to break the information asymmetry, and they are expected to play an important role in global environmental governance [18–20]. Moreover, the low-cost, high-efficiency digital economy industry has witnessed constant development; as a consequence, many new industries have appeared. The transformation and upgrading of traditional industries have been accelerated, particularly as the Chinese government has been making efforts to coordinate environmental protection and economic development. At the national level, the digital economy is becoming increasingly important for societal development. According to China's Digital Economy Development White Paper [21], the digital economy grew by 15.6% annually to 35.8 trillion yuan in 2019 or 36.2% of the gross domestic product (GDP). Societies worldwide are moving toward rapid optimal allocation and regeneration of resources through the digital industry. This is reflected, for example, in the 'Made in China 2025' strategy and the 'Industrial Internet' in the United States. The influence of emerging industries on environmental governance can be analysed through the identification, selection, filtering, storage, and use of big data.

Whether digital technology can improve environmental pollution is related to whether digitalisation can help reduce both energy consumption and the cost of environmental governance. The previous literature has studied the overall association between economy-wide energy consumption and information and communication technologies (ICTs). Some scholars argue that ICT has reduced the demand for energy through energy efficiency and sectoral changes. Schulte et al. [22] found that in the Organisation for Economic Cooperation and Development (OECD) countries, 'a 1% increase in ICT capital results in a 0.235% reduction in energy demand'. This is not due to a decrease in electricity consumption but a decline in other non-electric energy sources, possibly arising from the direct impact of ICTs and services on electricity and the indirect impact on non-electric energy carriers in other parts of the economy. ICTs can enrich environmental quality through dematerialisation of production, thereby supporting a less resource-intensive and lightweight economy [23,24]. Ren et al. [25] used the provincial data, systematic GMM method, and intermediate effect model of China from 2006 to 2017 to demonstrate that the relationship between Internet development and energy consumption structure has a negative impact. However, some scholars believe that ICT application will increase energy consumption due to the 'rebound effect' [26]; Zhou et al. [27] analysed the carbon emissions at the industry level in China by using the input–output method; the ICT sector can induce a large amount of emissions by requiring carbon-intensive intermediate inputs from non-ICT sectors. In other words, the application of ICT does not significantly improve the environment and may even worsen environmental problems. Some scholars believe that this influence is not good or bad. Noussan and Tagliapietra [28] forecasted the future European scenario and analysed the potential impact of digital technologies such as the Internet of Things on energy consumption and carbon dioxide emissions in the transportation field. The impact on green sustainability depends on user behaviour, economic conditions, transport, and environmental policies.

Information asymmetry is another challenge in environmental governance. It not only increases environmental governance costs and weakens the effectiveness of environmental policies, but it also leads to a lack of regulatory bodies in environmental governance and reduces the public's enthusiasm for environmental governance. In 2016, China launched an ecological and environmental protection big data service platform as part of the Belt and Road Initiative. 'Internet +', big data, remote sensing satellites, and other information technologies provide environmental information support to China and other countries along the initiative. The Internet's openness, interactivity, and real-time nature make public participation in environmental governance both possible and convenient [29]. Moreover, the Internet promotes environmental supervision, management, intelligence, accurate services, and rectifies previous environmental governance deficiencies [30,31]. Zuo et al. [32] made recommendations to adopt IOT technology to dynamically collect real-time product data related to energy consumption to improve energy efficiency and large-scale utilisation of clean energy. Li et al. [33] empirically concluded that digital technology promotes environmental sustainability in Chinese manufacturing.

Simultaneously, the digital economy is reshaping the global value chain. According to the 'smiling curve' theory, high added value is located at both tails of the curve, representing the upstream (pre-production research and development) and downstream (post-production services) of the value chain. Processing and assembly activities are located at the midpoint of the curve, indicating little added value [34]. In the past, China's manufacturing sector embraced economies of scale for profitability with high volume, low-value production that also created severe air pollution. As the energy factor shifts from the industrial to the service sector, growth in the more energy-efficient sectors will reduce emissions; consequently, the overall economy will be more energy efficient [35–37]. Original elements and resources are transferred from industries with low distribution efficiency to technology-intensive industries with high distribution efficiency [38]. Thus, upgrades to the industrial structure would have a substantial impact on pollution.

Additionally, a characteristic of the digital economy is the physical sharing of information. Spatial changes have completely overhauled logistics links, resulting in the emergence of new industries, such as e-commerce, which is witnessing rapid growth due to the high penetration of the Internet and the large numbers of mobile users [39]. E-commerce can improve environmental pollution, as it significantly reduces information search costs and product prices and does a better job of matching. Thus, these supply–demand resources significantly reduce transportation and distribution costs, require less energy consumption, and reduce carbon dioxide emissions compared to in-person shopping [40]. E-commerce can also significantly optimise the corporate structure and management, thereby improving production efficiency [41]. The digital economy changes the smile curve, reconstructs the industrial value chain, and realises green development under the value chain sharing economy.

By reviewing the previous literature, we find that, first, the existing literature discusses the impact of digitisation on carbon emissions, SO_2 emissions and energy consumption through the use of the Internet, output value proportion of the tertiary industry, and investment in the ICT industry as proxy indicators. It is worth noting that the digital economy has received more and more attention, while little empirical research has been conducted to explore whether the development of the digital economy can improve air pollution in China. Second, previous studies have always carried out regression analysis on ordinary panels or dynamic panels, ignoring the spatial correlation and spatial spillover effect of haze pollution. In reality, the diffusion of haze between different regions will lead to spatial correlation and spatial dependence. In spatial econometrics, neglecting spatial effects may lead to errors in estimation and analysis. In a digital environment, search costs are lower, which increases the potential scope and quality of the search. Digital products are often not competitors; that is, they can be replicated at zero cost. As the cost of transporting digital goods and information approaches zero, the role of geographical distance is also expected to change. Digital technology makes it easier to track behaviour [32], and the digital economy containing the above characteristics undoubtedly brings a new perspective for environmental governance. Therefore, we must ask, what impact does the digital economy have on haze pollution? Moreover, what are the channels through which this influence is generated? To answer the above questions, we empirically test the effects of the digital economy on haze pollution and its spatial spillover using data from 30 provinces in China. This study aims to provide insights into the potential impact of the digital economy on future environmental governance. It argues that to take full advantage of the digital economy in environmental sustainability, it is necessary to adopt appropriate policies, support efficient deployment, and shape the digital process politically and socially [42].

This study's main contributions are as follows: First, we construct the second-level indicators of digital infrastructure (representing digital technology) and digital industry (representing emerging industries) and evaluate the development of the digital economy using the entropy weight method. Second, from the perspective of spatiotemporal evolution characteristics of the digital economy and haze pollution, the relationship between them is discussed using the spatial model, filling the gap between the digital economy and ecological geography. Third, accurately solving the two-way causality between haze pollution and the digital economy leads to endogeneity problems. Two methods were used to test the robustness: the replacement of spatial matrix, and the construction of instrumental variables. The number of telephones per 10,000 people per city in 1984 further confirms the robust results of our quantitative research. Finally, this study discusses the mechanism of digital economy influencing haze pollution through industrial structure change using the threshold model.

2. Methodology and Data

2.1. Construction of the Spatial Weight Matrix

The spatial weight matrix reflects the spatial interaction between different regional research samples. Spatial statistical analysis begins with the establishment of a spatial weight matrix. In this study, we set up a spatial weight matrix with sample size n. All elements of W_{ij} are i, j = 1, $\cdots$, n, and the 0–1 adjacency weight matrix (W1) is expressed as

$$W_{ij} = \begin{vmatrix} W_{11} & W_{12} & \cdots & W_{1n} \\ W_{21} & W_{22} & \cdots & W_{2n} \\ \cdots & \cdots & \cdots & \cdots \\ W_{n1} & W_{n2} & \cdots & W_{nn} \end{vmatrix} \tag{1}$$

where $W_{ij} = W_{ji}$; at $W_{ij} = 0$, location j is not a neighbour of location I; and at $W_{ij} = 1$, location j is the neighbour of location i, where i = 1, 2, ..., n; j = 1, 2, ..., n. At $W_{ij} = 1$, province i has a common boundary with province j; otherwise, $W_{ij} = 0$.

Then, we constructed the weight matrix of geographical distance (W2), which is expressed as follows:

$$W_{ij} = \begin{cases} W_{ij} = \frac{1}{d_{ij}} \\ W_{ij} = 0 \end{cases} \tag{2}$$

Let the distance between the geographic centres of province i and province j be d_{ij}; the latitude and longitude of geographic centre point A of province i be β_1 and α_1, respectively; and the latitude and longitude of geographic centre point B of province j be β_2 and α_2, respectively. The Earth's radius is:

$$d_{ij} = R \cdot arc\ cos[cos\beta_1 cos\beta_2 cos(\alpha_1 - \alpha_2) + sin\beta_1 sin\beta_2] \tag{3}$$

2.2. Spatial Autocorrelation Analysis

For a comprehensive investigation of the spatial spillover effect of haze pollution and the digital economy, we use the global and local spatial correlation indexes. First, we test whether the research object has a spatial effect by conducting a spatial autocorrelation test for the development index of the digital economy and haze pollution. Spatial correlation analysis can measure the spatial effect of each year in the geographical distance matrix. We calculate the global Moran's index (Moran's I) as

$$I = \frac{n \sum_{i=1}^{n} \sum_{j=1}^{n} W_{ij} |x_i - \bar{x}||x_j - \bar{x}|}{\sum_{i=1}^{n} \sum_{j=1}^{n} W_{ij} \sum_{i=1}^{n} (x_i - \bar{x})^2} \tag{4}$$

The value range of Moran's I is [−1,1]. When I > 0, a positive autocorrelation exists between the two regions. Haze pollution or the development of the digital economy is characterised by spatial agglomeration. When I < 0, a negative correlation exists between the two regions or spatial discreteness. When I = 0, the distribution of haze pollution is random, and no spatial autocorrelation exists.

Global spatial correlation analysis examines the aggregation of the entire space. Local spatial correlation analysis is used to understand the development of the digital economy within each region or the degree of correlation between the haze pollution level in the focal region and nearby regions. The local Moran's I is calculated as

$$I = \frac{n|x_i - \bar{x}| \sum_{j=1}^{n} W_{ij}|x_j - \bar{x}|}{\sum_{i=1}^{n}(x_i - \bar{x})^2} \qquad (5)$$

Here, a positive I represents some areas with high (low) values surrounded by other regions with high (low) values—either high–high (H–H) or low–low (L–L). Moreover, a negative I represents an area with high (or low) values surrounded by other areas with a low (or high) value—either high–low (H–L) or low–high (L–H).

2.3. Econometric Methodology

The following model was established:

$$\ln PM2.5_{i,t} = \alpha_0 + \alpha_1 \ln DIGE_{i,t} + \alpha_2 X_{control\,i,t} + \mu_i + \delta_t + \varepsilon_{i,t} \qquad (6)$$

Here, $DIGE_{i,t}$ is an indicator of the development level of the digital economy in province i in period t; $X_{control\,i,t}$ is a series of control variables: population structure (PS), fixed assets (FA), energy situation (ES), and degree of innovation (IN) in Equation (6); μ_i, refers to the individual fixed effect of province i that is time-invariant; δ_t controls the time fixed effect; and $\varepsilon_{i,t}$ is a random perturbation term.

2.3.1. Spatial Autoregressive Model

Spatial correlation existed between our variables, and OLS may lead to inconsistencies in the parameter estimates. Therefore, this study introduced a spatial econometric model and analysed the influence of the digital economy on haze pollution in depth from both the space and time perspective. We selected the spatial autoregressive model (SAR) and spatial error model (SEM). The SAR is

$$Y = \alpha + \rho W y + X\beta + \varepsilon, \varepsilon \sim N[0, \sigma^2 I] \qquad (7)$$

A variable is affected not only by its explanatory variable but also by variables in other spaces. Here, Y is the explained variable, X is the independent variable, α is the constant term, W is the spatial weight matrix, WY is a vector of the spatial lag dependent variable, ρ denotes a spatial regression coefficient reflecting the spatial dependence of the sample observations, and ε is a random perturbation term. Substituting Equation (6) into the test of Equation (7), we obtain the following spatial econometric model:

$$\ln PM2.5_{i,t} = \alpha + \rho W \ln PM2.5_{i,t-1} + \beta_1 \ln DIGE_{i,t} + \beta_2 X_{control\,i,t} + \varepsilon \qquad (8)$$

2.3.2. Spatial Error Model

Equation (9) represents the SEM. The space disturbance term is related to the space population, and the disturbance term of a particular space affects and other spaces via the space effect.

$$Y = \alpha + X\beta + \varepsilon, \varepsilon = \lambda W\varepsilon + \mu, \mu \sim N[0, \sigma^2 I] \qquad (9)$$

where Y is the explained variable; X is the independent variable of exogenous influencing factors; α is the constant term; ε is a random error term; β represents the influence of the independent variable on the dependent variable; λ is the unevaluated coefficient of the spatial autocorrelation error term (also known as the spatial autocorrelation coefficient); and μ is an error term. Substituting Equation (6) into the test of Equation (9), we obtain the following spatial econometric model:

$$\ln PM2.5_{i,t} = \alpha + \beta_1 \ln DIGE_{i,t} + \beta_2 X_{control\,i,t} + \lambda W\varepsilon + \mu \qquad (10)$$

2.3.3. Threshold Model

This study tested whether the industrial structure mediates the relationship between the digital economy and haze pollution measured as particulate matter (PM2.5). The specific steps are as follows: in the digital economy development index (DIGE), the coefficient of α_1 is significant throughout the analysis in the linear regression model (6) for haze pollution of PM2.5, based on the construction of DIGE for the mediating variable IS in the linear regression equation of the industrial structure and DIGE for IS in the regression equation of PM2.5 by $\beta_1, \gamma_1, \gamma_2$. The significance of the regression coefficient determines whether a mediation effect exists. The specific form of the regression model is

$$\ln IS_{i,t} = \beta_0 + \beta_1 \ln DIGE_{i,t} + \beta_2 X_{control\ i,t} + \mu_i + \delta_t + \varepsilon_{i,t} \tag{11}$$

and

$$\ln PM2.5_{i,t} = \gamma_0 + \gamma_1 \ln DIGE_{i,t} + \gamma_2 \ln IS_{i,t} + \gamma_3 X_{control,t} + \mu_i + \delta_t + \varepsilon_{i,t} \tag{12}$$

In addition to the mediating effect model, the empirical test for the indirect transmission mechanism should consider Metcalfe's law—the value of the Internet is proportional to the square of the number of users. The development level and industrial structure upgrading of the digital economy may also indirectly reduce the non-linear dynamic spillover of haze pollution in the digital economy. Therefore, in order to study whether the digital economy has a non-linear impact on haze pollution through the intermediary mechanism of industrial structure change, the following panel threshold model is set:

$$\begin{aligned}\ln PM2.5_{i,t} = \phi_0 + \phi_1 \ln DIGE_{i,t} \times I(Adj_{i,t} \leq \theta) + \\ \phi_2 \ln DIGE_{i,t} \times I(Adj_{i,t} > \theta) + \phi_3 X_{controli,t} + \mu_i + \delta_t\end{aligned} \tag{13}$$

In Equation (13), $Adj_{i,t}$ is a threshold variable such as the digital economy and industrial structure, and $I(\cdot)$ represents indicator functions valued at 1 or 0, which meet the conditions in the parentheses—namely 1; otherwise 0. Equation (13) considers a single threshold case.

2.4. Data Source

2.4.1. Explained Variable

PM2.5 (ug/m^3). To address the lack of historical data on PM2.5 concentration levels, we used raster data from the atmospheric composition analysis group based on the annual average of global PM2.5 concentrations monitored by satellites [43]. Using ArcGIS software, we analysed the specific value of the annual mean PM2.5 concentration in Chinese provinces from 2011 to 2018. Using these data, the difficulty in using surface monitoring data based on point source data to measure the PM2.5 concentration of an area accurately was addressed.

2.4.2. Core Explanatory Variable

The core explanatory variable is the DIGE. With regards to the measurement of the digital economy's development level, as officials have not yet disclosed a comprehensive index of concrete information for it, the calculation faces certain difficulties and challenges [44]. Based on the method of Huang et al. [45], the present study adopted the indicators of Internet penetration rate, relevant practitioners, relevant output, and mobile phone penetration rate. Based on the 2011–2018 panel data of 30 provinces, to build the digital infrastructure and digital industry variable, this study developed secondary indices where the secondary index of digital infrastructure corresponds to mobile telephone exchange capacity (10,000 families), optical fibre cable line length (km), number of Internet broadband access ports (10,000 units), number of websites (10,000 units), popularisation rate of mobile telephones (unit/100), and number of Internet broadband access users (10,000 units). The secondary index of digital industry is number of computers per 100 people in the enterprise, number of websites per 100 enterprises, proportion of enterprises with e-commerce transaction activities on the Internet per 100 enterprises, and proportion of e-commerce sales in the GDP. Using the entropy method, the data of these 10 indicators were processed to obtain the DIGE.

2.4.3. Intermediate Variable

Industrial structure (IS) is the intermediate variable. The proportion of the tertiary sector's output value indicates whether an economy has an advanced industrial structure [46]. The larger the value, the smaller the negative impact on haze pollution. Therefore, the sign of the coefficient is expected to be negative.

2.4.4. Control Variables

The control variables include the following:

Population structure (PS). Owing to livelihood pressures, young people are more willing to risk high pollution emissions to earn higher incomes, and an increase in the proportion of the labour population aggravates haze pollution [47]. In this study, the proportion of people aged 15–64 years in the total population was used to measure the influence of total regional population distribution on haze pollution. Therefore, this study expected the coefficient sign to be positive.

Fixed assets (FA). Following Li et al. [48], FA is expressed as the total investment in fixed assets. FA investment is positively correlated with digital economy development and is an essential source of funds for promoting technological innovation. Therefore, this study expected a negative coefficient sign.

Energy situation (ES). Burning fossil fuels, especially coal, is regarded as an important source of haze pollution [49], and China is among the few countries whose energy consumption structure is dominated by coal. Therefore, the total amount of energy consumption (tons of standard coal) is used. The higher the proportion of coal consumption, the less likely it is to decrease haze concentration. We expected a positive coefficient sign.

Innovation degree (IN). IN is the number of patents granted by each province. The larger its value, the stronger the technological innovation ability, which helps improve the factor utilisation efficiency and reduce pollution emission intensity. Therefore, we expected a negative coefficient sign.

The index data for the core explanatory variables are available from the *China Statistical Yearbook* [50]. The index data for the intermediary and control variables are from the WIND and China Stock Market Accounting Research databases.

2.5. Data Description

Table 1 shows the descriptive statistics. To reduce errors and heteroscedasticity caused by different units, each variable was treated logarithmically. The results show that haze pollution varies significantly among different regions. The development index of the digital economy (lnDIGE) has a small mean and large standard error, while the standard error of the industrial structure (the mediating variable) is relatively small. Clear differences among provinces exist in terms of PS, FA, ES, and IN.

Table 1. Descriptive statistics.

Type of Variable	Variable	Obs	Mean	SD	Min	Max
Dependent Variable	lnPM2.5	240	3.500	0.478	2.164	4.426
Independent Variable	lnDIGE	240	−1.80	0.665	−3.565	−0.114
Intermediate Variables	lnIS	240	3.804	0.190	3.391	4.419
	lnPS	240	−0.306	0.048	−0.409	−0.176
Control Variables	lnFA	240	9.451	0.782	7.219	10.941
	lnES	240	9.421	0.646	7.378	10.568
	lnIN	240	8.061	1.398	4.248	10.882

3. Results and Discussion

3.1. Spatio-Temporal Evolution of China's PM2.5 Concentration and Digital Economy

This study selected three cross sections of time—2011, 2014, and 2018. The spatial clustering characteristics of the digital economy development and haze pollution distribution in 30 Chinese provinces were analysed using the natural fracture method.

As illustrated in Figure 1a–c, for PM2.5 pollution, the 30 provinces showed an overall decline over the 8 years of the haze index. Maximum PM2.5 concentration by region was found in east-central China and the provinces of Shandong, Henan, Anhui, and Jiangsu, among others, in 2011, 2014, and 2018. The PM2.5 concentration in these regions was three times the smog concentration in the next highest echelon. In Hubei, Shanxi, Guangdong, Guizhou, and Chongqing provinces, PM2.5 pollution levels improved significantly, while they deteriorated in Xinjiang, Liaoning, and Gansu. This result was affected not only by geographical location and meteorological conditions but also by the provinces' social and economic development [51]. The possible reasons are as follows: (1) Most economically developed provinces have relatively high PM2.5 levels and have consequently witnessed greater efforts to control air pollution. (2) The industrial division of labour in the provinces is changing. An increase in the proportion of the tertiary sector improves air quality, while the transfer of the industrial structure aggravates haze pollution in the receiving province. (3) In the central and western regions, which have low population density, PM2.5 pollution is not quite as severe, and little attention is paid to, or investments made for, mitigating air pollution, causing a continuous deterioration of air quality.

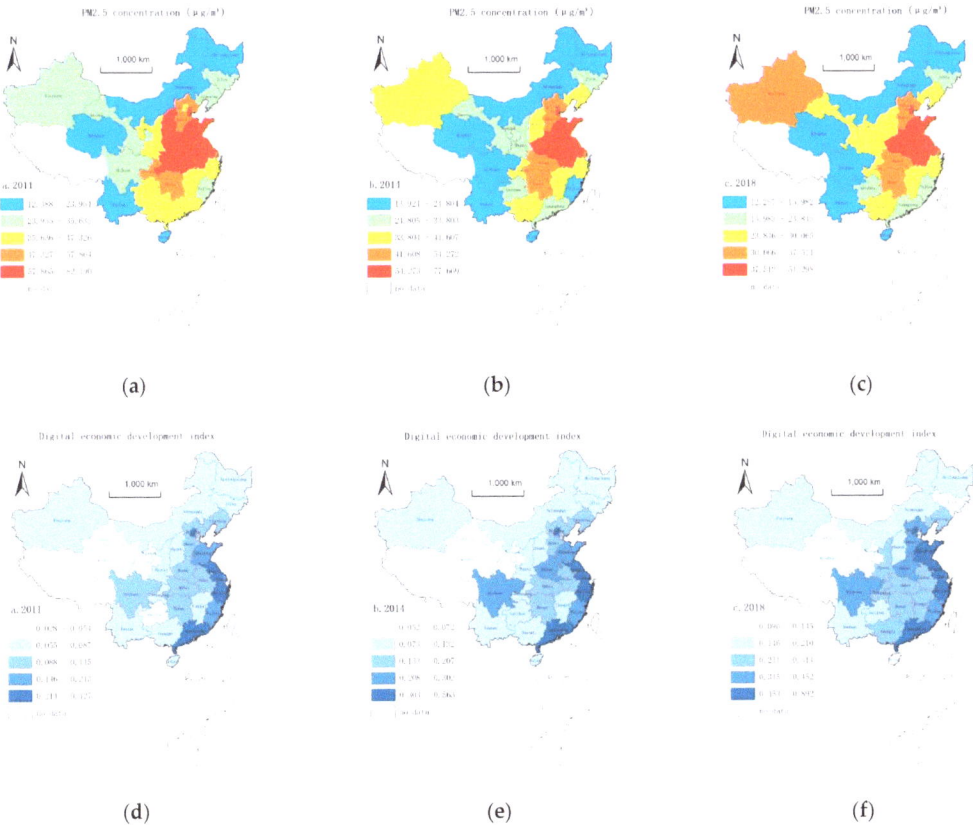

Figure 1. Spatio-temporal evolution of China's PM2.5 concentration levels (**a–c**) and digital economy (**d–f**) in 2011, 2014, and 2018.

Figure 1d–f show that, from 2011 to 2018, digital economy development was on the rise in all 30 provinces. China's three major economic belts are the bay area of the Yangtze River Delta, Guangdong province, and the Beijing–Tianjin–Hebei region, which witnessed substantial digital economy development in the first phase. Combined with other areas of the country, these form a clear core–periphery model wherein the eastern region's digital economy development index has leading areas, such as Guangdong, Jiangsu, Beijing, Shanghai, Zhejiang, Shandong, Shanxi, Shaanxi, and Guizhou. Moreover, Sichuan, Jiangxi, Anhui, Hubei, and other mid-west cities are catching up. Comparative advantage is implemented by digital economy development rotation. Simultaneously, the digital economy index reflects the imbalance and insufficiency among various regions in China [51]. The digital economy in Xinjiang, Gansu, Ningxia, and other regions in more remote areas is developing slowly, forming the bottom of the index. Thus, strengthening the Internet infrastructure construction in these areas is necessary.

3.2. Spatial Autocorrelation Analysis

To accurately understand the provincial-level digital economy and haze pollution agglomeration in the country, this study analysed the variables for the provinces with PM2.5 air pollution and digital economy development. Figure 2 shows the two indicators in the global Moran's I calculation formula: the 2011–2018 global Moran's I of the PM2.5 index, which is between 0.22 and 0.39 (p-value is 0.000–0.010, significant at 1%), with Z (I) 2.6–3.4 (Z >+ 2.58); and the global Moran's I of the digital economy, which is between 0.28 and 0.37 (p-value is 0.000–0.004, significant at 1%), with Z(I) 2.6–3.4 (Z >+ 2.58). Thus, the distribution of haze pollution and the digital economy presented significant spatial autocorrelation and had a geographical agglomeration feature. The more severe the haze pollution in the focal province, the higher the haze pollution in the neighbouring provinces. Moreover, the more advanced the digital economy in the focal province, the higher the degree of digital economy development in the neighbouring provinces.

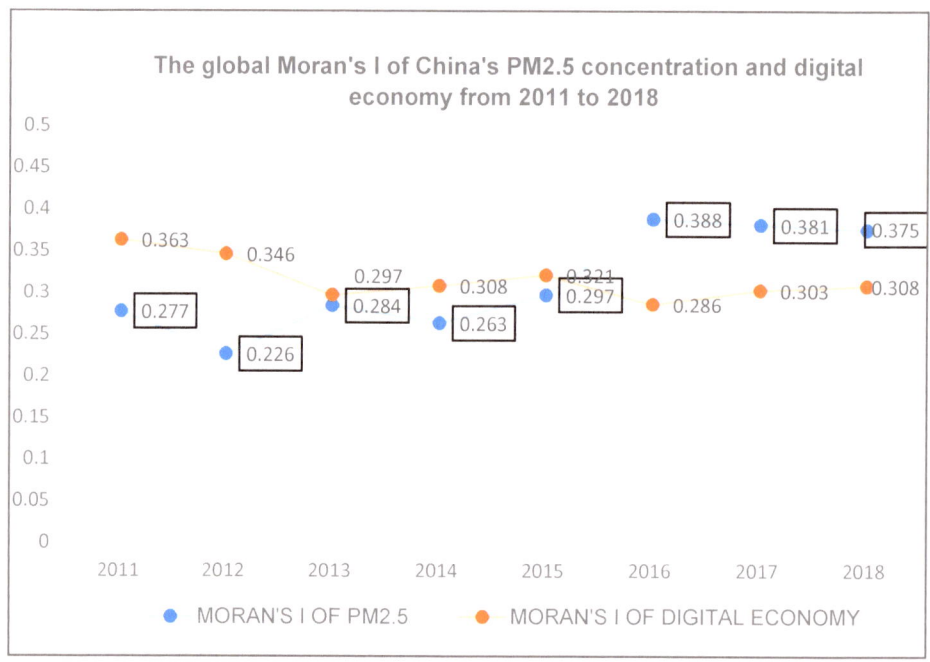

Figure 2. Global Moran's I of China's PM2.5 concentration levels and digital economy from 2011 to 2018.

Figure 3a–c show that haze pollution in 2011, 2014, and 2018 in the H–H agglomeration areas was mainly distributed in the Beijing–Tianjin–Hebei region and Yangtze River Delta; this includes the provinces that encompass Beijing, Tianjin, Shanghai, Jiangsu, and Zhejiang. The economically developed eastern region is the most populous province in China in terms of population density, urbanisation, industrialisation, new technology industry, and heavy industry base. Yunnan, Guizhou, and Sichuan in the west and Hunan, Hubei, and Hebei in the central region show an L–L agglomeration trend, while Gansu, Hainan, Hebei, Heilongjiang, and Jilin show an H–L agglomeration pattern. Haze pollution is geographically dispersed due to high pollution in this region [52].

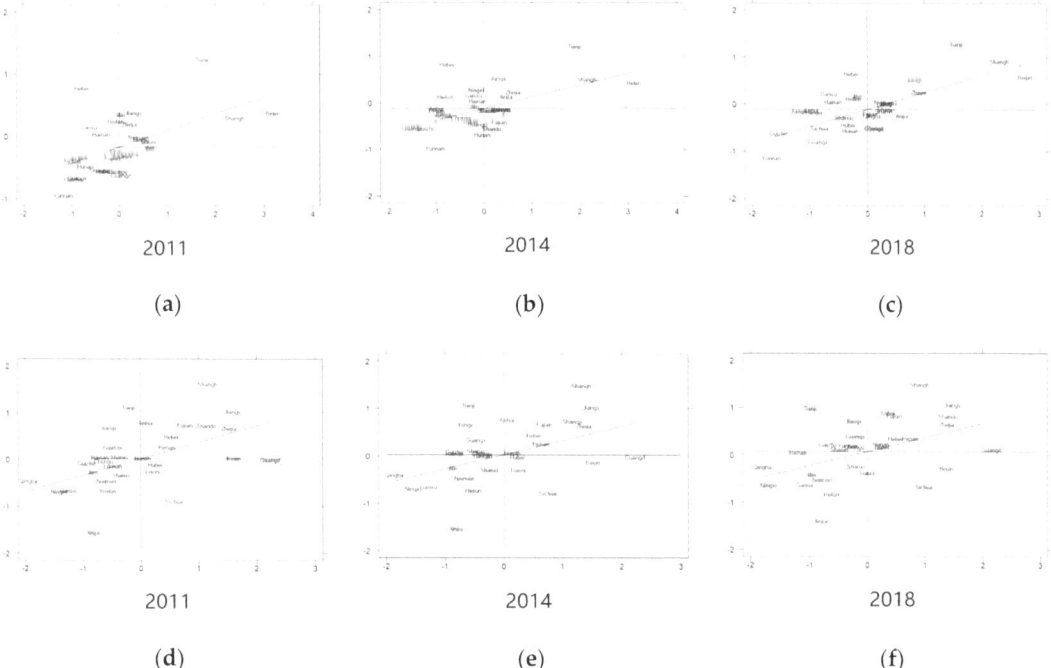

Figure 3. Local Moran's I of China's PM2.5 concentration levels (**a**–**c**) and digital economy (**d**–**f**) in 2011, 2014, and 2018.

Figure 3d–f show the relationship between the local space of the digital economy in 2011, 2014, and 2018. The Yangtze River Delta (Shanghai and Zhejiang in the east and Jiangsu), Shandong, Fujian, and Henan in central China, along with Hebei, all show H–H features. These are economically developed eastern coastal areas, with greater employment opportunities, strong talent agglomeration, and good development prospects, strengthening the development of the digital economy. However, the central region follows the existing trend, with an emerging digital economy. Tianjin, Jiangxi, and Guangxi all show the characteristics of the H–L agglomeration. Sichuan, Liaoning, and Hubei have an L–H agglomeration pattern. Compared with these three provinces, their neighbouring provinces are more attractive for developing the digital economy. Further, the remote western areas of Qinghai, Ningxia, Gansu, Xinjiang, Heilongjiang, and Inner Mongolia and the central areas of Shanxi and Hunan form an L–L agglomeration due to their low level of digital economy development and lack of a driving force in terms of digital economy development in neighbouring areas.

3.3. Spatial Panel Model Analysis

3.3.1. LM Test

Moran's I passed the significance test. The classical OLS regression had a significant spatial correlation; therefore, a spatial econometric model should be used for parameter estimation. As presented in Table 2, both LM-Lag and LM-Error passed the 1% significance level of the spatial dependence test. According to the criteria, LM-Lag and LM-Error should pass the significance test, and the lag and robust LM-Error should pass the 1% significance test. Thus, the spatial lag model and the SEM were used to estimate the regression. We introduced the neighbouring weighting matrix to the model and analysed the regression results.

Table 2. LM test.

Spatial Autocorrelation Text	Z-Value	p-Value
LM-lag	77.7777	0.000
Robust LM-lag	21.5651	0.000
LM-Error	105.2038	0.000
Robust LM-Error	48.9912	0.000

3.3.2. Regression Results and Discussion

As shown in Table 3, in the SAR estimation with a time fixed effect, the estimated value of ρ was 0.2, significant at the 5% level. This value indicates that neighbouring regions have a significant positive spatial spillover effect on PM2.5; an increase of 1% in PM2.5 concentration in neighbouring provinces leads to an increase of approximately 0.2% in PM2.5 concentration in the focal province. Thus, maintaining a province's particular approach to haze treatment cannot effectively solve inter-regional haze pollution. Consequently, transforming local treatment to regional joint prevention and control is necessary. In the SEM, the λ value was significant at the 10% level. This indicates that haze concentration is affected not only by observable factors such as population structure but also by observable factors in adjacent areas. The influencing factors are discussed below.

Table 3. Estimation results for different models (dependent variable is PM2.5).

	OLS (FE) (1)		SAR (2)		SEM (3)	
Variables	Estimate	T Value	Estimate	T Value	Estimate	T Value
(Intercept)	3.712 ***	5.55	0.010 ***	10.21	0.010 ***	10.21
lnDIGE	−0.263 **	−2.69	−0.216 ***	−3.44	−0.276 ***	−4.49
lnPS	0.064	0.10	−0.201	−0.36	−0.104	−0.19
lnFA	−0.021	−0.58	−0.030	−0.88	−0.038	−1.03
lnES	0.031 **	2.68	0.025 **	2.08	0.024 *	1.83
lnIN	−0.093 **	−2.32	−0.075 *	−1.91	−0.082 **	−2.08
ρ			0.200 **	2.39		
λ					0.172 *	1.78
R^2	0.663		0.666		0.662	

Note: ***, ** and * represent significance at 1%, 5% and 10% levels, respectively.

First, we focus on the effect and magnitude, of the core explanatory variable of the digital economy on haze pollution. The panel OLS, SAR, and SEM models showed that the digital economy development has a significantly negative effect on haze reduction, passing the significance test with a 99% confidence level. Every 1% increase in the development level of the digital economy reduces haze concentration in the region by approximately 0.2%. The possible reasons for this are as follows: (1) The digital economy promotes the construction of digital infrastructure through technological effects. (2) The digital economy, through structural effects, expands the proportion of digital industries, digitally empowers traditional industries, improves the energy efficiency and operational efficiency of traditional industries, promotes the rapid and efficient transformation and upgrading of traditional industries, and finally achieves low energy consumption and low emissions [53].

Second, from the perspective of energy, in the OLS and SAR models, the total energy consumption was significantly positive at 5%, which is consistent with expectations. This indicator shows a significant promoting effect on haze pollution [54]. The secondary sector includes industry and construction; the industrial consumption of coal, oil, non-ferrous metals, and other raw materials creating fine dust, which is the leading cause of haze pollution. The energy structure is the key factor responsible for aggravating haze pollution. Therefore, accelerating the transformation and upgrade of this structure is urgently necessary.

The degree of innovation variable in the OLS and SEM models was significantly negative at the 5% level. The early stage of economic development involves resource consumption to expand production and satisfy people's material needs. Thus, economic development neglects environmental protection to a certain extent. Moreover, although living standards are widely improved, natural resources become constrained. These aspects of early development highlight the importance of environmental protection in the reversed transmission of technology innovation, transforming economic development patterns, and optimising economic structure adjustment [55].

The population structure variable in the three models showed inconsistencies; its coefficient was positive in the OLS model, confirming that young people are willing to accept high pollution emissions in exchange for high income; thus, an increase in the labour population can increase or aggravate haze pollution [47]. However, in the SAR and SEM models, the coefficient was negative, possibly because labour population agglomeration significantly reverses the transmission of regional environment improvement to reduce smog pollution.

The coefficients of fixed assets were all negative, indicating that fixed asset investment is positively correlated with the development of the digital economy and is an important source of funds to promote technological innovation. Among the control variables, population structure and fixed assets were statistically significant.

3.4. Test for Threshold Regression Model

From the analysis of the theoretical model (8), we observed the mechanism through which the digital economy affects haze pollution. Considering that the development of the digital economy acts on haze reduction through structural effects, this study introduced the mediation variable index of industrial structure as the threshold for a threshold effect analysis to examine the influence of different intervals of the industrial structure on haze. The form of the panel threshold model was tested first. Subsequently, we followed Hansen [56] and used the bootstrap sampling method to simulate a likelihood ratio statistic of 200, estimating the threshold value and relevant statistics. The results show that a single threshold of F statistic was significant at 5%, while the double and triple thresholds were not significant. Thus, to analyse the effect of the digital economy on haze pollution, we considered the industrial structure to be a single threshold effect and assumed that the industrial structure is the threshold variable. As shown in Table 4, the negative influence of the digital economy on haze pollution continued to weaken, and the non-linear characteristics of the negative and diminishing 'marginal effect' of the digital economy remained. This trend shows that the dynamic influence of the digital economy on haze pollution is affected not only by its development level but also by the regulating influence of the industrial structure, which is reflected in the positive interaction between the digital economy and industrial structure. However, this effect gradually weakens with the change of industrial structure.

Table 4. Estimation results for the threshold regression model.

Variables		Intermediate Variable
		lnIS
Threshold	q_1	3.927
	q_2	
DIGE·I (Th ≤ q_1)		0.302 **
DIGE·I (q_1 < Th < q_2)		0.272 *
Control variables		YES
Number of periods		8
Number of provinces		30
R^2		0.6929

Note: **, * represent significance at 5%, 10% levels, respectively.

3.5. Heterogeneity Test

Owing to different resource endowments and stages of development, both the development level of the digital economy and haze pollution have noticeable heterogeneity in terms of their regional distribution. Therefore, regional differences may exist in the impact of the digital economy on haze pollution reduction, which are necessary to consider for an in-depth discussion.

First, a descriptive statistical explanation is provided for the differences in haze pollution and digital economy development levels in various regions. As shown in Table 5, in terms of PM2.5, the logarithmic mean value of haze pollution is the lowest in western China and highest in eastern China. Thus, the eastern region is significantly ahead of the central and western regions in terms of digital economic development. The mean difference between the eastern and the middle and western regions is approximately 0.574 and 0.89, respectively, reflecting a first-mover advantage. This result sets the foundation for the regional heterogeneity test of the effects of the digital economy on haze pollution. The regression analysis of regional heterogeneity is shown in Table 6. The results of models (1), (2), and (3) show that the digital economy in eastern China has a significant effect on reducing haze pollution, while the effect is not significant in central and western China. In other words, considering regional heterogeneity, the digital economy in eastern China has a higher positive effect on haze pollution reduction. This result is possibly because the digital economy in eastern China developed earlier and was at a higher level, causing the dividend of the impact of the digital economy on environmental governance to be released more fully.

Table 5. Descriptive statistics (different regions).

	lnPM2.5				
Region	Obs	Mean	Std. Dev.	Min	Max
East	88	3.639	0.447	2.618	4.426
Middle	64	3.657	0.439	2.629	4.409
West	88	3.247	0.430	2.164	4.046
	lnDIGE				
Region	Obs	Mean	Std. Dev.	Min	Max
East	88	−1.324	0.604	−2.760	−0.114
Middle	64	−1.898	0.454	−2.844	−0.929
West	88	−2.214	0.539	−3.565	−0.793

Table 6. Heterogeneity test (dependent variable is PM2.5).

	East (1)		Middle (2)		West (3)	
Variables	Estimate	T Value	Estimate	T Value	Estimate	T Value
(Intercept)	3.614 ***	6.09	2.795	0.99	4.545 ***	5.62

Table 6. Cont.

Variables	East (1)		Middle (2)		West (3)	
	Estimate	T Value	Estimate	T Value	Estimate	T Value
lnDIGE	−0.324 **	−2.81	−0.323	−0.97	−0.187	−1.76
lnPS	−0.256	−0.27	−1.568	−0.58	0.759	0.51
lnFA	0.002	0.11	0.010	0.09	−0.117 **	−2.26
lnES	0.045	2.70	0.034	1.33	0.020	1.00
lnIN	−0.103 **	−1.76	−0.081	−0.59	−0.083	−2.18
Number of periods	8		8		8	
Number of provinces	11		8		11	
R^2	0.692		0.587		0.723	

Note: ** represent significance at 5% levels, respectively.

4. Robustness Test

4.1. Changing the Spatial Matrix

A spatial econometric model, which is highly dependent on the spatial weight matrix, was used to study the influence of the digital economy on haze pollution. First, the neighbouring spatial weight matrix (W1) was used to determine whether the provinces are adjacent. Adjacency was set to 1, and non-adjacency to 0. Second, the robustness of the regression results was tested using the geographical distance spatial weight matrix (W2), which was constructed using the reciprocal of the square deviation of the distance between provinces.

From the test results in Table 7, we inserted the weight matrix (W2) into the spatial lag model (4) and SEM (5). We observed that the coefficient of the core variable of the digital economy was significantly negative in the SEM model. The lnDIGE regression coefficient was the largest and was significant at 1%, indicating that the digital economy's spatial influence on haze pollution is more likely to be in the error term of undetectable than the spatial correlation between the two in time. Therefore, the development of the digital economy can effectively reduce haze pollution, which is consistent with the main research results and proves that the regression results are robust.

Table 7. Estimation results for different models (dependent variable is PM2.5).

Variables	SAR (4)		SEM (5)	
	Estimate	T Value	Estimate	T Value
(Intercept)	0.010 ***	10.21	0.010 ***	10.21
lnDIGE	−0.132 *	−1.83	−0.250 ***	−3.88
lnPS	−0.344	−0.62	−0.103	−0.19
lnFA	−0.042	−1.23	−0.052	−1.41
lnES	0.022 *	1.87	0.017 *	1.23
lnIN	−0.064	−1.64	−0.085 **	−2.11
ρ	0.415 ***	3.21		
λ			0.450 **	2.98
R^2	0.665		0.659	

Note: ***, ** and * represent significance at 1%, 5% and 10% levels, respectively.

4.2. Use of Instrumental Variable

Selecting appropriate instrumental variables for the core explanatory variables can resolve endogeneity problems. Following Huang et al. [45], this study adopts the 1984 volume of each province's post and telecommunications business as the core explanatory variable and the instrumental variable of the comprehensive index of digital economy development. The instrumental variables must satisfy exogeneity and correlation. On the one hand, with the continuous development of traditional communications technology, previous levels of the local telecommunications infrastructure affect the subsequent stage of application of Internet technology from the technical level perspective and usage habits.

On the other hand, the impact of the use of traditional telecommunications tools, such as the use of fixed-line telephones, on economic development should meet the exclusivity. As their usage frequency gradually declines with social and economic development, the instrumental variable must satisfy the conditions.

As the original data of the selected tool variable are in cross-sectional form, they cannot be used directly in the econometric analysis of panel data. Based on Nunn and Qian [57], a variable that changes over time is introduced to construct the panel tool variable. The interaction term is constructed by the number of Internet users in the last year and the number of telephones per 10,000 people in each province in 1984. This statistic is used as the instrumental variable of the digital economy index of the province in that year. The results in columns (1) and (2) of Table 8 show that the effect of the digital economy on reducing haze pollution remains valid after considering endogeneity, and the results are all significant at 1%. For the test of the null hypothesis of insufficient identification of instrumental variables, the LM statistic P values are all 0.000, which significantly rejects the null hypothesis. In the test for weak identification of instrumental variables, the Wald F statistic is greater than the threshold value above 10% of the weak identification test. In general, the tests illustrate the rationality of choosing the cross-term, between the historical postal and telecommunications volumes of various provinces, and the number of Internet users in China as the instrumental variable of digital economy development.

Table 8. Test of the instrumental variables.

Variable	Instrumental Variable	
	(1)	(2)
lnDIGE	−0.424 ***	−0.239 ***
	(−18.70)	(−3.01)
Control variables	NO	YES
Province fixed effect	YES	YES
Year fixed effect	YES	YES
LM statistic	196.588	122.964
	[0.0000]	[0.0000]
Wald F statistic	308.97	289.62
	{16.38}	{16.38}
Number of periods	8	8
Number of provinces	30	30
R^2	0.6420	0.6634

Note: *** represents significance at 1% levels, respectively.

5. Conclusions

First, this study constructs an evaluation system for developing the digital economy at the provincial level in China from the two aspects of digital infrastructure (representing digital technology) and the digital industry (representing emerging industries). It calculates the development level of the digital economy in each province using the entropy weight method. Second, the spatial spillover effects of haze pollution and digital economy development are tested with the global and local spatial correlation indexes. Third, using the data of 30 provinces in China from 2011 to 2018, OLS regression and spatial SAR and SEM models were used to analyse the impact of digital economy development on haze pollution. Fourth, using the threshold model, the study discusses how the digital economy mechanism affects haze pollution through industrial structure change. Finally, the study divides the research samples into three regions (eastern, central, and western regions) to study the regional heterogeneity impact of digital economic development on haze pollution.

The findings of the present study are as follows: First, both haze pollution and digital economy distribution present significant global positive spatial spillover effects and local characteristics. Second, the digital economy has a positive impact on reducing smog. The development of the digital economy in neighbouring provinces has a significant positive spillover effect on reducing haze pollution in key provinces. The change of energy structure and innovation degree can effectively restrain the aggravation of haze pollution, and the conclusion is still valid in the robustness test using the instrumental variable method and adjusting the spatial matrix. Third, the results of the transmission mechanism show that the development of the digital economy can affect haze pollution by changing the industrial structure, showing the non-linear feature that the influence of haze reduction continues to weaken. Finally, in terms of regional differences, the impact of the digital economy on haze pollution is most significant in eastern China, while not significant in central and western China. Based on this study, the following policy recommendations are put forward.

First of all, the penetration and application of digital technology in environmental governance should be accelerated. We would increase investment in digital technologies; pay attention to the breadth and depth of applications in advanced fields such as the Internet, 5G, artificial intelligence, and big data; promote the circulation and sharing of resources, knowledge, and capital; and promote the improvement of digital economy in environmental governance, such as energy conservation and emission reduction. Second, the transformation and upgrading of industrial structure should be promoted, encouraging enterprises to vigorously develop cutting-edge technologies and promoting the continuous progress of digital industry and digitization of industry. Third, it is necessary to understand further the positive impact of the digital economy on reducing haze pollution in central and western China, indicating that a dynamic and differentiated digital economy strategy should be implemented. Finally, haze reduction policies should take into account spatial spillover and decomposition boundaries of administrative areas.

Although this study supplements the relevant research on the impact of the digital economy on haze and provides some theoretical reference for the digital economy on environmental governance, there is still room for further research. First, this paper measures the digital economy from two aspects: digital infrastructure and digital industry. Because of the existing data, it may have measurement errors. The evaluation is conducted at the provincial level, and the sample size is limited. In the future, it can be more detailed and micro, which may be more accurate in exploring the relationship between the two from the city level. Second, this study empirically analyses the spatial impact of digital economy development on haze pollution. The mechanism part is only carried out from the perspective of industrial structure, and subsequent studies should further explore the multi-dimensional impact of different mechanisms on haze. Finally, the development of the digital economy is cyclical, and each stage has a different impact on haze levels. This should be further investigated in future studies.

Author Contributions: Conceptualisation, J.Z. (Jie Zhou) and H.L.; methodology, J.Z. (Jianping Zhou); software, J.Z. (Jianping Zhou); validation, C.Z., J.Z. (Jianping Zhou) and H.L.; formal analysis, H.L.; investigation, C.Z. and H.L.; resources, C.Z. and J.Z. (Jie Zhou); data curation, J.Z. (Jianping Zhou); writing—original draft preparation, J.Z. (Jie Zhou); writing—review and editing, J.Z. (Jie Zhou); visualisation, J.Z. (Jie Zhou); supervision, H.L.; project administration, H.L.; funding acquisition, C.Z. and H.L. All authors have read and agreed to the published version of the manuscript.

Funding: The National Natural Science Foundation of China and the Natural Science Foundation of Zhejiang Province funded this project under grant numbers 61902349 and 2021C03144, respectively.

Institutional Review Board Statement: Not applicable.

Informed Consent Statement: Not applicable.

Data Availability Statement: Not applicable.

Conflicts of Interest: The authors declare no conflict of interest.

References

1. Peters, A.; Dockery, D.W.; Muller, J.E.; Mittleman, M.A. Increased particulate air pollution and the triggering of myocardial infarction. *Circulation* **2001**, *103*, 2810–2815. [CrossRef]
2. Chang, T.; Graff Zivin, J.; Gross, T.; Neidell, M. Particulate pollution and the productivity of pear packers. *Am. Econ. J. Econ. Policy* **2016**, *8*, 141–169. [CrossRef]

3. Ebenstein, A.; Fan, M.; Greenstone, M.; He, G.; Zhou, M. New evidence on the impact of sustained exposure to air pollution on life expectancy from China's Huai river policy. *Proc. Natl. Acad. Sci. USA* **2017**, *114*, 10384–10389. [CrossRef]
4. Chen, S.; Oliva, P.; Zhang, P. Air pollution and mental health: Evidence from China. In *NBER Working Paper 24686*; National Bureau of Economic Research, Inc.: Cambridge, MA, USA, 2018; Available online: https://www.nber.org/system/files/working_papers/w24686/w24686.pdf (accessed on 3 May 2021).
5. Grossman, G.M.; Krueger, A.B. Economic growth and the environment. *Q. J. Econ.* **1995**, *110*, 353–377. [CrossRef]
6. Apergis, N.; Ozturk, I. Testing environmental Kuznets curve hypothesis in Asian countries. *Ecol. Indic.* **2015**, *52*, 16–22. [CrossRef]
7. Birdsall, N. Another look at population and global warming. In *Policy Research Working Paper Series, 1020*; The World Bank: Washington, DC, USA, 1992; Available online: https://www.researchgate.net/profile/Nancy-Birdsall/publication/23721567_Another_Look_at_Population_and_Global_Warming/links/546f64ab0cf24af340c089e4/Another-Look-at-Population-and-Global-Warming.pdf (accessed on 2 May 2021).
8. Dietz, T.; Rosa, E. Rethinking the environmental impacts of population, affluence and technology. *Human Ecol. Rev.* **1994**, *1*, 277–300.
9. O'Neill, B.C.; Liddle, B.; Jiang, L.; Smith, K.R.; Fuchs, R. Demographic change and carbon dioxide emissions. *Lancet* **2012**, *380*, 157–164. [CrossRef]
10. Menz, T.; Welsch, H. Population aging and carbon emissions in OECD countries: Accounting for life-cycle and cohort effects. *Energy Econ.* **2012**, *34*, 842–849. [CrossRef]
11. Ohlan, R. The impact of population density, energy consumption, economic growth and trade openness on CO_2 emissions in India. *Nat. Hazards* **2015**, *79*, 1409–1428. [CrossRef]
12. Kumar, A.; Srivastava, S.; Goudar, R.H. Efficient operating system switching using mode bit and hibernation mechanism. *CSI Transact. ICT* **2013**, *1*, 67–74. [CrossRef]
13. Lv, B.; Zhang, B.; Bai, Y. A systematic analysis of PM2.5 in Beijing and its sources from 2000 to 2012. *Atmos. Environ.* **2016**, *124*, 98–108. [CrossRef]
14. Qerimi, D.; Dimitrieska, C.; Vasilevska, S.; Rrecaj, A.A. Modeling of the solar thermal energy use in urban areas. *Civ. Eng. J.* **2020**, *6*, 1349–1367. [CrossRef]
15. Abdulrazzaq, L.R.; Abdulkareem, M.N.; Yazid, M.R.M.; Borhan, M.N.; Mahdi, M.S. Traffic congestion: Shift from private car to public transportation. *Civ. Eng. J.* **2020**, *6*, 1547–1554. [CrossRef]
16. Gibergans-Baguena, J.; Hervada-Sala, C.; Jarauta-Bragulat, E. The quality of urban air in barcelona: A new approach applying compositional data analysis methods. *Emerg. Sci. J.* **2020**, *4*, 113–121. [CrossRef]
17. Benaissa, F.; Bendahmane, I.; Bourfis, N.; Aoulaiche, O.; Alkama, R. Bioindication of urban air polycyclic aromatic hydrocarbons using petunia hybrida. *Civ. Eng. J.* **2019**, *5*, 1305–1313. [CrossRef]
18. Wu, J.; Song, G.; Jie, L.; Deze, Z. Big data meet green challenges: Big data toward green applications. *IEEE Syst. J.* **2016**, *10*, 888–900. [CrossRef]
19. Hong, J. Causal relationship between ICT R&D investment and economic growth in Korea. *Technol. Forecast. Soc. Chang.* **2017**, *116*, 70–75. [CrossRef]
20. Latif, Z.; Mengke, Y.; Latif, S.; Ximei, L.; Pathan, Z.H.; Salam, S.; Jianqiu, Z. The dynamics of ICT, foreign direct investment, globalisation and economic growth: Panel estimation robust to heterogeneity and cross-sectional dependence. *Telemat. Inform.* **2018**, *35*, 318–328. [CrossRef]
21. China's Digital Economy Development White Paper. 2020. Available online: http://www.caict.ac.cn/kxyj/qwfb/bps/202007/t20200702_285535.html (accessed on 19 April 2021).
22. Schulte, P.; Welsch, H.; Rexhaeuser, S. ICT and the demand for energy: Evidence from OECD countries. *Environ. Resour. Econ.* **2016**, *63*, 119–146. [CrossRef]
23. Fuchs, C. The implications of new information and communication technologies for sustainability. *Environ. Dev. Sustain.* **2008**, *10*, 291–309. [CrossRef]
24. Ishida, H. The effect of ICT development on economic growth and energy consumption in Japan. *Telemat. Inform.* **2015**, *32*, 79–88. [CrossRef]
25. Ren, S.; Hao, Y.; Xu, L.; Wu, H.; Ba, N. Digitalization and energy: How does internet development affect China's energy consumption? *Energy Econ.* **2021**, *98*, 105220. [CrossRef]
26. Lange, S.; Pohl, J.; Tilman, S. Digitalization and energy consumption. Does ICT reduce energy demand? *Ecol. Econ.* **2020**, *176*, 106760. [CrossRef]
27. Zhou, X.; Zhou, D.; Wang, Q.; Su, B. How information and communication technology drives carbon emissions: A sector-level analysis for China. *Energy Econ.* **2019**, *81*, 380–392. [CrossRef]
28. Noussan, M.; Tagliapietra, S. The effect of digitalization in the energy consumption of passenger transport: An analysis of future scenarios for Europe. *J. Clean. Prod.* **2020**, *258*, 120926. [CrossRef]
29. Chong, M.; Choy, M. The Social Amplification of haze-related risks on the internet. *Health Commun.* **2016**, *33*, 14–21. [CrossRef] [PubMed]
30. Granell, C.; Havlik, D.; Schade, S.; Sabeur, Z.; Delaney, C.; Pielorz, J.; Havlik, F.N.; Bodsberg, R.; Berre, A.; Mon, J.L. Future Internet technologies for environmental applications. *Environ. Model. Softw.* **2016**, *78*, 1–15. [CrossRef]

31. Zhu, J.; Xie, P.; Xuan, P.; Jin, Z.; Yu, P. Renewable Energy Consumption Technology under Energy Internet Environment. In Proceedings of the 2017 IEEE Conference on Energy Internet and Energy System Integration (EI2), Beijing, China, 26–28 November 2017; Available online: http://www.researchgate.net/publication/322351391_Renewable_energy_consumption_technology_under_energy_internet_environment (accessed on 24 March 2021).
32. Zuo, Y.; Tao, F.; Nee, A.Y.C. An internet of things and cloud-based approach for energy consumption evaluation and analysis for a product. *Int. J. Comput. Integr. Manuf.* **2017**, *31*, 337–348. [CrossRef]
33. Li, Y.; Jing, D.; Li, C. The impact of digital technologies on economic and environmental performance in the context of Industry 4.0: A moderated mediation model. *Int. J. Prod. Econ.* **2020**, *229*, 107777. [CrossRef]
34. Shin, N.; Kraemer, K.L.; Dedrick, J. Value capture in the global electronics industry: Empirical evidence for the 'smiling curve' concept. *Indus. Innov.* **2012**, *19*, 89–107. [CrossRef]
35. Li, K.; Lin, B. The non-linear impacts of industrial structure on China's energy intensity. *Energy* **2014**, *69*, 258–265. [CrossRef]
36. Tian, X.; Chang, M.; Shi, F.; Tanikawa, H. How does industrial structure change impact carbon dioxide emissions? A comparative analysis focusing on nine provincial regions in China. *Environ. Sci. Policy* **2014**, *37*, 243–254. [CrossRef]
37. Chang, N. Changing industrial structure to reduce carbon dioxide emissions: A Chinese application. *J. Clean. Prod.* **2015**, *103*, 40–48. [CrossRef]
38. Qin, J.; Liu, Y.; Grosvenor, R. Data Analytics for Energy Consumption of Digital Manufacturing Systems Using Internet of Things Method. In Proceedings of the 2017 13th IEEE Conference on Automation Science and Engineering (CASE 2017), Xi'an, China, 20–23 August 2017; Available online: http://www.researchgate.net/publication/322565613_Data_analytics_for_energy_consumption_of_digital_manufacturing_systems_using_Internet_of_Things_method (accessed on 12 March 2021).
39. Oláh, J.; Kitukutha, N.; Haddad, H.; Pakurár, M.; Máté, D.; Popp, J. Achieving sustainable e-commerce in environmental, social and economic dimensions by taking possible trade-offs. *Sustainability* **2019**, *11*, 89. [CrossRef]
40. Romm, J. The Internet and the new energy economy. *Resour. Conserv. Recycl.* **2002**, *36*, 197–210. [CrossRef]
41. Bloom, N.; Sadun, R.; Van Reenen, J. Americans do it better: US multinationals and the productivity miracle. *Am. Econ. Rev.* **2012**, *102*, 167–201. [CrossRef]
42. Santarius, T.; Pohl, J.; Lange, S. Digitalization and the decoupling debate: Can ICT help to reduce environmental impacts while the economy keeps growing? *Sustainability* **2020**, *12*, 7496. [CrossRef]
43. van Donkelaar, A.; Martin, R.V.; Brauer, M.; Boys, B.L. Use of satellite observations for long-term exposure assessment of global concentrations of fine particulate matter. *Environ. Health Perspect.* **2015**, *123*, 135–143. [CrossRef]
44. Liu, J.; Yang, Y.-J.; Zhang, S.-F. Research on the measurement and driving factors of China's digital economy. *Shanghai J. Econ.* **2020**, *6*, 81–96.
45. Huang, Q.; Yu, Y.-Z.; Zhang, S.-L. Internet development and productivity growth in manufacturing industry: Internal mechanism and China experiences. *China Ind. Econ.* **2019**, *8*, 5–23.
46. Cheng, Z.; Li, L.; Liu, J. Industrial structure, technical progress and carbon intensity in China's provinces. *Renew. Sustain. Energy Rev.* **2018**, *81*, 2935–2946. [CrossRef]
47. Buehn, A.; Farzanegan, M.R. Hold your breath: A new index of air pollution. *Energy Econ.* **2013**, *37*, 104–113. [CrossRef]
48. Li, G.; Guo, F.; Di, D. Regional competition, environmental decentralisation, and target selection of local governments. *Sci. Total Environ.* **2021**, *755*, 142536. [CrossRef]
49. Ma, L.-M.; Zhang, X. The spatial effect of China's haze pollution and the impact from economic change and energy structure. *China Ind. Econ.* **2014**, *4*, 19–31.
50. China Statistical Yearbook, 2011–2018. Available online: http://www.stats.gov.cn/tjsj/ndsj/ (accessed on 18 May 2021).
51. Zhao, X.; Zhou, W.; Han, L.; Locke, D. Spatiotemporal variation in PM2.5 concentrations and their relationship with socioeconomic factors in China's major cities. *Environ. Int.* **2019**, *133*, 105145. [CrossRef]
52. Zhao, T.; Zhang, Z.; Liang, S. Digital economy, entrepreneurship, and high-quality economic development: Empirical evidence from urban China. *Manag. World* **2020**, *10*, 65–76.
53. Li, X.-Z.; Chen, H.-L.; Zhang, X.-D. Research on the industrial performance of the convergence of information industry with manufacturing industry based on the data of Zhejiang province. *China Soft Sci.* **2017**, *1*, 22–30.
54. Shao, S.; Li, X.; Cao, J.; Yang, L. China's economic policy choices for governing smog pollution based on spatial spillover effects. *Econ. Res. J.* **2016**, *9*, 73–88.
55. Ren, Y.; Zhang, G. Can city innovation dispel haze? *China Popul. Resour. Environ.* **2020**, *30*, 111–120.
56. Hansen, B.E. Threshold effects in non-dynamic panels: Estimation, testing, and inference. *J. Econom.* **1999**, *93*, 345–368. [CrossRef]
57. Nunn, N.; Qian, N. US food aid and civil conflict. *Am. Econ. Rev.* **2014**, *104*, 1630–1666. [CrossRef]

MDPI AG
Grosspeteranlage 5
4052 Basel
Switzerland
Tel.: +41 61 683 77 34

Sustainability Editorial Office
E-mail: sustainability@mdpi.com
www.mdpi.com/journal/sustainability

Disclaimer/Publisher's Note: The statements, opinions and data contained in all publications are solely those of the individual author(s) and contributor(s) and not of MDPI and/or the editor(s). MDPI and/or the editor(s) disclaim responsibility for any injury to people or property resulting from any ideas, methods, instructions or products referred to in the content.

www.ingramcontent.com/pod-product-compliance
Lightning Source LLC
LaVergne TN
LVHW072332090526
838202LV00019B/2408